International Nuclear Science and Technology Conference

(INST 2014)

Journal of Physics: Conference Series Volume 611

Bangkok, Thailand
28 – 30 August 2014

ISBN: 978-1-5108-0359-6
ISSN: 1742-6588

Printed from e-media with permission by:

Curran Associates, Inc.
57 Morehouse Lane
Red Hook, NY 12571

Some format issues inherent in the e-media version may also appear in this print version.

Printed by Curran Associates, Inc. (2015)

For permission requests, please contact the Institute of Physics at the address below.

Institute of Physics
Dirac House, Temple Back
Bristol BS1 6BE UK

Phone: 44 1 17 929 7481
Fax: 44 1 17 920 0979

techtracking@iop.org

Additional copies of this publication are available from:

Curran Associates, Inc.
57 Morehouse Lane
Red Hook, NY 12571 USA
Phone: 845-758-0400
Fax: 845-758-2634
Email: curran@proceedings.com
Web: www.proceedings.com

TABLE OF CONTENTS

Journal of Physics: Conference Series **611** (2015) 011001 doi:10.1088/1742-6596/611/1/011001

The 1st International Nuclear Science and Technology Conference 2014 (INST2014)

Announcement

Nuclear technology has played an important role in many aspects of our lives, including agriculture, energy, materials, medicine, environment, forensics, healthcare, and frontier research. The International Nuclear Science and Technology Conference (INST) aims to bring together scientists, engineers, academics, and students to share knowledge and experiences about all aspects of nuclear sciences.

INST has evolved from a series of national conferences in Thailand called Nuclear Science and Technology (NST) Conference, which has been held for 11 times, the first being in 1986.

INST2014 was held in August 2014 and hosted by Thailand Institute of Nuclear Technology (TINT). The theme was "Driving the future with nuclear technology". The conference working language was English.

The proceedings were peer reviewed and considered for publication.

The topics covered in the conference were:

- Agricultural and food applications [AGR]
- Environmental applications [ENV]
- Radiation processing and industrial applications [IND]
- Medical and nutritional applications [MED]
- Nuclear physics and engineering [PHY]
- Nuclear and radiation safety [SAF]
- Other related topics [OTH]
- Device and instrument presentation [DEV]

Awards for outstanding oral and poster presentations will be given to qualified students who present their work during the conference.

Journal of Physics: Conference Series **611** (2015) 011001 doi:10.1088/1742-6596/611/1/011001

Chair's Welcome Address

Dear colleagues and friends,

On behalf of the organizing committee and Thailand Institute of Nuclear Technology, I would like to welcome you to International Nuclear Science and Technology Conference 2014 (INST2014) in Bangkok, Thailand, from 28-30 August 2014.

INST2014 is the first international conference that covers current topics in nuclear science, technology and applications being organized by Thailand Institute of Nuclear Technology. Colleagues from all breadths of nuclear and radiation research will present and share their work at the conference. This year, there are over 300 participants from Thailand and 13 other countries, with 41 oral presentations and 54 poster presentations.

Through INST2014, the committee hopes to provide a stage where new ideas are generated, innovation created, knowledge discussed and exchanged, friends made and strengthened, and collaboration initiated and followed through.

The organizing committee thanks Ministry of Science and Technology, Thailand, and Thailand Institute of Nuclear Technology for hosting this event and thanks Nuclear Society of Thailand for co-hosting the event and Korean Association for Radiation Application (KARA) for sponsoring this conference. We welcome you to Bangkok.

With warmest regards,

N. Poramatikul

Nipavan Poramatikul
Chair, INST2014
Director, Nuclear Research and Development Division, TINT

Journal of Physics: Conference Series **611** (2015) 011001 doi:10.1088/1742-6596/611/1/011001

List of Committees

Host Organization: Thailand Institute of Nuclear Technology (TINT), Public Organization, Ongkharak, Nakorn Nayok, Thailand

Organizing Committee

Nipavan Poramatikul (Chairperson)
Ajaya Malakrong
Panita Ruenbanterng
Kanchalika Dechates
Kanokporn Boonsirichai
Kasinee Hemvichian
Roppon Picha
Archara Phattanasab
Dhanaj Seangchantr
Adisak Punyanut
Saensuk Wetchagarun
Sudkanung Phumkem
Sorada Chanintayuttavong
Wannipa Phianphak (Secretary)

Academic Committee

Kanokporn Boonsirichai (Chairperson)
Phiriyatorn Suwanmala
Suchada Segsanviriya
Roppon Picha
Moleephan Dangprasert
Sarinrat Wonglee
Pimpon Uttayarat
Kampanart Silva
Naruemon Neramitmansook (Secretary)

Editorial Team

Phiriyatorn Suwanmala
Roppon Picha
Amares Chatt - Dalhousie University, Canada
Olgun Guven - Hacettepe University, Turkey
Kanokporn Boonsirichai
Moleephan Dangprasert
Sarinrat Wonglee
Pimpon Uttayarat
Kampanart Silva
Naruemon Neramitmansook

Oral Presentations

Venue: Centara Grand at Central Plaza Ladprao, Bangkok, Thailand
Dates: 28-30 August 2014

Thursday, 28 Aug

Time	Speaker	Organization	Title
9:50-10:20	R. Mongkolnavin	Chulalongkorn University, Thailand	Nuclear Research and Development in the AEC Era ** Keynote Presentation **
10:30-11:00	R. Szymczak	Tradewinds, Australia	Marine ecosystem impacts of radioactive releases from Fukushima Daiichi - past, present and future
11:00-11:30	A. Chatt	Dalhousie University, Canada	Simultaneous Multielement Species Analysis using Neutron Activation
11:30-12:00	S. Mankhetkorn	Chiang Mai University, Thailand	Visualization of Inflammation at Early Stage of Lung Cancer in Xenografted Temporally Immunocompromised Rats by Ferrioxamine Magnetic Resonance Imaging
14:30-15:00	O. Güven	Hacettepe University, Turkey	Past, Present and Future of Radiation Processing of Polymers
15:00-15:30	M. Aritomi	Tokyo Institute of Technology, Japan	Learning from The Accidents at Fukushima Daiichi Nuclear Power Plants
15:30-15:50	D. Ruffolo	Mahidol University, Thailand	Research in Thailand Using Neutron Monitors to Study Solar Storms and Solar Effects on Our Space Radiation Environment
16:00-16:20	K. Sukrod	Thailand Institute of Nuclear Technology	Improvement of Portable Computed Tomography System for On-field Applications
16:20-16:40	P. Seeprasert	Kyoto University, Japan	Role of Saprotrophic Fungi to Enhance the Retention Cesium and Strontium: Toxicity and Accumulation
16:40-17:00	P. Kessaratikoon	Thaksin University, Thailand	Specific Activity of Natural and Anthropogenic Radionuclides and Radiological Hazard Assessment in Surface Soil Samples Collected along the Andaman Sea Coast in Southern Thailand

Friday, 29 Aug

Time	Speaker	Organization	Title
9:00-9:20	A. Petchrak	Thailand Institute of Nuclear Technology	Preliminary Laboratory-scale Study for Leak Detection in Industrial Applications Using Radiotracers at TINT
9:20-9:40	T. Buasri	Tokyo Institute of Technology, Japan	Development of TiNiPd High Temperature Shape Memory Alloy for Sensor Application **Runner-up of the Student Oral Presentation Award**
9:40-10:00	K. Saenboonruang	Kasetsart University, Japan	Recent Developments in GEM-Based Neutron Detectors

9:00-9:20	T. Liamsuwan	Thailand Institute of Nuclear Technology	Characterization of Neutron Calibration Fields at the TINT's 50 Ci Americium-241/Beryllium Neutron Irradiator
9:20-9:40	P. Intharat	Thammasat University, Thailand	Investigation of L-H Transition in Tokamak Plasma Using BALDUR Code
9:40-10:00	K. Tiyapun	Thailand Institute of Nuclear Technology	Validation of the MCNP Computational Model for Neutron Flux Distribution with the Neutron Activation Analysis Measurement
10:30-10:50	M. Fukushima	Ishinomaki Senshu University, Japan	Japanese Fir Tree Leaves as Indicators of 137Cs Contamination in Oshika Peninsula, Miyagi Prefecture
10:50-11:10	K. Titipornpun	Suratthani Rajabhat University, Thailand	Measurements of Indoor Radon Concentrations in the Chaiya and the Tha Chana Districts, Surat Thani Province, Thailand
11:10-11:30	N. Baumann	Institute of Resource Ecology, Germany	Uranium Contents in Plants and Mushrooms Grown on a Uranium Mining Waste Tip Site in Former East Germany
11:30-11:50	S. Khuntong	Kasetsart University, Thailand	Assessment of Radionuclides in Marine Sediment along the Upper Gulf of Thailand
10:30-10:50	D. Y. Kim	Sungkyunkwan University, Korea	Simulation and measurement of characteristics of MICROMEGAS detector for alpha detection from 241Am source
10:50-11:10	W. Buangam	Sirindhorn International Institute of Technology, Thailand	Investigation of Impurity Behaviors using CORE modeling for L-mode Tokamak in BALDUR Code
11:10-11:30	F. Tabbakh	Nuclear Science and Technology Institute of Iran	An approach to efficiency enhancement in thermoelectric generators
11:30-11:50	D. Gihwala	Cape Peninsula University of Technology, South Africa	The Complementarity of PIGE and PIXE Spectrometry for Elemental Analysis
14:30-14:50	P. Charoenphun	Ramathibodi Hospital, Thailand	A Long-lived Radiotracer, 89Zr-oxine, for Cell Tracking Using PET
14:50-15:10	N. Juntong	Synchrotron Light Research Institute, Thailand	Linear Accelerator Development for Cancer Treatment at SLRI
15:10-15:30	L. Kovacs	Cidade Universitaria, Brazil	Iodine analysis in whole blood: an alternative for sport medicine
15:30-15:50	K. Boonsirichai	Thailand Institute of Nuclear Technology	Simulation and measurement of characteristics of MICROMEGAS detector for alpha detection from 241Am source
14:30-14:50	I. Kovalev	Lomonosov Moscow State University, Russia	The Mini Calorimeter for the NUCLEON Experiment
14:50-15:10	B. Chatthong	Sirindhorn International	Investigation of Toroidal Flow Effects on L-H transition in Tokamak Plasma Based on Bifurcation

		Institute of Technology, Thailand	Model **Winner of the Student Oral Presentation Award**
15:10-15:30	T. Lim	Satrec Initiative, Korea	Development of Advanced Mobile Radiation Survey System Using Gamma Spectrometry
15:30-15:50	S. H. Lee	Korean Association for Radiation Application, Korea	Presentation by Korean Association for Radiation Application
16:00-16:20	W. Sudprasert	Kasetsart University, Thailand	EPR Study of Free Radicals in Cotton Fiber for Its Potential Use as a Fortuitous Dosimeter in Radiological Accident
16:20-16:40	N. Ya-anant	Thailand Institute of Nuclear Technology	Safety and Security Management on Disused Sealed Radioactive Sources in Thailand
16:40-17:00	T. Kim	KEPCO International Nuclear Graduate School, Korea	Best Estimate Methodology for Realistic Safety Evaluation

Saturday, 30 Aug

Time	Speaker	Organization	Title
9:00-9:30	K. Okamoto	University of Tokyo, Japan	Post-Fukushima Research Activities for Nuclear Safety
9:30-10:00	R. Frew	IAEA	Stable Isotope Deltas – Nature's Own Traceability System for Verifying Food Authenticity
10:00-10:30	K. Klakhaeng	Rice Department, Thailand	Induced Mutations in Thai Rice
10:40-11:00	S. Sangaroon	Mahasarakam University, Thailand	Study of Collimated Neutron Flux Monitors for MAST and MAST Upgrade
11:00-11:20	N. Sanguansak	Suranaree University of Technology, Thailand	Present Status of BNCT Project in Thailand
11:20-11:40	W. Pasanphan	Kasetsart University, Thailand	A series of radiation processed nanostructural chitosan derivatives for biomedicine, agriculture, and bioplastics

Journal of Physics: Conference Series **611** (2015) 011001 doi:10.1088/1742-6596/611/1/011001

Panel discussion: Future trends of nuclear research

Speaker	Organization	Field of research
K. Okamoto	Tokyo University, Japan	Nuclear engineering
M. Aritomi	Tokyo Institute of Technology, Japan	Nuclear engineering and management
S. Chongkum	Thailand Institute of Nuclear Technology	(Host country)
R. Frew	Food and Environmental Protection Laboratory, IAEA	Food authenticity and traceability
O. Guven	Hacettepe University, Turkey	Radiation processing
R. Szymczak	Tradewinds, Australia	Environmental studies
T. Liamsuwan	Thailand Institute of Nuclear Technology	(Moderator)

Oral Sessions' Chairs

Chair	Organization
C. Cherdchu	Nuclear Society of Thailand, Thailand
S. Nilsuwankosit	Chulalongkorn University, Thailand
S. Wetchagarun	Thailand Institute of Nuclear Technology, Thailand
M. Sonsuk	Nuclear Society of Thailand, Thailand
T. Onjun	Sirindhorn International Institute of Technology, Thailand
W. Sudprasert	Kasetsart University, Thailand
S. Sangaroon	Mahasarakham University, Thailand
P. Uttayarat	Thailand Institute of Nuclear Technology, Thailand
S. Nilsuwankosit	Chulalongkorn University, Thailand
P. Ampornrat	Office of Atoms for Peace, Thailand
K. Silva	Thailand Institute of Nuclear Technology, Thailand
P. Uttayarat	Thailand Institute of Nuclear Technology, Thailand

Poster Presentations

Venue: Centara Grand at Central Plaza Ladprao, Bangkok, Thailand
Dates: 28-30 August 2014

Organization	Title	Authors
Panjab University, India	Role of Deformed Coulomb Potential in the Fusion Process at Different Orientations of the Colliding Nuclei	R. Kumari
Nuclear Science and Technology Institute of Iran	Carbohydrate based materials for gamma radiation shielding	F. Tabbakh, V. Babaee, and Z. Naghsh-nehzad
Chungnam National University, Korea	Indoor Particulate Matter and the Associated Elements at a Subway Station in Korea by Instrumental Neutron Activation Analysis	J.H. Lee, J.H. Jeong, J.M. Lim, and K.H. Kim
Chiang Mai University, Thailand	An ASEAN Ion Beam Analysis Center at Chiang Mai University, Thailand	U. Tippawan, T. Kamwann, L.D. Yu, S. Intarasiri, N. Puttaraksa, S. Unai, C. Thongleurm, and S. Singkarat
Chiang Mai University, Thailand	Low-energy Accelerator-based Nuclear Biotechnology for Applications in Agriculture and Biomedicine	L.D. Yu, S. Anuntalabhochai, B. Phanchaisri, W. Wongkham, and T. Vilaithong
Chulalongkorn University, Thailand	Innovative Neutron Shielding Material Composing of Natural Rubber-Styrene Butadiene Rubber Blends Diboron Trioxide	C. Jumpee and D. Wongsawae
Chulalongkorn University, Thailand	An Industrial Radiography Exposure Device Based on Measurement of Transmitted Gamma-Ray Intensity	C. Polee, N. Chankow, S. Srisatit, and D. Thong-Aram **Runner-up of the Student Poster Presentation Award**
Chulalongkorn University, Thailand	A simple laboratory system for diffusive radon flux measurements	C. Kranrod, S. Chanyotha, and W.C. Burnett
Kasetsart University, Thailand	Determination of Cesium-137 in the Oyster (Crassostrea belcheri) meat and seawater samples from the Lower Gulf of Thailand and Radiation Dose Assessment of the General Public.	P. Pakkong, N. Phaopeng, Y. Tumnoi
King Mongkut's Institute of Technology Ladkrabang, Thailand	Shielding Design for Storage of Radioactive Waste from TRR-1/M1 Pool Refurbishment	K. Songkaitiwong, T. Chaiya, S. Janhomkul, C. Tippayakul, and S. Lekchaum **Winner of the Student Poster Presentation Award**
King Mongkut's Institute of Technology Ladkrabang, Thailand	Preliminary validation of computational model for neutron flux prediction of Thai Research Reactor (TRR-1/M1)	S. Sabaibang, C. Tipayakul, and S. Lekchaum
King Mongkut's University of Technology Thonburi, Thailand	Application of Lab View on Ionization Chamber to Measurement Radiation	P. Kerdchockchai, T. Soodprasert, E. Hoonnivathana, P. Naemchnthara, P. Limsuwan, and K. Naemchanthara
Mahidol University, Thailand	Factors effecting adsorption of 137Cs in marine sediment samples from the upper gulf of Thailand	C. Saengkul, P. Sawangwong, and P. Pakkong
Office of Atoms for Peace, Thailand	Simulation on Radioactive Material Positioning System	T. Tularak, W. Rayangam, and S. Kiatisakwatana
Office of Atoms for Peace, Thailand	Analysis of Radon in Shallow wells: A Case Study at Pijit district in Songkhla, Thailand	A. Charoensri, S. Siriboonprapob, and N. Sastri
Office of Atoms for Peace, Thailand	Cs-137 Concentrations in Seawater from the Gulf of Thailand and the Andaman Sea and Radiation Doses in Marine biota after the Fukushima Dai-ichi NPP	Y. Tumnoi, N. Nakkaew, C. Khrautongkieo, and S. Udomsomporn

	Accident	
Prince of Songkhla University, Thailand	NAA Comparison of Nutrients in Egg Yolk and Egg White	S. Ruengdit, P. Maijan, J. Channuie, and R. Picha
Rajabhat Rajanagarindra University, Thailand	Application of the Optically Stimulated Luminescence (OSL) in Environmental Monitoring.	P. Nuthongkum and S. Monthonwattana
Ramathibodi Hospital, Thailand	Dose-Response Curves for Dicentrics and PCC Rings: Preparedness for Radiological Emergency in Thailand	B. Rungsimaphorn, B. Rerkamnuaychoke, and W. Sudprasert
Royal Thai Army Chemical Department, Thailand	Feasibility study on nitrogen in explosives using X-ray photoelectron spectroscopy: Chemical fertilizer	P. Dararutana
Sirindhorn International Institute of Technology, Thailand	Study of Pellet Injection Effects on Transport Barrier Formation and Pedestal Width Using Bifurcation Concept	K. Wihakhaphirom, B. Chatthong, R. Picha, N. Poolyarat, and T. Onjun
Sirindhorn International Institute of Technology, Thailand	Review on Design of Beam Shaping Assembly Based on the D-T Reaction for BNCT	M. Asnal, T. Onjun, and T. Liamsuwan
Sirindhorn International Institute of Technology, Thailand	Effects of Current on Behaviors of Saturated Magnetic Island in Tokamak	W. Kanjanaput, R. Picha, J. Promping, N. Poolyarat, and T. Onjun
Sirindhorn International Institute of Technology, Thailand	Studying Transport Barrier Formation using Bi-Stable Sandpile Model	B. Kitpitak, W. Kanjanaput, N. Poolyarat, R. Picha, and T. Onjun
Thailand Institute of Nuclear Technology	The Development of 3D Graphics for Simple Implementation of Photon and Neutron Transport Code	P. Siangsanan
Thailand Institute of Nuclear Technology	Gamma and Neutron Attenuation Properties of Barite-Cement Mixture	R. Picha, J. Channuie, S. Khaweerat, T. Liamsuwan, J. Promping, W. Ratanatongchai, K. Silva, and S. Wonglee
Thailand Institute of Nuclear Technology	Application of Gamma Radiation on Bio-oil Produced from Pyrolysis of Soybean Cake	P. Pichestapong, U. Injarean, P. Prapakornrattana, and K. Charoen
Thailand Institute of Nuclear Technology	99mTc-Ciprofloxacin for Diagnosis of Bacterial Infection	A. Aungurarat, T. Ngamprayad, M. Dangprasert, S. Phumkem, and B. Jowanaridhi
Thailand Institute of Nuclear Technology	Formulation of DOTATATE Cold Kit for Preparation of Ga-68 Radiopharmaceutical	W. Sriwiang, T. Duangta, P. Kaeopookum, and S. Wongsanit
Thailand Institute of Nuclear Technology	Application of gamma irradiation for sterilizing purpose in marl product (Dinsorepong)	N. Neramitmansook, K. Chahorm, W.Sriwichai, W. Pilamart, and W. Phianphak
Thailand Institute of Nuclear Technology	Gamma Irradiation Induces DNA Double-Strand Breaks in Fibroblasts: A Model Study for the Development of Biodosimetry	P. Uttayarat, K. Boonsirichai, and T. Tangtong
Thailand Institute of Nuclear Technology	Mutation Breeding for Acidic Soil Resistance in Rice	V. Puripunyavanich, K. Boonsirichai, S. Chookaew, and V. Khamvarn
Thailand Institute of Nuclear Technology	Evaluation and Improvement of Microbiological Qualities of Frozen Seafood by Gamma Irradiation.	K. Chahorm, N. Neramitmansook, and W. Phianphak
Thailand Institute of Nuclear Technology	Preliminary Study of 13C/12C Application for the Detection of Adulterated Honey	S. Loestanatjaroen
Thailand Institute of Nuclear Technology	Variations of 210Po Activity in Mussels (Perna viridis) of Samut Sakhon and Its Contribution of Radiation Dose	B. Porntepkasemsan, K. Srisuksawad, and V. Permnamthip
Thailand Institute of Nuclear Technology	Determination of Exposure Dose Error on Personal Dosimetry Irradiation Caused by the Standard Source (137Cs) Travelling Time	T. Rungseesamran, D. Saengchantr, N. Kamwang, and K. Chongkitivitya

Thailand Institute of Nuclear Technology	Estimation of the Indoor Radon and the Annual Effective Dose from Granite Samples	P. Sola, W. Srinuttakul, and P. Kewsuwan
Thailand Institute of Nuclear Technology	Determination of Self-absorption Corrections Factor for Radioactivity Measurement of Seawater and plankton samples by gross α/β counter	S. Wongsanit, P. Pakkong, Y. Tumnoi, and S. Udomsomporn
Thailand Institute of Nuclear Technology	Microorganisms in Fermented Bio-extract from Irradiated Silk Waste	N. Sermkiattipong and T. Tangthong
Thailand Institute of Nuclear Technology	Spatial distribution of 137Cs in surface soil under different land uses in Chao Phraya watershed: Potential used as sediment source tracing	K. Srisuksawad, B. Porntepkasemsan, N. Noipow, A. Omanee, W. Wiriyakitnateekul, R. Choybudha, and P. Srimawong
Thailand Institute of Nuclear Technology	Measurement of Q-values of Nuclear Reactions Using Gas-filled Neutron Detectors	J. Channuie, T. Liamsuwan, R. Picha, W. Ratanatongchai, S. Khaweerat, S. Wonglee, J. Promping, and K. Silva
Thailand Institute of Nuclear Technology	Effect of gamma radiation and electron beams on microbiological quality and protein patterns of 4 beans	S. Chookaew, J. Eamsiri, W. Pewlong, and S. Sajjabut
Thailand Institute of Nuclear Technology	Influence of gamma radiation and electron beam on microbial load of Pueraria mirifica at various storage time	J. Eamsiri, W. Pewlong, S. Sajjabut, and S. Chookaew
Thailand Institute of Nuclear Technology	Effects of Radiation on Mechanical Properties of Poly(butylene succinate) and Cassava Starch Blends	K. Hemvichian, K. Dechasasawat, W. Kangsumrith, and P. Suwanmala
Thailand Institute of Nuclear Technology	Decontamination of Clostridium perfringens and Salmonella spp. in Thai Fermented Fish (Pla-ra) by Gamma Radiation	P. Prakhongsil, W. Phianphak, A. Malakrong, and C. Komolamisra
Thailand Institute of Nuclear Technology	Hydrogen Determination using PGNAA by Small Neutron Source	W. Ratanatongchai, S. Khaweerat, J. Channuie, S. Wonglee, R. Picha, J. Promping, K. Silva, T. Liamsuwan, S.Limlai, P. Boonyamanee, and W. Boonnuch
Thailand Institute of Nuclear Technology	The Measurement of Stable Isotope Ratio by Cavity Ring-Down Spectroscopy (CRDS) and the Application on Hydrology and Groundwater Geochemistry in Thailand.	N. Noipow, K. Srisuksawad, K. Kamdee, M. Yongprawat, and B. Porntepkasemsan
Thailand Institute of Nuclear Technology	Neutron beam characterization for Neutron radiography facility at the Thai research reactor TRR-1/M1	S. Wonglee, C. Polee, S. Khaweerat, W. Ratanatongchai, and N. Chankow
Thailand Institute of Nuclear Technology	Improvement of Microbiological Quality on of Cordyceps militaris Products by Gamma Radiation	S. Sajjabut, W. Pewlong, J. Eamsiri, and S. Chookaew
Thailand Institute of Nuclear Technology	Simulation of Plasma Performance of the Tokamak DEMO Nuclear Fusion Reactors Design	J. Promping, W. Kanjanaput, W. Buangam, B. Chathong, R. Picha, N. Poolyarat, Y. Pianroj, and T. Onjun
Thailand Institute of Nuclear Technology	Environmental Effects of Monazite Contamination	K. Boonsirichai, Y. Tumnoi, P. Uttayarat, K. Silva, A. Phattanasab, and V. Puripunyavanich
Thailand Institute of Nuclear Technology	A Nuclear Method to Authenticate Buddha Images	S. Khaweerat, W. Ratanatongchai, J. Channuie, S. Wonglee, R. Picha, J. Promping, K. Silva, and T. Liamsuwan
Thammasat University, Thailand	Characteristic of TPF-I Current Signals	T. Kunamaspakorn, N. Poolyarat, R. Picha, J. Promping, and T. Onjun
Thammasat University, Thailand	Electrical Characterization of Dielectric Barrier Discharge in Atmospheric Air or Plasma Production to Improve Seed germination	T. Traikool, N. Poolyarat, R. Picha, and T. Onjun

Peer review statement

All papers published in this volume of *Journal of Physics: Conference Series* have been peer reviewed through processes administered by the proceedings Editors. Reviews were conducted by expert referees to the professional and scientific standards expected of a proceedings journal published by IOP Publishing.

Role of deformed Coulomb potential in the fusion process at different orientations of the colliding nuclei

Raj Kumari

Department of Physics, Panjab University, Chandigarh-160014

E-mail: rajkumari80pu@gmail.com

Abstract. We discuss the effect of adding deformation in the Coulomb potential in the fusion process using well tested proximity based nuclear potentials. This effect is analyzed at different orientations of the two colliding nuclei. Our study reveals that different (oblate/prolate) quadruple deformations in the Coulomb potential affect the total interaction potential in the inner region in different manner. In addition to nature of deformation (oblate/prolate), the depth of the fusion pocket also varies with change in orientations of colliding nuclei. A totally opposite trend of variation is observed in case of spherical-oblate and oblate-oblate systems as compared to spherical-prolate and prolate-prolate systems.

1. Introduction

The heavy-ion collisions produce a variety of phenomena that depend on the mass/charge of the colliding nuclei as well as on the incident energy and impact parameter of a reaction. At low incident energies, colliding matter leads to complete (incomplete) fusion [1, 2, 3, 4, 5, 6, 7, 8, 9, 10, 11, 12, 13, 14]. Whereas, with increase in the incident energy, rare phenomena such as multifragmentation [15, 16, 17, 18, 19] as well as associated phenomena such as collective flow and elliptic flow [20, 21, 22] can also emerge. Due to large excitation energy, the fine structures such as shell effects, shape of a nucleus as well as quantum features don't play significant role at higher incident energies where multifragmentation is a dominant channel. These effects, however, can play a role at low excited matter where fusion has a large share.

In recent years, one has seen renewed interest in the field of fusion. This renewed interest has been due to availability of neutron rich nuclei and synthesis of super heavy elements. In this direction, Puri and coworker [4, 5, 6, 7] studied the fusion barrier heights and positions by using various proximity potentials. Their studies revealed that out of different proximity potentials, Akyüz Winther (AW) 95, Bass 80 and Denisov DP reproduce the experimental data better than others. These potentials nicely explain the fusion probabilities at above barrier energies. In addition, many other proximity based potentials are also available in the literature [8, 9, 10, 11, 12, 13, 14]. In all above calculations, nuclei were assumed to be spherical, therefore, deformation was neglected in the nuclear as well as Coulomb potential. Though many calculations have been reported for deformed nuclear potentials, only few address the question of deformed Coulomb potential [23]. Our present aim is to use the best adjusted proximity based potentials along with deformed Coulomb potential to see if deformed Coulomb potential plays any role at different orientations of the colliding nuclei or not. In present study, the nuclear potential is assumed to be independent of deformation effects.

In present work, we employ AW 95, Bass 80 and Denisov DP to study the effect of oblate and prolate quadruple deformations (in the Coulomb potential) on the fusion process.

2. The Model

The nuclear part of the interaction potential $V_N(R)$ between two colliding surfaces, taken for present analysis, is derived using proximity approach and is mentioned in Refs. [4, 5, 6, 7]. The expression for the Coulomb interaction $V_C(R)$ between two deformed colliding nuclei is given by Wong [24] as

$$\begin{aligned} V_C(R,\theta) &= \frac{Z_1Z_2e^2}{R} + \sqrt{\frac{9}{20\pi}}\frac{Z_1Z_2e^2}{R^3}\sum_{i=1}^{2}[R_i(\theta_i)]^2\beta_{2i}P_2(cos\theta_i) \\ &+\frac{3}{7\pi}\frac{Z_1Z_2e^2}{R^3}\sum_{i=1}^{2}[R_i(\theta_i)]^2[(\beta_{2i}P_2(cos\theta_i)]^2. \end{aligned} \quad (1)$$

where Z_1 and Z_2 are their respective charges, R is the internuclear distance, θ_i is the angle of orientation (taken anticlockwise) and is defined as the angle between the axis of collision and symmetry axis as shown in figure 1. $R_i(\theta_i)$ is the nuclear radius for a deformed nucleus which is given by

$$R_i(\theta_i) = R_i[1 + \sqrt{\frac{5}{4\pi}}\beta_{2i}P_2(cos\theta_i)], \quad (2)$$

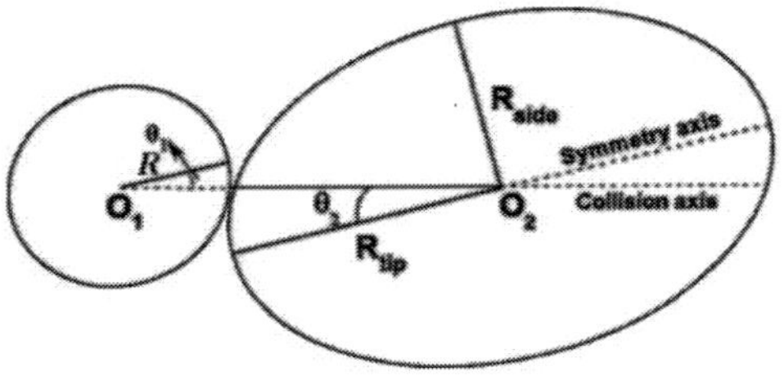

Figure 1. Collision of a spherical and prolate-deformed nucleus in the same plane (*i.e.* ϕ=0).

R_i being the effective sharp radius of *ith* nucleus. The different forms of effective sharp radius used in case of AW 95, Bass 80 and Denisov DP are given in Ref. [7]. The parameter β_{2i} corresponds to quadruple deformation for *ith* nucleus. In general, the quadrupole moment for a deformed nucleus is given by

$$\beta_2 = \frac{2}{5}Z(b^2 - a^2), \quad (3)$$

where Z represents atomic number of a nucleus, a and b in the equation represent semi-major and semi-minor axis, respectively. For spherically symmetric nucleus, $\beta_2 = 0$ (i.e. $b = a$). In a prolate deformed nucleus, $\beta_2 > 0$ (i.e. $b > a$), and in oblate deformed nucleus $\beta_2 < 0$ (i.e. $b < a$). The quantity $P_2(cos\theta_i)$ is the legendre polynomial for n=2. Since the colliding nuclei can have many possible orientations (θ_1, θ_2) during the experiment, therefore, we take different cases of orientations of two deformed colliding nuclei while studying the fusion between them. We restrict to the condition that both the deformed colliding nuclei are lying in the same plane (*i.e.* $\phi = 0$). Therefore, the Coulomb potential calculated above is independent of angle ϕ. As we know that the quadruple deformation (β_{2i}), higher order deformations (β_{3i}, β_{4i}), etc., satisfy the relation $\beta_2^{2i} \approx \beta_{l\geq 3}$, therefore, only the terms which are linear and quadratic in quadruple deformation of nuclei are considered and the terms including higher order deformations are neglected [25, 26]. The values of quadruple deformation parameter (β_{2i}) for present study are taken from Ref. [27].

Adding Coulomb potential to the nuclear part, we get the total interaction potential which, for zero value of orbital momentum, is given by

$$V_T(R) = V_N(R) + V_C(R, \theta). \tag{4}$$

These models are used to calculate fusion barrier heights and positions. For further details of these models, the reader is referred to Refs. [4, 5, 6, 7].

3. Results and Discussion

For the present study, the reactions of ^{30}Si+^{62}Ni, ^{30}Si+^{24}Mg, ^{27}Al+^{70}Ge and ^{24}Mg+^{24}Mg are considered. These reactions are considered because, the value of deformation parameter for the above mentioned colliding nuclei is large enough to see the effect of deformation on the total interaction potential.

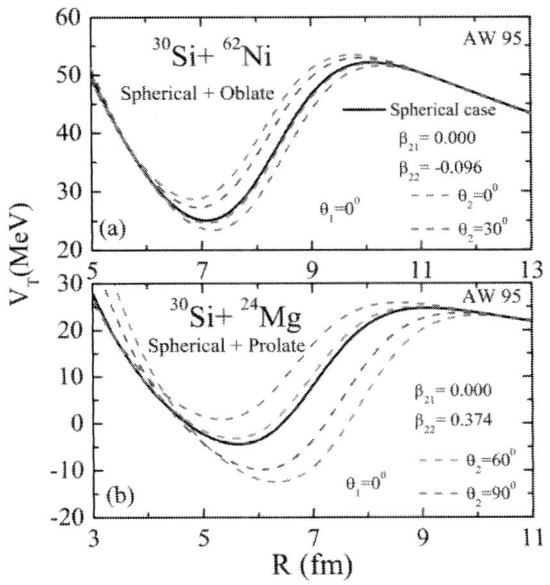

Figure 2. (Color online) The total interaction potential V_T(MeV) is displayed as a function of inter-nuclear distance R (fm) for the reactions of ^{30}Si + ^{62}Ni (see upper panel) and ^{30}Si + ^{24}Mg (see lower panel). Only AW 95 based calculations are displayed here. Similar trends are observed with Bass 80 and Denisov DP (not shown here). Further details are given in the text.

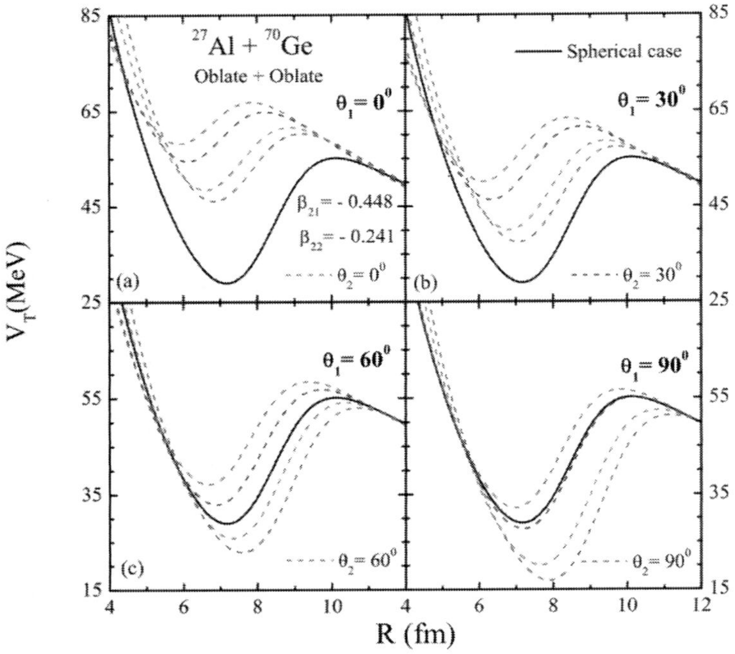

Figure 3. (Color online) Same as figure 2, but for the reaction of ^{27}Al + ^{70}Ge. Here both colliding nuclei are oblate-deformed. Various lines have same meaning as in figure 2.

In figure 2, we display the total interaction potential V_T(MeV) as a function of internuclear distance R (fm) for spherical (projectile)-deformed (target) systems. Here, black solid line corresponds to the total interaction potential calculated using AW 95, assuming both colliding nuclei to be spherical. The red, blue, magenta and olive dashed lines correspond to different orientations of the colliding nuclei (taking deformation into account), *i.e.* $(\theta_1 = 0^0, \theta_2 = 0^0)$, $(\theta_1 = 0^0, \theta_2 = 30^0)$, $(\theta_1 = 0^0, \theta_2 = 60^0)$ and $(\theta_1 = 0^0, \theta_2 = 90^0)$, respectively.

We find that the effect of adding deformation to Coulomb potential on total interaction potential in the inner region is significant, whereas, the effect in the tail region is negligible. It is clear from the figure 2(a) that the depth as well as width of the fusion pocket for ^{30}Si+^{62}Ni *i.e.* spherical projectile with oblate-deformed target ($R_x = R_y > R_z$) increase with increasing relative orientation of two deformed colliding nuclei. The maximum depth of fusion pocket is observed in case of $\theta_1 = 0^0, \theta_2 = 90^0$ orientation. Whereas, depth is minimum in case of $(\theta_1 = 0^0, \theta_2 = 0^0)$. This is because, an oblate-deformed target finds maximum height of fusion barrier while colliding along its longest axis with the spherical projectile and lowest barrier height while colliding along the shortest axis. However, in case of ^{30}Si+^{24}Mg (see figure 2(b)), *i.e.* spherical projectile with prolate-deformed target ($R_x = R_y < R_z$), the depth as well as width of the fusion pocket decrease with increasing relative orientation of two deformed colliding nuclei. The maximum depth of fusion pocket is observed in case of $(\theta_1 = 0^0, \theta_2 = 0^0)$. Whereas, depth is minimum in case of $(\theta_1 = 0^0, \theta_2 = 90^0)$. The reason is that a prolate-deformed target finds maximum height of fusion barrier while colliding along its shortest axis (side collision) with the

Journal of Physics: Conference Series **611** (2015) 012001 doi:10.1088/1742-6596/611/1/012001

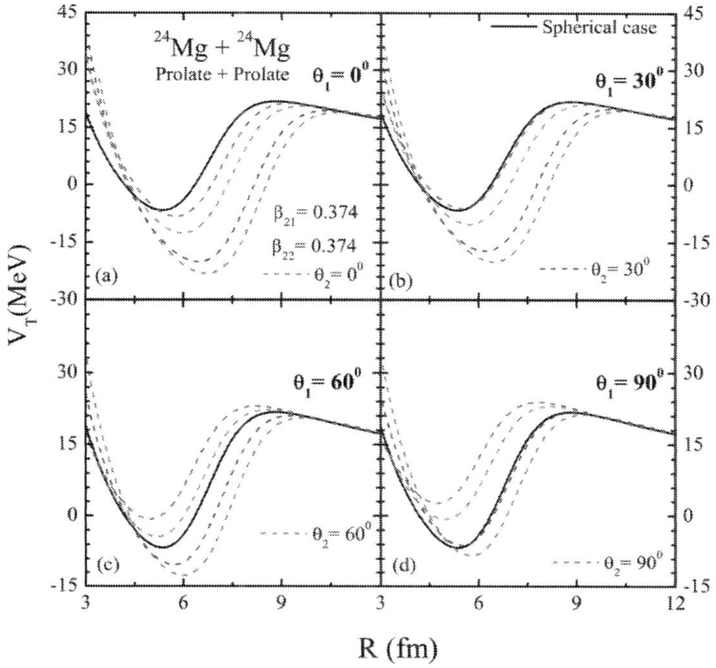

Figure 4. (Color online) Same as figure 2, but for the reaction of ^{24}Mg + ^{24}Mg. Here both colliding nuclei are prolate-deformed.

spherical projectile and lowest barrier height while colliding along the longest axis (tip collision).

In figure 3, we observe the behavior of total interaction potential as a function of internuclear distance for ^{27}Al + ^{70}Ge (oblate + oblate-deformed system) at different orientations of two colliding nuclei. A careful analysis of the four panels (3(a), (b), (c), and (d)), reveals that when oblate-deformed projectile is oriented at angles $\theta_1 = 0^0$ and $\theta_1 = 30^0$, the depth as well as width of interaction potential for all possible orientations ($\theta_1 = 0^0$ to $\theta_1 = 90^0$) of oblate-deformed target is less as compared to the spherical-spherical case. However, when oblate-deformed projectile is oriented at angles $\theta_1 = 60^0$ and $\theta_1 = 90^0$, the depth and width of fusion pocket start increasing with increasing orientation angles of oblate-deformed target and in most of the cases, these values are more as compared to that in spherical-spherical case. Out of different possibilities of orientation of two colliding nuclei, the lowest fusion barrier for the collision of oblate-deformed projectile with oblate-deformed target is observed at $\theta_1 = 90^0, \theta_2 = 90^0$ i.e. along the polar axis. The maximum repulsion for fusion is found for $\theta_1 = 0^0, \theta_2 = 0^0$ i.e. along the equatorial axis. Moreover, for a fixed orientation of oblate-deformed projectile (say $\theta_1 = 0^0$), the fusion barrier height decreases with increasing orientation angle of oblate-deformed target w.r.t. collision axis.

Similarly, the behavior of total interaction potential as a function of internuclear distance is studied for prolate+ prolate-deformed systems in figure 4. On carefully analyzing the four panels (4(a), (b), (c), and (d)), we find that when prolate-deformed projectile is oriented at angles $\theta_1 = 0^0$ *and* $\theta_1 = 30^0$, the depth as well as width of interaction potential for all possible

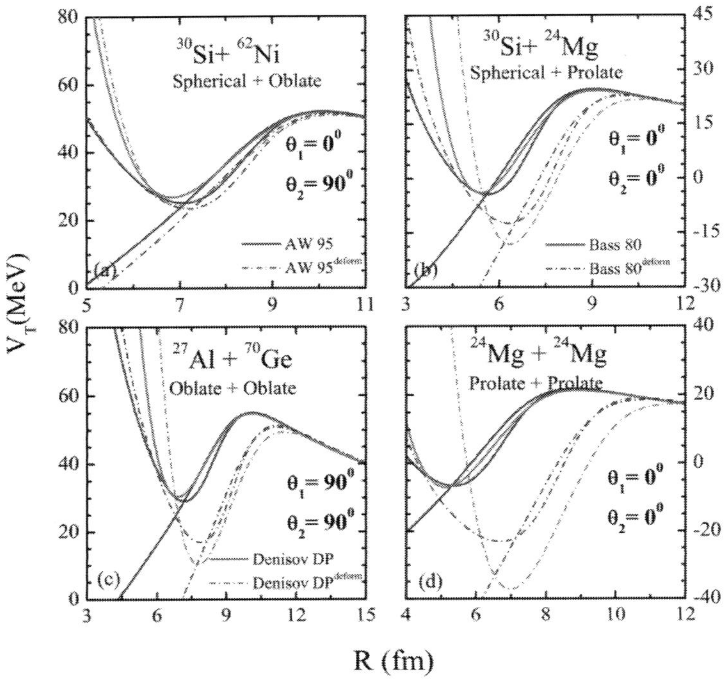

Figure 5. (Color online) Same as figure 2, but for the reactions of ^{30}Si + ^{62}Ni and ^{30}Si + ^{24}Mg in the upper panels and for ^{27}Al + ^{70}Ge and ^{24}Mg + ^{24}Mg in the lower panels.

orientations ($\theta_1 = 0^0$ *to* $\theta_1 = 90^0$) of prolate-deformed target is more as compared to the spherical-spherical case and this trend is exactly opposite to that observed in figure 3. Also, when oblate-deformed projectile is oriented at angles $\theta_1 = 60^0$ and $\theta_1 = 90^0$, the depth and width of fusion pocket start decreasing with increasing orientation angles of oblate-deformed target. Out of different possibilities of orientation of two colliding nuclei, the lowest fusion barrier for the collision of prolate-deformed projectile with prolate-deformed target is observed at $\theta_1 = 0^0, \theta_2 = 0^0$ i.e. tip collision. Maximum repulsion for fusion is found for $\theta_1 = 90^0, \theta_2 = 90^0$ i.e. along the equatorial axis. Moreover, for a fixed orientation angle of prolate-deformed projectile (say $\theta_1 = 0^0$), the fusion barrier height increases with increasing orientation angle of prolate-deformed target w.r.t. collision axis. We observe exactly opposite behavior of total interaction potential with internuclear distance for prolate + prolate-deformed systems compared to oblate + oblate-deformed systems.

In figure 5, a comparative study is conducted using AW 95, Bass 80 and Denisov DP. Here, purple, wine and pink solid lines represent the calculations for spherical-spherical cases using AW 95, Bass 80 and Denisov DP, respectively. Whereas, the calculations including deformations using AW 95, Bass 80 and Denisov DP are represented by purple, wine and pink dash-dotted lines, respectively. For different deformed shapes of the two colliding nuclei, the interaction potential is displayed at only those orientation angles at which fusion barrier are found to be minimum. From last three figures i.e. figures. 2, 3 and 4, we find that for spherical + oblate (see figure 2(a)), spherical + prolate (see figure 2(b)), oblate + oblate (see figure 3(d)) and prolate + prolate systems (see figure 4(a)), the minimum fusion barrier is at orientation angle ($\theta_1 = 0^0, \theta_2 = 90^0$), ($\theta_1 = 0^0, \theta_2 = 0^0$), ($\theta_1 = 90^0, \theta_2 = 90^0$) and ($\theta_1 = 0^0, \theta_2 = 0^0$), respectively. All three proximity based potentials i.e. AW 95, Bass 80 and Denisov DP show similar trend for interaction potential at critical orientations (corresponding to minimum depth of fusion pocket) of the colliding nuclei w.r.t. collision axis. In all the four cases (figs. 5(a), (b), (c) and (d)), the interaction potential becomes less repulsive or more attractive on including the deformed Coulomb potential (at above mentioned orientation angles) in the total interaction potential.

4. Summary

We studied the role of deformed Coulomb potential in fusion process by employing proximity based potentials due to AW 95, Bass 80 and Denisov DP. This detailed study reveals that shapes of the colliding nuclei affect the interaction potential in the inner region in different manner. In addition, the orientations of the nuclei participating in a reaction significantly affect the fusion barrier. The prolate-deformed colliding nuclei find minimum repulsion along tip collision, whereas, the oblate-deformed colliding nuclei find lowest fusion barrier along polar axis. These conclusions can be helpful in deciding the conditions for the fusion of deformed nuclei or synthesis of super-heavy elements during experiments.

Acknowledgement

This work is supported by financial grant from Department of Science and Technology (DST), Government of India. R. K. also acknowledges research grant from Council of Scientific and Industrial Research (CSIR), New Delhi. The author is thankful to Professor Rajeev K. Puri for insightful discussions on the present work.

References

[1] Li G.-Q. and Xu G.-O. 1989 *Phys. Rev.* C **39** 276
[2] Li G.-Q. and Xu G.-O. 1990 *Phys. Rev.* C **41** 169
[3] Zhang J *et al* 2013 *Nucl. Phys.* A **909** 36
[4] Dutt I and Puri R K 2010 *Phys. Rev.* C **81** 044615
[5] Dutt I and Puri R K 2010 *Phys. Rev.* C **81** 047601
[6] Dutt I and Puri R K 2010 *Phys. Rev.* C **81** 064608
[7] Dutt I and Puri R K 2010 *Phys. Rev.* C **81** 064609
[8] Puri R K, Chattopadhyay P and Gupta R K 1991 *Phys. Rev.* C **43** 315
[9] Puri R K and Gupta R K 1992 *Phys. Rev.* C **45** 1837
[10] Puri R K and Gupta R K 1995 *Phys. Rev.* C **51** 1568
[11] Puri R K, Sharma M K and Gupta R K 1998 *Eur. Phys. J* A **3** 277
[12] Arora R, Puri R K and Gupta R K 2000 *Eur. Phys. J* A **8** 107
[13] Puri R K and Dhiman N K 2005 *Eur. Phys. J* A **23** 429
[14] Puri R K, Arora R and Gupta R K 1999 *Phys. Rev.* C **60** 054619
Kumari R 2013 *Nucl. Phys.* A **917** 85
[15] Vermani Y K, Goyal S and Puri R K 2009 *Phys. Rev.* C **79** 064613
[16] Vermani Y K and Puri R K 2009 *Europhys. Lett.* **85** 62001
[17] Vermani Y K *et al* 2010 *J. Phys. G: Nucl. Phys.* **37** 015105
[18] Vermani Y K and Puri R K 2009 *J. Phys. G: Nucl. Phys.* **36** 105103
[19] Kaur S and Puri R K 2013 *Phys. Rev.* C **87** 014620
Kaur S and Puri R K 2014 *Phys. Rev.* C **89** 057603
Kumar R, Gautam S and Puri R K 2014 *Phys. Rev.* C **89** 064608
[20] Kumar S, Sharma M K and Puri R K 1998 *Phys. Rev.* C **58** 3494
[21] Gautam S *et al* 2010 *J. Phys. G: Nucl. Phys.* **37** 085102

[22] Bansal R *et al* 2013 *Phys. Rev.* C **87** 061602
Bansal R *et al* 2014 *J. Phys. G : Nucl. Phys.* **41** 035103
[23] Phookan C K and Kalita K 2013 *Nucl. Phys.* A **899** 29
[24] Wong C Y 1973 *Phys. Rev. Lett.* **31** 766
[25] Denisov V Yu and Pilipenko N A 2007 *Phys. Rev.* C **76** 014602
[26] Denisov V Yu and Pilipenko N A 2008 *Ukr. J. Phys.* **53** 845
[27] Moller P and Nix J R 1995 *Atomic and Nuclear Data Tables* **59** 185

Gamma and neutron attenuation properties of barite-cement mixture

R Picha[1], J Channuie, S Khaweerat, T Liamsuwan, J Promping, W Ratanatongchai, K Silva, and S Wonglee

Thailand Institute of Nuclear Technology, Ongkharak, Nakorn Nayok, Thailand

E-mail: aeroppon@gmail.com

Abstract. For the neutron radiography facility renovation plan at Thai Research Reactor, mixed barite-concrete blocks of different compositions were tested for their photon and neutron radiation attenuation properties. ^{60}Co and ^{137}Cs isotopes were used as the gamma sources; ^{241}Am-Be was used as the neutron source. For detection, a scintillation counter and a BF_3 tube were used. The intensities at various energies were measured and attenuation coefficients were calculated. Samples of barite mixture were analyzed with X-ray. The results involving the effects of barite are reported and discussed.

1. Introduction

Nuclear radiation can be generated from many sources, natural and man-made, such as radioisotopes, reactors, accelerators. Such radiation is widely used in many fields such as nuclear medicine, medical diagnosis, industrial tracing and radiography. Besides its beneficial effects, radiation can also cause harm. Adequate protection is thus the first priority when handling radiation sources.

Neutron radiography (NR) experiment at Thai Research Reactor 1/Modification 1 (TRR-1/M1) is one of several beam experiment facilities at Thailand's only research reactor. NR has been in operation for more than 20 years. Its current wall setup is shown in figure 1. Due to dislocation of concrete blocks, there has been an initiation to renovate the facility with several improvements including new concrete walls.

When the beam shutter is opened, neutrons from the reactor core travel along the neutron beam tube (left side of the figure) inside a wall. The beam is then incident on a sample before being absorbed into the beam stopper embedded in the concrete wall. For the purpose of increasing its potential to conduct future research, the facility is currently undergoing renovation stages, which involve replacing the beam shutter, the shielding walls, and the imaging system. Based on recent literature [1-3], barite concrete blocks were tested for their radiation shielding properties. Another available aggregate, copper slag, which seemed to possess desirable characteristics in terms of strength and gamma shielding [4-5], was also used for comparison.

In this work, the process using these concrete samples as shielding material for gamma and neutron radiation is discussed. The content is organized in the following manner. In section 2, the methods of study are described. Section 3 goes through the results on concrete properties. Finally, the summary and conclusion are given in section 4.

[1] To whom any correspondence should be addressed.

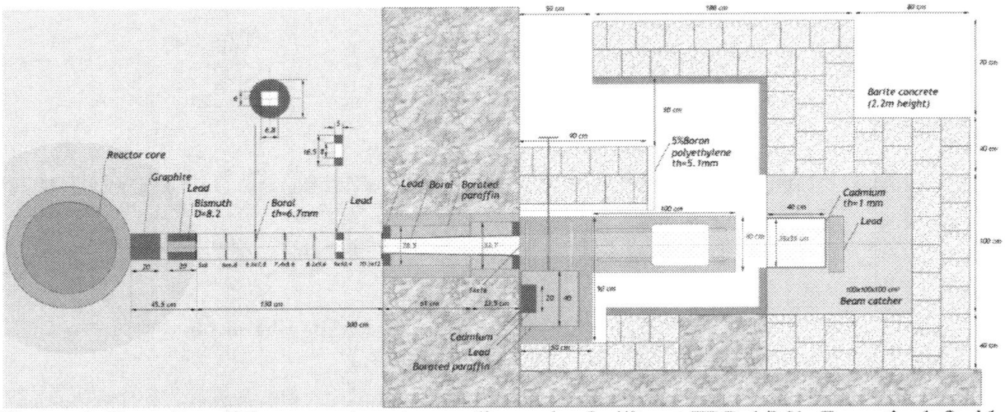

Figure 1. Top view of the current neutron radiography facility at TRR-1/M1. From the left side to the right side is the reactor core inside a water pool, the beam tube consisting of graphite, lead, bismuth, and boral, the reactor wall with a beam tube inside, the beam shutter, the sample staging area, beam stopper, and the concrete shielding walls.

2. Materials and Methods

The reactor neutron radiography system consists of six main parts: 1) the neutron source, 2) the beam shutter, 3) the collimator, 4) the sample stage, 5) the image capture system, and 6) the shielding walls. In the overall renovation plan, several improvements are for each of these components. The neutron source in this setup comes is the fission reaction of uranium-235. The beam shutter, when in the "Closed" position, is used as the first line of defense against neutron beam. The collimator is used for focusing the beam, and can be used as secondary safety equipment. It is one of the parts not currently being used but planned for the future. The sample stage is also planned to be improved so that it can be remotely controlled. For the image capture system, there are several options: X-ray films, digital imaging plates, or a camera. Finally, shielding walls are planned to be built with a known composition to replace the current walls. To design and evaluate the composition for the new shielding walls, Monte Carlo simulation and actual measurements were conducted. For the shielding wall blocks, barite powder and copper slag was mixed with cement at different ratios. The blocks were tested with gamma and neutron sources.

For characterization, samples were also tested using X-ray sources via fluorescence and diffraction techniques.

2.1. Concrete composition

Barite ($BaSO_4$) powder was formed into cement mortar. For the first batch, four different barite concrete (BaCon) formulae with different barite-to-cement ratios were used. The measured densities are between 2.7-2.9 g cm^{-3}. The compositions are shown in table 1.

Table 1. Weight compositions of barite concrete (BaCon) samples.

formula	cement	fine barite	water	density (g/cm^3)
BaCon 1	280	3090	435	2.91
BaCon 2	310	3050	430	2.72
BaCon 3	350	3000	420	2.91
BaCon 4	400	2940	420	2.78

Another batch of concrete was mixed using copper slag as aggregate, along with a smaller fraction of barite powder. Compositions of copper slag concrete (CuslCon) are shown in table 2.

Table 2. Weight compositions of copper slag concrete samples.

formula	cement	fine barite	agg Cu slag	water	density (g/cm^3)
CuslCon 1	400	400	2050	200	2.68
CuslCon 2	400	800	1700	200	3.04
CuslCon 3	400	1200	1350	200	3.06
CuslCon 4	400	1600	1000	200	2.92

2.1.1. X-ray analysis

Concrete samples were analyzed with X-ray fluorescence (XRF) and X-ray diffraction (XRD) in powder forms. XRF was done using PANalytical Epsilon 5 spectrometer with energy range 25-100 keV (atuo adjust). XRD was done using a Bruker device, operated at 30 mA and 30 kV. From XRD spectra (figure 2), $BaSO_4$ orthorhombic peaks were well identified ($a{:}b{:}c$ = 8.792:5.430:7.110, α=90°, β=90°, γ=90°). XRF results are listed in table 3. Composition of copper slag aggregate is shown in table 4.

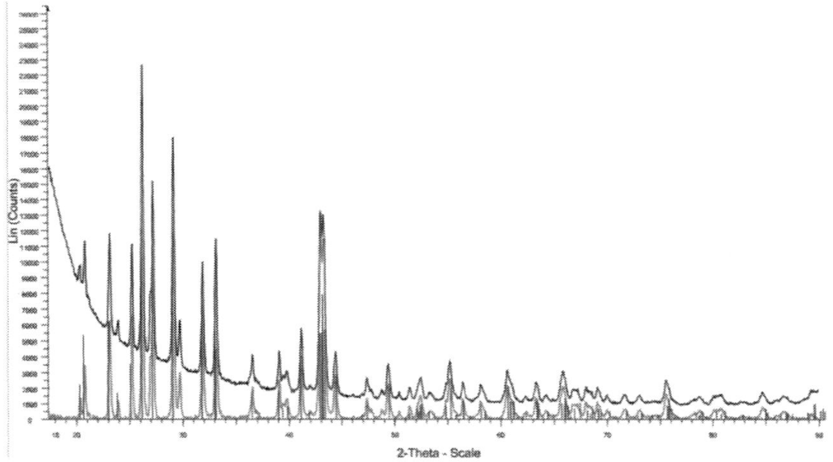

Figure 2. XRD spectrum of one of the BaCon samples (formula 1). Top spectrum is raw measurement. Lower spectrum has background subtraction applied. Vertical lines are barite peaks.

Table 3. Molecular concentration of barite concrete (BaCon) from X-ray fluorescence analysis. The values are normalized (total 100%).

formula	BaO (%)	SO_3 (%)	Sum (%)	Statistical error (%)
1	50.8	23.5	74.2	0.4
2	50.0	23.0	73.0	0.4
3	49.1	22.2	71.4	0.4
4	48.0	21.5	69.5	0.4

Table 4. Major molecular constituents of copper slag aggregate from X-ray florescence analysis. The values are normalized (total 100%).

compound	Fe_2O_3	SiO_2	Al_2O_3	ZnO	CuO	CaO
concentration (%)	74.752	13.396	2.644	2.275	1.609	1.324

2.2. Shielding tests
Monte Carlo simulation was performed along with actual measurements. In the measurements, analyses were carried out using two gamma sources providing know energies. The sources used were Cs-137 and Co-60. After energy spectra were obtained, attenuation coefficients (μ) at three energies were calculated using the available 0.661, 1.173, and 1.333 MeV photopeaks.

2.2.1. Monte Carlo simulation. Simple simulation setup was studied using PHITS (Particle and Heavy Ion Transport code System), a general purpose Monte Carlo particle transport code created by Japan Atomic Energy Agency (JAEA). The setup is shown in figure 3. Two types of sources were used to study the shielding capability of the materials. First one is a 4-MeV gamma source and the second is a 2-MeV neutron source. These energies are assessed to be the maximum energies in the actual radiography setup. Source cross-section area is 10x10 cm^2. Density between barite concrete and copper slag concrete are kept the same in the simulation (3.35 g/cm^3). In actual experiment, the densities are likely different. For the weight composition, copper-slag concrete consists of 45% copper slag (about 60% of it is Fe_2O_3), 40% barite, 10% cement, and 5% water, whereas barite concrete consists of approximately 46% Ba and 5% Fe.

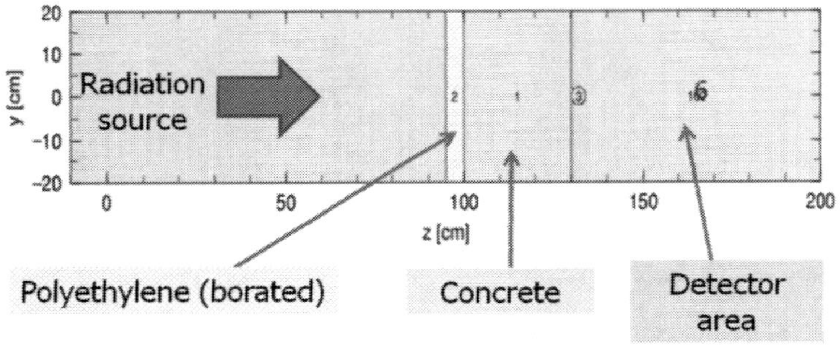

Figure 3. Geometry and location of radiation source, materials, and detector used in Monte Carlo simulation. The concrete volume is changed between different concrete compositions.

2.2.2. Gamma measurement. Lanthanum bromide scintillation detector (Canberra LABR-1.5x1.5) was used to measure gamma signals. A simple linear setup with the source and the detector separated by 40 cm was used, as shown in figure 2. Partial collimation using lead blocks was applied. The counting time was set at 5 minutes per sample. The measured count rates and percent transmission between different concrete compositions are shown in figure 5. It can be seen that the gamma shielding properties of the new concrete samples are comparable to the old one.

Journal of Physics: Conference Series **611** (2015) 012002 doi:10.1088/1742-6596/611/1/012002

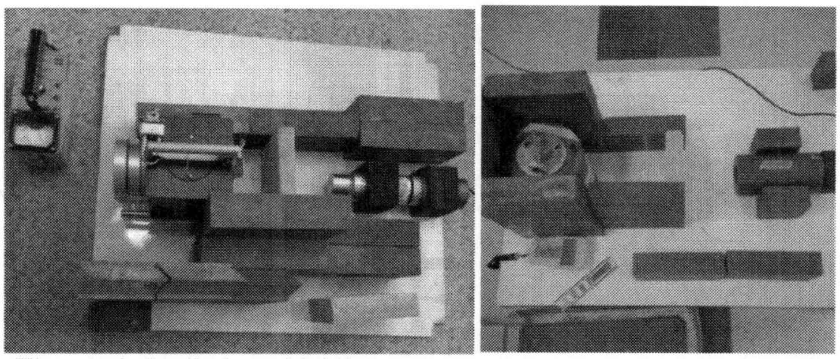

Figure 4. Radiation shielding test setup for Cs-137 source (left) and Co-60 source (right) and LABR setup.

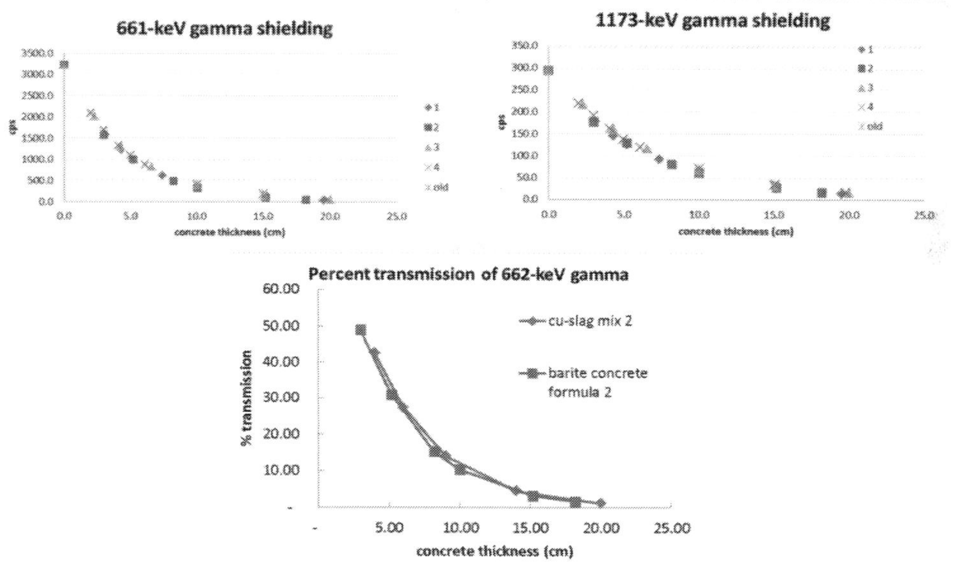

Figure 5. (Top row) Gamma count rates for different BaCon compositions at different values of concrete thickness, and (Bottom row) percent transmission of gamma ray through BaCon formula 2 and CuslCon formula 2.

The gamma intensity is related to the material thickness according to Beer-Lambert law:

$$I(x) = I(0)e^{-\mu x} \quad (1)$$

where I is the intensity as a function of thickness x and μ is the energy-dependent attenuation coefficient.

2.2.3. Neutron measurement. Boron trifluoride (BF_3) detector operated with a high voltage supply of 1,800 V was used to detect neutrons. A 50-Ci Am-241/Be neutron source with a collimator was used as the neutron source, with the setup shown in figure 6. This newly arranged facility at TINT, Nakorn Nayok, is aimed for nuclear analytical techniques and irradiation applications. Source-to-detector distance was set at 60 cm. The counting time was 1 minute. Beer-Lambert law is also applied to the data to find out the macroscopic cross section Σ.

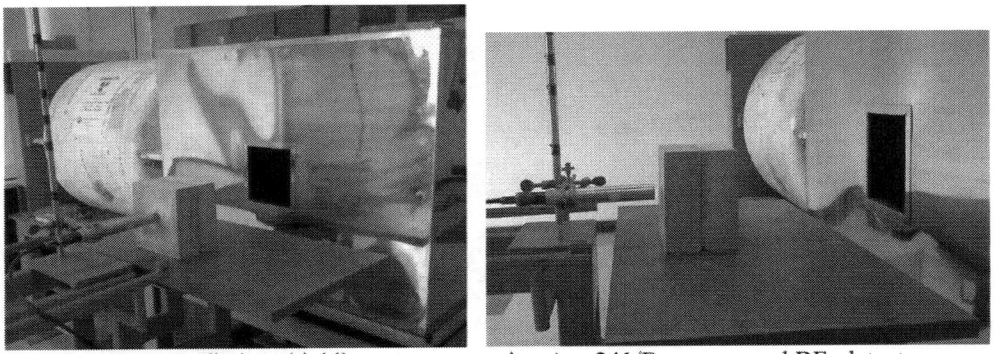

Figure 6. Radiation shielding test setup using Am-241/Be source and BF_3 detector.

3. Results and Discussions

Simulation results using the Monte Carlo method are shown in table 5. Two different concrete thicknesses and two radiation sources were used. Compared to the barite concrete, the Cu-slag concrete is predicted to yield similar results in gamma shielding and much better results in neutron shielding.

Table 5. Fluence ratio of Cu-slag concrete to barite concrete from simulation.

concrete thickness	radiation source	fluence ratio of Cu-slag concrete to barite concrete, computed in air space behind concrete	
		neutron	gamma
30 cm	4-MeV gamma	-	1.107
40 cm	4-MeV gamma	-	1.117
30 cm	2-MeV neutron	0.533	1.220
40 cm	2-MeV neutron	0.383	0.951

From the shielding tests, gamma ray attenuation coefficients decrease as the gamma energy increases. For both gamma and neutron shielding properties, there is not a large variation in the coefficient values across different concrete formulae. The coefficients are summarized in table 6. The calculated energy-specific gamma attenuation coefficients are similar across all concrete formulae. With the current neutron setup, the measurements are inconsistent with the simulation predictions. A future experiment setup incorporating high-density polyethylene sleeves for neutron thermalization is planned.

Table 6. Radiation attenuation coefficients of different concrete formulae. Subscript of μ indicates the gamma energy.

formula	*gamma*			*neutron*
	$\mu_{0.661}$ (cm^{-1})	$\mu_{1.173}$ (cm^{-1})	$\mu_{1.333}$ (cm^{-1})	Σ (cm^{-1})
BaCon 1	0.223±0.001	0.157±0.001	0.146±0.001	0.113±0.031
BaCon 2	0.229±0.001	0.156±0.001	0.147±0.001	0.112±0.032
BaCon 3	0.212±0.001	0.145±0.001	0.136±0.001	0.101±0.024
BaCon 4	0.209±0.001	0.144±0.001	0.140±0.002	0.113±0.034
current	0.200±0.011	0.150±0.001	0.141±0.001	0.100±0.006
CuslCon 1	0.185±0.002			0.025±0.002
CuslCon 2	0.222±0.001			0.024±0.002
CuslCon 3	0.212±0.002			0.025±0.002
CuslCon 4	0.206±0.001			0.023±0.002

4. Conclusions

For the purpose of shielding neutron and gamma rays in the neutron radiography area at research reactor TRR-1/M1, concrete blocks, formed by mixing in barite and copper slag were formulated and constructed. Characterization was done using XRF and XRD techniques. The results from the newly formulated concrete (except BaCon 3 and 4) are comparable to that of the current concrete in terms of gamma shielding. Similar gamma shielding properties are observed for the two concrete types. Results agree qualitatively with the prediction from Monte Carlo simulation. Copper slag concrete is theoretically more effective than barite concrete in terms of neutron shielding. However, using the existing experimental setup, the neutron measurements yield the opposite results. New measurements using high-density polyethylene sleeves, which were shown to thermalize neutrons well, are planned for the future study. For the shielding of gamma rays, barite concrete is an effective material. However, the brittleness of the aggregate-free formulae keeps them from practical applications. Experiments with other compositions and larger aggregate size will be conducted at the next stage. Besides using material with good shielding properties, it is equally essential to have good safety equipment such as dose monitoring detectors clearly displaying the dose rate in the area.

Acknowledgments

Authors acknowledge assistance from Office of Atoms for Peace for providing a Cs-137 source and TINT Nuclear Technology Service Center for a Co-60 source and XRF analysis of the samples. This ongoing project is part of MOU between TINT and Siam City Cement Company (SCCC) for radiation-shielding concrete development.

References

[1] Akkurt I, Basyigit C, Kilincarslan S, and Mavi B 2005 The shielding of gamma-rays by concretes produced with barite. *Prog. Nucl. Energ.* **46** 1-11

[2] Akkurt I, Akyildirim H, Mavi B, Kilincarslan S, and Basyigit C 2010 Gamma-ray shielding properties of concrete including barite at different energies *Prog. Nucl. Energ.* **52** 620-623

[3] Esen Y and Yilmazer B 2011. An investigation of x-ray and radio isotope energy absorption of heavyweight concretes containing barite *Bull. Mater. Sci.* **34** 1 169-175

[4] Al-Jabri K S, Hisada M, Al-Oraimi S K, and Al-Saidy A H 2009 Copper slag as sand replacement for high performance concrete *Cement. Concrete. Comp.* **31** 483-488

[5] Shi C, Meyer C, and Behnood A 2008 Utilization of copper slag in cement and concrete *Resour. Conserv. Recy.* **52** 1115-1120

Investigation of toroidal flow effects on *L-H* transition in tokamak plasma based on bifurcation model

B Chatthong[1] and T Onjun
School of Manufacturing Systems and Mechanical Engineering, Sirindhorn International Institute of Technology, Thammasat University, Pathum Thani, Thailand

E-mail: boonyarit.chatthong@gmail.com

Abstract. The impacts of toroidal flow on the *L-H* transition phenomenon in tokamak plasmas based on bifurcation concept are investigated. A set of pressure and density transport equations with both neoclassical and anomalous effects included is considered. The anomalous transport suppression mechanism considered is only from a flow shear, which is calculated based on a force balance equation with both pressure gradient and toroidal velocity components included. The toroidal velocity can be calculated using four different models. The first model is an empirical model in which the velocity is dependent on a local ion temperature. The second model is based on neoclassical toroidal viscosity theory in which the velocity is driven by ion temperature gradient. In the third model, the velocity is dependent on current density flow in plasma. The fourth model is essentially the third model including effect of an intrinsically generated bootstrap current. It is found that inclusion of toroidal velocity can substantially increase the plasma pressure and density, mainly due to an increase of the pedestal width. It is also found that the pedestal for pressure tends to form first. After the pedestal forms, it expands inwards with the characteristic of super-diffusive nature in initial state and become sub-diffusive nature in final state before reaching steady state. The expansion speed depends sensitively on the strength of flow shear effect. It is also found that the time required plasma to reach steady state after the *L-H* transition is much longer than that for *L-H* transition.

1. Introduction

Discovery of high confinement mode (*H*-mode) in tokamak plasma is one of the greatest achievement in fusion research [1]. Unlike low confinement mode (*L*-mode), the *H*-mode can provide significant improvement in plasma performance i.e. high central temperature, high density and pressure, and sufficient long enough confinement time. Consequently, a fusion output power is largely increased. For that reason, a modern fusion project, such as International Thermonuclear Experimental Reactor (ITER), is designed based on *H*-mode operation [2]. Evidently, the *H*-mode plasma occurs as a result of formation of an edge transport barrier (ETB), which is defined as a local narrow region near the edge of plasma, called a "pedestal" [3]. The pedestal has several key characteristics, such as a transport reduction and relatively strong gradients. The sharp gradient at plasma edge causes the plasma profiles at its core to rise up because of its stiffness property. The success of ITER depends largely on the achievement of *H*-mode plasma and its performance level relies on both the pedestal values and pedestal width. Historically, the *H*-mode plasma was first discovered over 30 years ago in

[1] To whom any correspondence should be addressed.

ASDEX, Germany [1]. Since then, several tokamaks around the world have also been able to achieve the mode. Experimentally, it is known that the plasma can reach *H*-mode if two criteria are satisfied; firstly, the operation is equipped with either a divertor or a limiter, and secondly, plasma heat injection, regardless of heating method, surpasses a threshold value. When these two criteria are met, the plasma makes a sudden transition from *L*-mode to *H*-mode, called *L-H* transition. Recently, an intermediate confinement mode (*I*-mode) occurring during the transition was discovered and it has been extensively investigated [4]. The mechanism of this mode is still not clear. Therefore, it will not be considered in this work.

Even though the *H*-mode was found several decades ago and many experiments have successfully operated in this mode on a daily basis, the comprehensive understanding of physics behind *L-H* transition is still not completely developed. Many theoretical and empirical concepts revolve around the idea that the transition is caused by radial electric field in the plasma. Generally, it is known that the plasma transport is composed of neoclassical and anomalous contributions; the latter being much more dominant except possibly near plasma center. It was found from experiments that the total plasma transport is reduced to the predictable neoclassical value inside the transport barrier [5], which could imply that the anomalous transport is quenched. Theoretically, it is explained that the anomalous transport can be stabilized by flow velocity shear because of a convection cell breaking up[6]. Moreover, there are some experimental evident which show that the anomalous transport fluxes can be suppressed by a sheared flow within the local region of transport barrier [7-8]. Therefore, many ideas propose that the transport barrier forms as a result of anomalous transport suppression by magnetic shear or/and flow shear [7], which is related to the radial electric field. In general, the flow effect can have different impacts on an unstable mode, such as strong effect on ITG, but no effect on ETG.

Bifurcation concept can be used to explain formation of ETB and physics of *L-H* transition. Several previous works used this approach to either qualitatively or quantitatively model the transition and study barrier mechanics. Bifurcation concept is based on the idea that the *L-H* transition phenomenon is an intrinsic property of the plasma. Namely, there exists a regime in the plasma where two different plasma states can co-exist, i.e. low and high confinement states. This allows the plasma to abruptly bifurcate from one to another. The model is frequently illustrated as an *s*-curve, non-monotonic, in fluxes (heat/particle) versus gradients (pressure/density) space composing of one unstable and two stable, low and high, branches, as can be seen in reference. [9]. When the heat or particle flux reaches a certain threshold the plasma gradient jumps from low to high branch, implying *L-H* transition. The radial location of an ETB is represented where the plasma has high value of gradient. The work in this paper is based on the two-field bifurcation model analyzed in reference. [9] where heat and particle transport equations are studied simultaneously. The setback of the original model was the assumption that both neoclassical and anomalous diffusivities to be constant. Also, in calculation of radial electric field the toroidal and poloidal rotation terms are neglected. In this work, the anomalous diffusivities are calculated based on profiles stiffness model, which means that in the low gradients regimes, the plasma transport is governed solely by neoclassical effect, while in the high gradients regimes, the dominating anomalous transport is driven by the gradients [10]. Moreover, it was found that in some regimes of the plasma, toroidal rotation can significantly dominates the calculation of radial electric field as can be seen in reference. [11]. In addition, it is known that the toroidal flow can play significant role in plasma performance improvement [12]. So in this paper, effects of toroidal rotation are included for investigation.

In this work, the pressure and density transport equations are numerically solved simultaneously for spatiotemporal profiles of plasma pressure and density, *L-H* transition, pedestal width and its dynamics. Each equation is composed of neoclassical and anomalous transports, source term and the flow shear as suppression mechanism. The suppression only affects on the anomalous channel for the reason mentioned previously. Three calculations of toroidal velocity are used; the first one is based on empirical observation in which toroidal velocity is proportional to plasma temperature [12], the second one is based on neoclassical toroidal viscosity (NTV) physics [11], and the last one is based on empirical approximation [13]. The three models are based on different physical regimes in the plasma which are not necessary mutually exclusive so they can partially coexist. The effects of each model toward this bifurcation picture of *L-H* transition are studied and results are compared.

This paper is organized as follows: brief descriptions of bifurcation model as well as toroidal velocity models are presented in section 2; numerical results and analysis are discussed in section 3; and the conclusion is given in section 4.

2. Bifurcation concept and toroidal velocity models

This section introduces the models used in the work which include bifurcation and toroidal velocity models. The main transports equations are solved numerically using discretization method. The code is developed in MATLAB environment based on the built-in "*pdepe*" function which is appropriate for solving initial-boundary problems for parabolic-elliptic partial differential equations in one dimension.

2.1. Two-field bifurcation model

The model used in this work is based on the framework of bifurcation concept introduced in reference. [9]. It conceptually explains that the *L-H* transition is an intrinsic property of the plasma where its confinement mode can abruptly change providing that some criteria are satisfied, i.e. ratio of anomalous over neoclassical diffusivities and heat or particle fluxes must exceed threshold values [14]. Graphically, the model is depicted as bifurcation diagram illustrating an *s*-curve shape graph of pressure/density gradients against heat/particle fluxes, respectively. A jump of pressure and density gradients in the graph indicates that the plasma has entered *H*-mode. The model consists of coupled pressure and density transport equations which are expressed in slab geometry as shown:

$$\frac{3}{2}\frac{\partial p}{\partial t}-\frac{\partial}{\partial x}\left[\chi_{neo}+\frac{\chi_{ano}}{1+\alpha v_E'^2}\right]\frac{\partial p}{\partial x}=H(x)\,, \tag{1}$$

$$\frac{\partial n}{\partial t}-\frac{\partial}{\partial x}\left[D_{neo}+\frac{D_{ano}}{1+\alpha v_E'^2}\right]\frac{\partial n}{\partial x}=S(x)\,, \tag{2}$$

where p is the plasma pressure, n is the plasma density, χ_{neo} and D_{neo} represent thermal and particle neoclassical transport coefficients, respectively, χ_{ano} and D_{ano} represent thermal and particle anomalous transport coefficients, respectively, v_E' is the flow shear suppression, α is a positive constant representing strength of the suppression, and H(x) and S(x) are the thermal and particle sources. Essentially, these two equations represent conservation of energy and mass, respectively. Note that throughout this work, the thermal and particle sources are estimated to be localized at plasma center and edge, respectively, using Gaussian distribution. The slab geometry is justified because tokamak plasma is assumed to have toroidal and poloidal symmetries in circular shape limit. The stabilizing mechanic for anomalous transport is the flow shear v_E', which accounts for the known reduction of turbulent transport by sheared radial electric field [6]. It couples the two transport equations based on force balance calculation:

$$v_E' \propto E_r' = \left(\frac{p'}{n}-v_\theta B_\phi + v_\phi B_\theta\right)', \tag{3}$$

where v_θ and v_ϕ are the poloidal and toroidal velocities, respectively, and B_θ and B_ϕ are poloidal and toroidal magnetic fields, respectively. The poloidal rotation term, which is a result from neoclassical calculation, is neglected in this work. Whereas, the toroidal rotation term was found to be dominating in some plasma regime as illustrated in reference. [11] so its effect will be studied in this work because it was ignored in previous bifurcation related works. Models for toroidal velocity prediction are discussed in the next section. The neoclassical transport coefficients are simply set to be constant while the anomalous transport coefficients follows critical gradient transport model similar to that described in reference. [10], with the functional forms:

$$\chi_{ano} = c_{\chi} (p' - p'_c) \mathrm{H}(p' - p'_c), \tag{4}$$

$$D_{ano} = c_{D} (n' - n'_c) \mathrm{H}(n' - n'_c), \tag{5}$$

where c_{χ} and c_D are proportional constants, p'_c and n'_c are the critical gradients for pressure and density fields, respectively, and H represents a Heaviside step function. This local anomalous transport model realizes the physical plasma transport in which at low gradient, near plasma center, the neoclassical effect dominates the transport, while at high gradient, near plasma edge or transport barrier location, the anomalous transport is driven by the gradient itself.

2.2. Toroidal velocity models

This section discusses the toroidal velocity models used in this paper. Four different toroidal velocity models are employed, each governs different physics in tokamak plasma. These models provide explicit calculations of toroidal velocity which means the toroidal momentum equation is not included. This requires extensive modelling and analysis, and is left for future work. Model a) is based on an empirical observation which assumes that the toroidal flow is proportional to local plasma temperature T [12]:

$$v_{\phi,a} \propto T. \tag{6}$$

This model is applicable in tokamak which is directly driven by on-axis neutral beam injection (NBI) heating, for example JET. Model b) is based on neoclassical toroidal viscosity (NTV) physics which applies to plasma with symmetry breaking via the application of a non-axisymmetric field [11]. It has the form:

$$v_{\phi,b} \propto \frac{\partial T}{\partial r}. \tag{7}$$

So, the toroidal velocity in this model is proportional to the gradient of local plasma temperature gradient. Model c) is an empirical estimation discussed in Ref. [13], where the toroidal velocity is assumed to be proportional to plasma toroidal current density:

$$v_{\phi,c} \propto j_\phi = j_0 \left(1 - \frac{r^2}{a^2}\right)^{\nu}, \tag{8}$$

where j_ϕ and j_0 are toroidal current density and its value at plasma center, r/a is normalized minor radius and ν is a constant to control current shaping. This form of plasma current is a typical estimation for on-axis current drive system [15]. Model d) is similar to model c) except the extra bootstrap current j_b term. This bootstrap current is an intrinsic current generated in the plasma it has the form [15]:

$$j_b = -T \frac{\partial n}{\partial r}. \tag{9}$$

So, the last model has the form:

$$v_{\phi,d} \propto j_\phi + j_b, \tag{10}$$

3. Numerical results and discussions

The numerical simulations in this work are results of solving transport equations (1) and (2) simultaneously and self-consistently. In accordance with physical observations, the anomalous transports are assumed to be one order of magnitude over the neoclassical transports, while particle diffusivities are assumed to be one third of the thermal diffusivities. The heat and particle sources are assured to be large enough so the plasmas can access the *H*-mode.

3.1. Plasma response on profile and pedestal width

Effects of toroidal rotation on plasma profiles are investigated in this section. This can be seen in figure 1, which shows the simulation results for plasma pressure and density at plasma core and edge as a function of normalized minor radius (r/a) at steady state. The four toroidal velocity models are used for simulations in order to compare with results without toroidal flow included ($v_{\phi,0}$). In general, inclusion of toroidal flow tends to increase plasma pressure and density except at its edge where transport barrier is formed. This is not surprising because the plasma in the edge region is governed solely by neoclassical effect, which remains unaffected by the toroidal term. Moreover, the toroidal term facilitates the enhancement of transport reduction, thus higher plasma profiles. The top panels show plasma profiles near its center, it can be estimated that the increases are approximately 0.04%, 1.72%, 3.09% and 3.96% of center pressure, and 0.03%, 1.58%, 2.60% and 3.39% of center density for simulations using model a), b), c) and d), respectively.

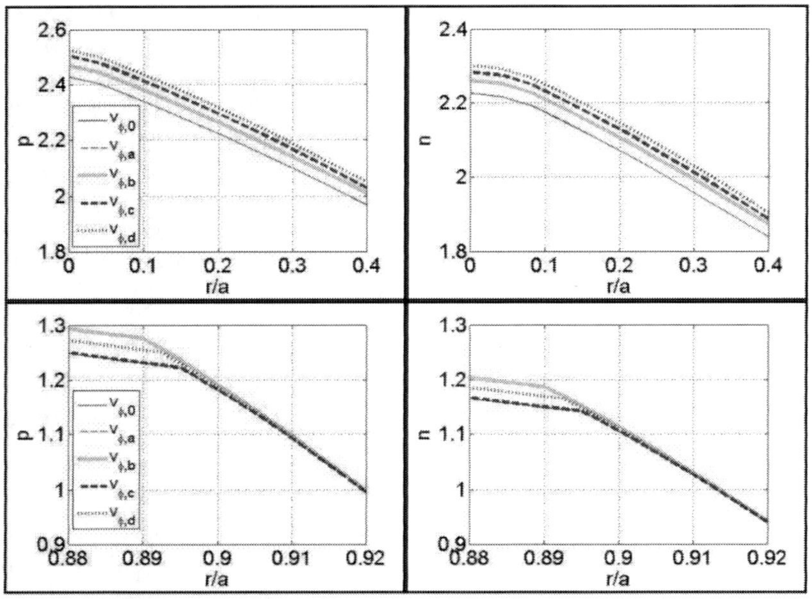

Figure 1. Plasma core profiles for pressure (top left) and density (top right), and plasma edge profiles for pressure (bottom left) and density (bottom right) at steady state. Each line represents simulation results without toroidal flow (thin solid), using model a (thin dashed), model b (thick solid), model c (thick dashed) and model d (dotted).

The toroidal effects yield somewhat different results at pedestal area (area with high gradients) as seen in the bottom panels of figure 1. First of all, the results show that the pedestal width remains unchanged from simulation without toroidal effect when using model a) and c) with pedestal width of 0.105. On the other hand, the simulations using model b) and d) yield pedestal widths of 0.110 and 0.108, respectively. Note that the pedestal with appears to be the same for pressure and density channels regardless of the model used. The values at the top of pedestal (p_{ped}, n_{ped}) are also increased by inclusion of toroidal flow. The increases are approximately 0.09%, 4.40%, 0.04% and 2.29% of p_{ped}, and 0.08%, 4.07%, 0.04% and 2.12% of n_{ped} for simulations using model a), b), c) and d), respectively. The reasons for the increase being difference from model to model at plasma center and edge can be seen in figure 2 which shows radial electric field, pressure gradient and toroidal flow contributions as a function of normalized minor radius. First of all, equation (3) implies that the shear or gradient of toroidal velocity profile is an essential ingredient for transport reduction calculation. It can be see that in model a), the profile is relatively flat throughout the whole plasma so it does not affect the flow shear much. In model b), the profile appears to have slight gradient at plasma center and strong gradient at plasma edge. Model c) has somewhat high gradient near plasma center and becomes flat near plasma edge. And model d) has both high gradients at plasma center and plasma edge. These results agree with the increase of plasma values at its center and pedestal.

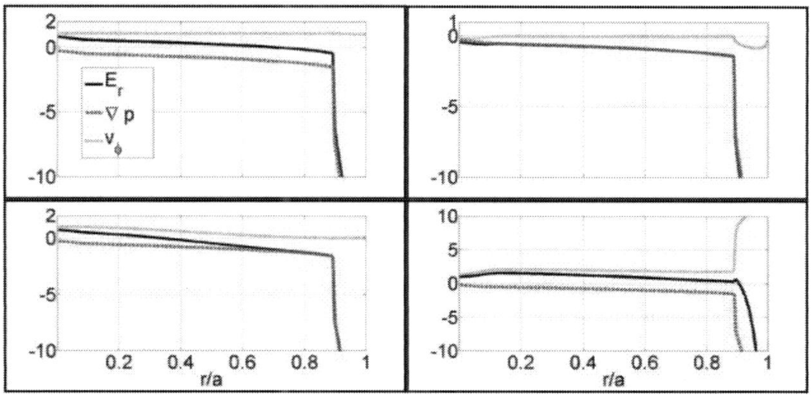

Figure 2. Profiles of radial electric field (solid), pressure gradient term (dashed) and toroidal rotation term (dotted) at steady state for simulations using model a(top left), b (top right), c (bottom left) and d (bottom right).

3.2. Pedestal dynamics

This section illustrates the pedestal dynamics once *L-H* transition occurs. Figure 3 demonstrates the change of pedestal width after it is formed. Initially, the transport barrier only forms at pressure channel. After forming, it starts to expand. It appears that the pedestal growth is categorized as superdiffusive behaviour ($\Delta_{ped} \propto t^b, b > 0.5$), agreeing with the turbulent nature of the plasma because in this phase the suppression effect is still not too high, especially at the particle channel suppression, thus turbulent transport still plays a considerable role. The speed of pedestal width expansion appears to be related to the gradient of toroidal velocity profile near pedestal area. Left panel of figure 3 shows that the barrier expands fastest for simulations using model b) with expansion rate of 2.9 s^{-1}, then model d) with rate of 1.4 s^{-1}, their toroidal velocity gradients near the pedestal are 3.2 and 2.4, respectively. The correlation is not clear in model a) and c) which yield very small negative value of toroidal velocity gradients while the expansion rate is slightly less than that of model d). At some time later, the transport reduction effect becomes large enough to suppress the density channel so the transport barrier in density channel starts to form. Note that the pedestal width of both pressure and density channels are about the same regardless of the model used. The onset of transport barrier formation in density channel also improves density profiles significantly. This actually reduces the first contribution in the radial electric field calculation of equation (3) resulting in the flow shear being less effective. As a result, both barrier widths slowly decrease or collapse until the plasma reaches steady state. The decrease is slower than the width expansion because now the plasma is governed solely by the neoclassical transport which is a slower process than turbulent transport. In this process the width dynamics becomes subdiffusive or even slower. There is an interesting point worth mentioning here, which is the time it takes the plasma to evolve during H-mode is around one order of magnitude slower than the time it takes for the plasma to evolve from L to H mode.

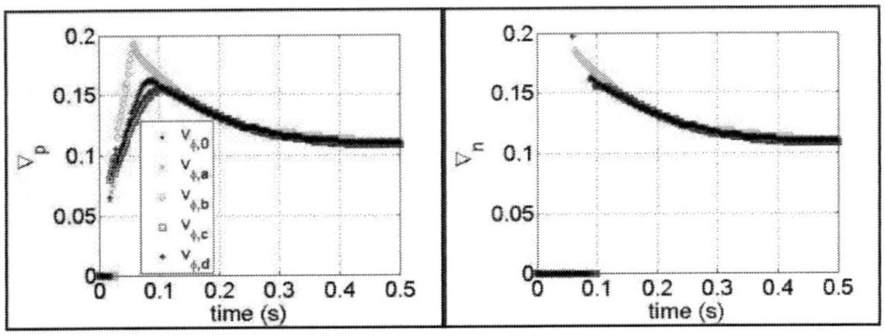

Figure 3. Time evolution of pedestal width of pressure (left) and density (right) channels.

4. Conclusion

Numerical method is used to solve the coupled pressure and density transport equations based on bifurcation concept. The transport effects included are neoclassical transport which is assumed to be constant and anomalous transport which is inspired by the critical gradient transport model. The suppression mechanism is the flow shear calculating from the force balance equation which includes toroidal flow effect. Four simple toroidal velocity models are considered in the calculation of radial electric field based on the force balance equation. It is found that inclusion of toroidal velocity can substantially increase the plasma pressure and density, mainly due to an increase of the pedestal width. It is also found that the pedestal for pressure tends to form first. After the pedestal forms, it expands inwards with the characteristic of super-diffusive nature in initial state and become sub-diffusive nature in final state before reaching steady state. The expansion speed depends sensitively on the strength of flow shear effect. It is also found that the time required plasma to reach steady state after the *L-H* transition is much longer than that for *L-H* transition.

Acknowledgments

This work is supported by the Commission on Higher Education (CHE) and the Thailand Research Fund (TRF) under contract No.RSA5580041. B Chatthong thanks the Royal Thai Scholarship and IRFM CEA Cadarache and acknowledges the discussion at the 3rd APTWG, Korea 2013.

References

[1] Wagner F *et al.*, 1982 *Phys. Rev. Lett.* **49** 1408
[2] Aymar R, Barabaschi P and Shimomura Y, 2002 *Plasma Phys. Control. Fusion* **44** 519
[3] Hubbard A E, 2000 *Plasma Phys. Control. Fusion* **42** A15
[4] Xu Y *et al.*, 2015 *Plasma Phys. Control. Fusion* **57** 014028
[5] Wagner F, 2007 *Plasma Phys. Control. Fusion* **49** B1
[6] Biglari H, Diamond P H and Terry P W, 1990 *Physics of Fluids B: Plasma Physics* **2** 1-4
[7] Burrell K H, 1997 *Phys. Plasmas* **4** 1499-1518
[8] Connor J W, Fukuda T, Garbet X, Gormezano C, Mukhovatov V, Wakatani M, Group t I D and Physics t I T G o T a I B, 2004 *Nucl. Fusion* **44** R1
[9] Malkov M A and Diamond P H, 2008 *Phys. Plasmas* **15** 122301
[10] Garbet X *et al.*, 2004 *Plasma Phys. Control. Fusion* **46** 1351
[11] Chatthong B and Onjun T, 2013 *Nucl. Fusion* **53** 013007
[12] Chatthong B, Onjun T and Singhsomroje W, 2010 *Nucl. Fusion* **50** 064009
[13] Chatthong B and Onjun T, 2014 *Songklanakarin J. Sci. Technol.* **36** 375-387
[14] Chatthong B, Sarazin Y, Onjun T, Wongsrisujarit N, Imbeaux F, Strugarek A, Picha R and Poolyarat N, 2011 38th EPS Conference on Plasma Physics (Strasbourg) P4.097
[15] Wesson J, 2004 Clarendon Press ; Oxford University Press

Characterization of neutron calibration fields at the TINT's 50 Ci americium-241/beryllium neutron irradiator

T Liamsuwan[1], **J Channuie and W Ratanatongchai**
Nuclear Research and Development Division, Thailand Institute of Nuclear Technology (Public Organization), 9/9 Moo 7 Saimoon, Ongkharak, Nakorn Nayok, 26120 Thailand

E-mail: thiansinl@tint.or.th

Abstract. Reliable measurement of neutron radiation is important for monitoring and protection in workplace where neutrons are present. Although Thailand has been familiar with applications of neutron sources and neutron beams for many decades, there is no calibration facility dedicated to neutron measuring devices available in the country. Recently, Thailand Institute of Nuclear Technology (TINT) has set up a multi-purpose irradiation facility equipped with a 50 Ci americium-241/beryllium neutron irradiator. The facility is planned to be used for research, nuclear analytical techniques and, among other applications, calibration of neutron measuring devices. In this work, the neutron calibration fields were investigated in terms of neutron energy spectra and dose equivalent rates using Monte Carlo simulations, an in-house developed neutron spectrometer and commercial survey meters. The characterized neutron fields can generate neutron dose equivalent rates ranging from 156 μSv/h to 3.5 mSv/h with nearly 100% of dose contributed by neutrons of energies larger than 0.01 MeV. The gamma contamination was less than 4.2-7.5% depending on the irradiation configuration. It is possible to use the described neutron fields for calibration test and routine quality assurance of neutron dose rate meters and passive dosemeters commonly used in radiation protection dosimetry.

1. Introduction

Reliable measurement of neutron radiation is important for monitoring and protection in workplace where neutrons are present. In Thailand, neutrons are used in research, non-destructive testing, gemstone irradiation, and neutron activation analysis. Available neutron sources are radioactive sources, neutron generators and the Thai Research Reactor.

Despite a variety and increasing interest of neutron source applications, there is no standard calibration facility for neutron measuring devices available in Thailand and Southeast Asia. Therefore, neutron area monitoring devices need to be sent to Japan, Europe or the USA for annual calibration, while personal dosemeters for neutrons are calibrated once by manufacturers upon purchase and the received dose is routinely evaluated using computer software.

Recently, Thailand Institute of Nuclear Technology (TINT) has set up a multi-purpose neutron irradiation facility equipped with a 50 Ci ^{241}AmBe neutron source. Among other applications, the facility was developed for generating neutron calibration fields for radiation protection devices. Unlike recommended calibration facilities for neutron measuring devices [1, 2], the facility at TINT is not a

[1] To whom any correspondence should be addressed.

low-scatter type and the room size is not as large as most of the existing facilities. Moreover, since it is a multi-purpose facility, the neutron source is kept inside a cask and neutron irradiation is done through a collimator.

Due to scattering and absorption of neutrons by the facility's components, a simple set of correction factors for the determination of neutron dose is not achievable. Instead, Monte Carlo simulations, accompanied by measurements of neutron energy spectra and neutron dose, are required for characterization of the neutron calibration fields.

In this paper, the design of the neutron irradiation facility is described. The neutron calibration fields were characterized using Monte Carlo simulations accompanied by measurements with an in-house developed neutron spectrometer and commercial dose rate meters. The computational and experimental approaches served as validation tools for each other.

2. Material and Methods

2.1. Neutron irradiation facility

Figure 1 shows the layout of the TINT's neutron irradiation facility. The room is 6.5 m in width, 8.5 m in length, and 3.5 m in height. The neutron irradiator consists of the source storage container and the collimator enclosure, both sitting on an iron table. The collimator aperture is centered at 0.9 m above the concrete floor. To irradiate samples, a long handle is used for pulling the source from the storage container towards the collimator aperture. A warning light is activated as soon as the source is moved away from the centre of the storage container.

The storage container is a cylindrical tank with a diameter of 68.6 cm and a length of 94.8 cm. It is filled with neutron shielding materials. The 50 Ci ^{241}AmBe neutron source is housed in a cavity of 7.6 cm in diameter inside the storage container. The source has a cylindrical double encapsulation made of stainless steel. The outer dimensions are 5 cm in diameter and 8.52 cm in length. The wall thickness of each encapsulation is 0.2 cm. The neutron emission rate is 10^8 n/s (±10%) according to the manufacturer (QSA Global Inc.).

The collimator enclosure is a stainless steel box filled with a mixture of paraffin and boric acid. The dimensions of the enclosure are 69 cm in width, 90 cm in length, and 66 cm in height. The collimator duct of a rectangular pyramid shape sits inside the enclosure with the aperture of 10 × 10 cm^2 inwards and 15 × 15 cm^2 outwards. It is possible to place a neutron absorbing material inside the collimator duct to reduce the neutron flux. For that purpose, a graphite block of 14 cm thickness was constructed to fit the collimator duct.

For radiation safety reasons, the shielding wall of 150 cm height consisting of 5 cm thick borated polyethylene sheets and 40 cm thick heavy concrete was constructed behind and at one end of the irradiator. The shielding wall was designed to minimize the total dose equivalent rates to comply with the dose limit of 10 μSv/h outside the irradiation area.

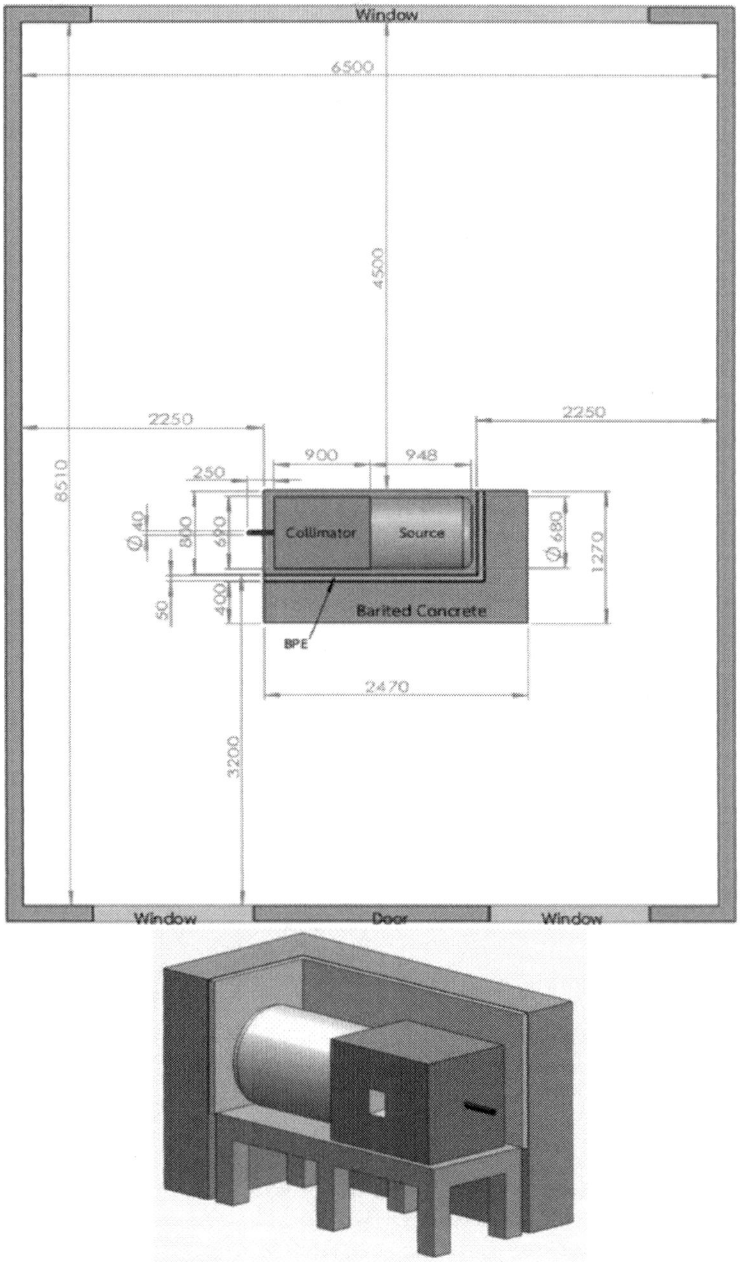

Figure 1. The layout of the TINT's neutron irradiation facility. The upper figure shows the dimensions (unit: mm) of the room and the neutron irradiator. The lower figure displays the setup of the neutron irradiator, which consists of the storage container and the collimator enclosure with the aperture of 15 × 15 cm^2, both sitting on the iron table. It is possible to insert a graphite absorber inside the collimator duct to reduce the neutron flux. The neutron irradiator is surrounded by the shielding wall, which is made of 5 cm thick borated polyethylene sheets and 40 cm thick heavy concrete.

2.2. Neutron field characterization

2.2.1. Monte Carlo simulations. The general purpose Monte Carlo Particle and Heavy Ion Transport code System (PHITS) version 2.64 [3] was used for simulating the neutron irradiation facility. The simulation took into account the source encapsulation, storage container, collimator, iron table, concrete floor, ceiling and shielding materials. Interactions of neutrons, photons and charged particles were considered. The Event-Generator mode implemented in PHITS was used for transporting neutrons with energy below 20 MeV. The cut-offs for radiation transport were set at 1 keV for scattered photons and 1 keV/u for secondary ions. The neutron energy spectrum of the bare source (without encapsulation) was taken from the ISO recommendation [4].

Spherical virtual detectors with the diameter of 1 cm filled with air were used for tallying neutrons and gammas, which were incident at selected positions of the room. Neutron dose equivalent (H_N) was calculated using the ICRP conversion method [5] according to

$$H_N = \sum_i k_i(E)\phi_{i,N}(E) \qquad (1)$$

where $k_i(E)$ is the ICRP's ambient dose equivalent per unit neutron fluence ($H^*(10)/\phi$) (Figure 2) [5], E is the neutron energy, and $\phi_{i,N}(E)$ is the calculated neutron fluence in the energy interval between E and $E+\Delta E$. For scattered gammas, the kerma approximation implemented in PHITS was used for determination of scattered photon dose. The simulation results were normalized to the neutron emission rate (10^8 neutrons/s).

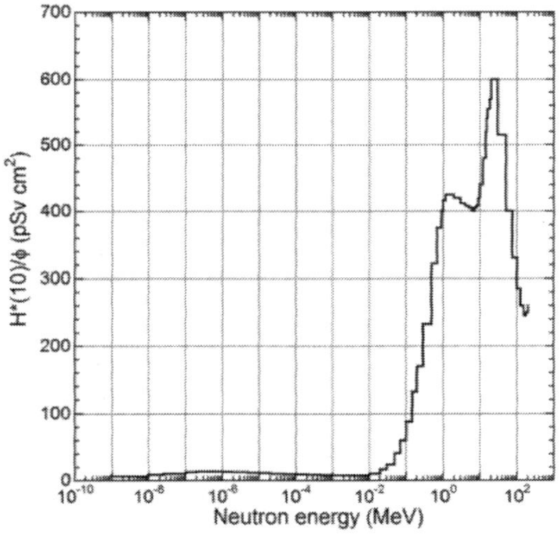

Figure 2. The ICRP's ambient dose equivalent per unit neutron fluence [5].

2.2.2. Radiation measurements. Measurements of neutron and gamma dose rates and neutron energy spectra were carried out to accompany the Monte Carlo calculations. Commercial survey meters were used for measurements of neutron and gamma dose equivalent rates (Ludlum model 12-4 of Ludlum Measurements, Inc. and Radiagem™ 2000 of Canberra Industries, Inc., respectively). For neutron

Journal of Physics: Conference Series **611** (2015) 012004 doi:10.1088/1742-6596/611/1/012004

spectrometry, a neutron spectrometer was developed in-house with a similar design as the Nested Neutron Spectrometer of Dubeau et al [6] (patented and commercialized by Detec) and the same working principles as well-known Bonner sphere spectrometers [7].

The neutron spectrometer consists of a small cylindrical ^{3}He thermal neutron detector of the model LND251 (LND, Inc.) and 7 cylindrical moderator shells made of high density polyethylene. The outer diameters of the moderator shells range from 7.5 to 20 cm. The smallest shell has an inner cavity that fits the neutron detector. Each shell was designed to fit into the next larger shell. By sequentially placing each shell to the detector and to the smaller shells, the neutron count rates associated with eight moderator thicknesses were measured. The neutron count rates (C) were unfolded to obtain the neutron energy spectra (ψ) for 50 energy bins by solving a system of linear equations:

$$C = R\psi \tag{2}$$

where R is the spectrometer's response matrix obtained using Monte Carlo simulations. Equation (2) was solved by the Levenberg-Marquardt method implemented in MATLAB (The MathWorks, Inc.).

3. Results and Discussion

Figure 3 shows the calculated and measured neutron energy spectra at the source-to-detector distances of 100 and 200 cm, without the graphite absorber inside the collimator duct. The Monte Carlo simulation and the spectrometer measurement showed good agreement in the neutron energy range of 0.1-10 MeV, covering the energy region in which the ICRP fluence-to-dose conversion factors are relatively large (Figure 2). Discrepancies between the two data sets are observed at lower energies, but neutrons of this energy range have a small contribution to the total neutron dose equivalent. At energies above 10 MeV, the spectrometer overestimated the simulation results. Since the ICRP conversion factor maximizes in this energy range (Figure 2), the neutron dose equivalent determined using the spectrometer are expected to be higher than those obtained using the Monte Carlo simulations. Comparing the neutron energy spectra at 100 and 200 cm source-to-detector distances, the magnitude of the neutron flux at the closer distance is about 4-4.5 times as high as the neutron flux at the larger distance, in close agreement to the inverse square law.

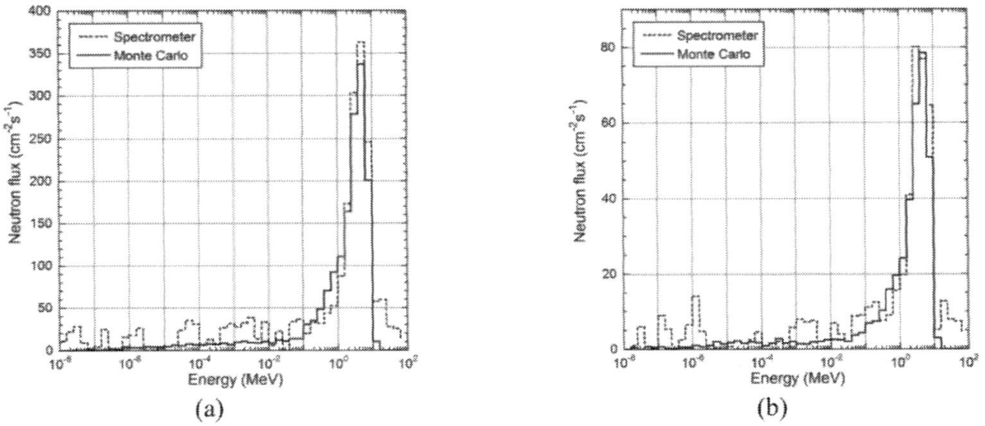

(a) (b)

Figure 3. The neutron energy spectra at the source-to-detector distances of (a) 100 and (b) 200 cm, without the graphite absorber inside the collimator duct. The solid lines were obtained using the Monte Carlo simulations and the dashed lines were determined by the measurement with the in-house developed neutron spectrometer.

Table 1 and Table 2 show the evaluated neutron and gamma dose equivalent rates at the source-to-detector distances 75, 100 and 200 cm, with and without the graphite absorber inside the collimator duct, respectively. At closer distances (<75 cm) the field size would be too small (smaller than 20 × 20 cm^2), while at larger distances (>200 cm) the detector would be too close to the concrete wall and the room scatter contribution appreciably increased. The final dose equivalent rates were averaged over the results obtained by the Monte Carlo simulations, the survey meters and, in some cases, also the neutron spectrometer. For neutrons, the standard deviations of the means were relatively small without the graphite absorber (from less than 1% up to 18%), and increased up to 30% when the graphite absorber was used. The calculated and measured gamma dose rates differed by 1-8.4 folds, but in all cases the gamma contributions were less than 4.2-7.5% of the neutron dose equivalent rates.

The maximum neutron dose equivalent rate was about 3.5 mSv/h without the graphite absorber. By placing the graphite absorber inside the collimator duct, it was possible to reduce the neutron dose equivalent rate to 156 μSv/h at the source-to-detector distance of 200 cm. In all cases, nearly 100% of the neutron dose equivalent rates were contributed by neutrons with energies larger than 0.01 MeV. Despite the small contribution of thermal neutrons to the total neutron dose, the thermal component can be important for calibration of detectors that are sensitive to thermal neutrons e.g. TLD, BF_3 and ^3He detectors.

Table 1. Neutron and gamma dose equivalent rates at different source-to-detector distances (without the graphite moderator).

Distance (cm)	Neutron dose equivalent rates (μSv/h)				Gamma dose equivalent rates (μSv/h)		
	MC	Survey meter	Spectrometer	Average	MC	Survey meter	Average
75	3,537	3,500	-	3,518.7	75	148	111.7
100	1,896	2,000	2,243	2,046.3	59	85	72.1
200	445	375	534	451.3	5	21	13.0

Table 2. Neutron and gamma dose equivalent rates at different source-to-detector distances (with the graphite moderator).

Distance (cm)	Neutron dose equivalent rates (μSv/h)			Gamma dose equivalent rates (μSv/h)		
	MC	Survey meter	Average	MC	Survey meter	Average
75	1,463	1,000	1,231.6	80	85	82.5
100	786	525	655.5	49	49	48.9
200	177	135	156.0	1	12	6.5

Apart from generating neutron dose equivalent rates for calibration of dose rate meters, the investigated neutron fields can be used for irradiation of integrating dosemeters e.g. passive dosemeters. It is possible to adjust the irradiation time according to the required neutron dose, the source-to-detector distance and the configuration with or without the graphite absorber. Table 3 shows examples of the irradiation time required for neutron dose equivalents of 0.2-3 mSv at the source-to-detector distance of 100 cm, with and without the graphite absorber. This range of neutron dose equivalents is required for the routine quality assurance of the TLD and OSL personal dosemeters serviced by TINT.

For relatively low dose, it is possible to choose between the configurations with and without the graphite absorber. In both cases, the normalized neutron energy spectra were not significantly different. For applications that require the time uncertainty to be minimal, longer irradiation time (with the graphite absorber) is preferred. For relatively large dose (e.g. 2-3 mSv), due to time constraints, the configuration without the graphite absorber might be the only alternative.

4. Conclusions

The neutron fields at the TINT's multi-purpose neutron irradiation facility were characterized using Monte Carlo simulations, an in-house developed neutron spectrometer and commercial survey meters. The facility is equipped with a 50 Ci ^{241}AmBe neutron irradiator with the collimator of 15 × 15 cm^2 aperture. It is possible to place a neutron absorbing material inside the collimator duct to reduce the neutron flux. The investigated neutron fields generated neutron dose equivalent rates ranging from 156 µSv/h to 3.5 mSv/h. In all cases, the largest dose component was due to neutrons of energies above 0.01 MeV. Depending on the irradiation setup, the gamma contamination was at maximum 4.2-7.5% of the neutron dose. Irradiation time for integral dose measurement can be varied by changing the irradiation configuration such as the source-to-detector distance and the neutron absorbing material inside the collimator duct. The characterized neutron fields can facilitate calibration test and quality assurance of neutron dose rate meters and personal dosemeters commonly used in radiation protection dosimetry. The work has paved the way for setting up a calibration facility for neutron measuring devices in Thailand.

Table 3. Required neutron irradiation time and expected gamma dose equivalents at the source-to-detector distance of 100 cm.

Neutron dose equivalent (mSv)	Without graphite absorber		With graphite absorber	
	Irradiation time (min)	Expected gamma dose equivalent (mSv)	Irradiation time (min)	Expected gamma dose equivalent (mSv)
0.2	5.9	0.01	18.3	0.01
0.4	11.7	0.01	36.6	0.03
0.6	17.6	0.02	54.9	0.04
0.8	23.5	0.03	73.2	0.06
1.0	29.3	0.04	91.5	0.07
1.5	44.0	0.05	137.3	0.11
2.0	58.6	0.07	183.1	0.15
3.0	88.0	0.11	274.6	0.22

Acknowledgments

The authors wish to thank Department of Nuclear Engineering, Chulalongkorn University and Nuclear Technology Service Center and Safety Unit at TINT for supporting some of the experimental equipment.

References

[1] IAEA 1988 *Guidelines on Calibration of Neutron Measuring Devices (Technical Report Series No. 285)* (Vienna, Austria: International Atomic Energy Agency)
[2] ISO 2000 *Reference neutron radiations - Part 2: Calibration fundamentals of radiation protection devices related to the basic quantities characterizing the radiation field* (Geneva, Switzerland: International Organization for Standardization)
[3] Sato T *et al* 2013 Particle and Heavy Ion Transport code System, PHITS, version 2.52 *J. Nucl. Sci. Technol.* **50** 913–23
[4] ISO 2001 *Reference neutron radiations - Part 1: Characteristics and methods of production* (Geneva, Switzerland: International Organization for Standardization)
[5] ICRP 1996 Conversion Coefficients for use in Radiological Protection against External Radiation (ICRP Publication 74) *Ann. ICRP* **26** 3–4
[6] Dubeau J, Witharana S S H, Atanackovic J, Yonkeu A and Archambault J P 2012 A neutron spectrometer using nested moderators *Radiat. Prot. Dosimetry* **150** 217–22

[7] Thomas D J and Alevra A V 2002 Bonner sphere spectrometers—a critical review *Nucl. Instr. Meth. Phys. Res. A.* **476** 12–20

International Nuclear Science and Technology Conference 2014 (INST2014) IOP Publishing
Journal of Physics: Conference Series **611** (2015) 012005 doi:10.1088/1742-6596/611/1/012005

The simulation of *L-H* transition in tokamak plasma using MMM95 transport model

P Intharat[1,4], B Chatthong[2], T Onjun[2], N Poolyarat[1] and R Picha[3]

[1] Department of Physics, Faculty of Science and Technology,Thammasat University, Pathum Thani, Thailand
[2] School of Manufacturing Systems and Mechanical Engineering, Sirindhorn International Institute of Technology, Thammasat University, Pathum Thani, Thailand
[3] Thailand Institute of Nuclear Technology, Bangkok, Thailand

Email: pat_int@yahoo.com

Abstract. BALDUR integrative predictive modelling code together with a Multimode (MMM95) anomalous transport model is used to simulate the evolution profiles, including plasma current, temperature, density and energy in a tokamak reactor. It is found that a self-transition from low confinement mode (*L*-mode) to high confinement mode (*H*-mode) regimes can be achieved once a sufficient auxiliary heating applied to the plasma is reached. The result agrees with experimental observations from various tokamaks. A strong reduction of turbulent transport near the edge of plasma is also observed, which is related to the formation of steep radial electric field near the edge regime. From transport analysis, it appears that the resistive ballooning mode is the dominant term near the plasma edge regime, which is significantly reduced during the transition.

1. Introduction

Harvesting nuclear fusion energy from tokamak plasma becomes one of the great challenges among the physicists and engineer. Nuclear fusion reaction requires high energy to overcome the Coulomb's repulsive force between nuclei. In 1979, it was found in ASDEX upgrade tokamak that when sufficient external heating was applied, the plasma temperature, density and confinement time increased significantly. This new operational regime is called "high confinement mode (*H*-mode)" [1-7]. In *H*-mode, the plasma yields higher performance, which is often measured by quantities such as temperature, density and stored energy. Higher temperature and density can increase the nuclear fusion reactions. The difference between the *H*-mode and a low confinement mode (*L*-mode) is the presence of the steep gradient at the edge region, called edge transport barrier (ETB) [5, 8]. It is most interesting to investigate the *L-H* transition in tokamak plasma.

BALDUR integrated predictive modelling code [9] was developed to simulate time-evolution of plasma current, density and temperature profiles. It includes various phenomena in tokamak experiments, such as sources, sinks, transports of thermal energy and particles, etc. The BALDUR code has been intensively tested against experimental data from various tokamaks, in which the temperature and density profiles yield an RMS about 10% [4, 10]. In this work, both thermal and

[4] To whom any correspondence should be addressed.

Journal of Physics: Conference Series **611** (2015) 012005 doi:10.1088/1742-6596/611/1/012005

particles transports are described by using a Multimode 95 anomalous core transport (MMM95) model. The external heating is applied to the plasma as electron cyclotron resonance heating (ECRH) and ion cyclotron resonance heating (ICRH). Once critical threshold is reached, the plasma makes a transition from *L*-mode to *H*-mode, called "*L-H* transition". This work studies the *L-H* transition by using the MMM95 anomalous core transport. The *H*-mode access can be referred from the experiment results using the threshold scaling law and the reduction of anomalous transport near the edge regime of plasma. This work is organized as follows; chapter 2 describes description of simulation, chapter 3 presents simulation results and the conclusion is given in chapter 4.

2. Description of the simulation

2.1. MMM95 Anomalous core transport model

In this work, MMM95 core transport is used to predict the core and edge transport in BALDUR simulation [10-11, 13] model. It is implemented in BALDUR to predict the profiles of temperature and density. MMM95 consists of Weiland models, Guzdar-Drake models and its kinetic ballooning model. The Weiland model described the Ion Temperature Gradient (ITG) and Trapped Electron Modes (TEM), it is a theory-based model derived from the fluid concepts. Generally, the Weiland model is largest contribution in the plasma core regime. On the other hand, the Guzdar-Drake model described both resistive and kinetic ballooning modes (RB and KB, respectively). Their calculation is proportional to the pressure gradient and collision. The resistive ballooning mode has high contribution near the edge of plasma. Whereas, the kinetic ballooning model usually provides a small contribution to total diffusivity throughout the entire plasma, except near the magnetic axis. All of the anomalous transport contributions from the MMM95 transport model are multiplied by κ^{-4}, since the models were originally derived for circular plasmas [11]. Parameter κ is plasma elongation which describes the geometry of plasma cross section shape. Both thermal and particle transport parameters in MMM95 core transport model are described below:

$$\chi_i = 0.8\chi_{i,(ITG,TEM)} + 1.0\chi_{i,(RB)} + 1.0\chi_{i,(KB)} , \tag{1}$$

$$\chi_e = 0.8\chi_{e,(ITG,TEM)} + 1.0\chi_{e,(RB)} + 1.0\chi_{e,(KB)} , \tag{2}$$

$$D_H = 0.8D_{H,(ITG,TEM)} + 1.0D_{H,(RB)} + 1.0D_{H,(KB)} , \tag{3}$$

$$D_z = 0.8D_{z,(ITG,TEM)} + 1.0D_{z,(RB)} + 1.0D_{z,(KB)} , \tag{4}$$

where χ_i and χ_e are ion and electron diffusivities, respectively (m/s^2), D_H is particle diffusivity and D_z is impurity diffusivity. ITG and TEM represent contributions from Weiland model. RB is transport from resistive ballooning model and KB is transport from kinetic ballooning model.

2.2. The H-mode power threshold model

It is found that the *L-H* transition occur when the total applied heating power exceeds a threshold power . The standard working hypothesis, supported by many observations, is that *H*-mode occurs when the total power or the power transported across the separatrix (P_{sep}) exceeds a threshold value ($P_{sep} > P_{thr}$) [1]. The P_{sep} can be calculated as follows:

$$P_{sep} = P_{ohm} + P_{aux} + P_{\alpha} - \frac{dw}{dt} - P_{loss} , \tag{5}$$

where P_{ohm} is power in ohmic stage, P_{aux} is auxiliary heating, P_{α} is alpha power, $\frac{dw}{dt}$ is derivative of plasma stored energy with respect to time and P_{loss} is loss power. The threshold power P_{thr} can be calculated based on scaling law as follows [1]:

$$P_{thr} = 2.15\kappa n_{e20}^{0.58} B^{0.82} R^{1.00} a^{0.81} 2 M_{amu}^{-1} , \tag{6}$$

where κ is the elongation, n_{e20} is the electron density (10^{-20} m^{-3}), B is the toroidal magnetic field, R and a are the plasma major and minor radii, respectively, and the M is the mass of ion species.

2.3. The radial electric field

The physics of edge transport barrier is important for the understanding of the evolution of turbulent transport. At the onset of the *L-H* transition, the turbulent transport is found to decrease near the edge region of the plasma. The key for understanding of the ETB formation is $\omega_{E\times B}$ (flow shear) stabilization. The flow shear is known to be able to quench the turbulent transport. It depends on plasma electric field on radial direction. Theoretically and experimentally the flow shear stabilization [7, 13] is calculated as:

$$\omega_{E\times B} = \frac{(RB_\theta)^2}{B_\varphi} \frac{\partial}{\partial \psi} \frac{E_r}{RB_\theta}, \qquad (7)$$

where $\omega_{E\times B}$ is the flow shear rate (s^{-1}), R is the major radius, B_θ is the poloidal magnetic field, B_ϕ is the toroidal magnetic field, and E_r is the radial electric field for the main plasma ions. The E_r can be used to indicate the plasma mode of the tokamak experiments [7] because the E_r profile is dependent on the pressure gradient at the onset of *L-H* transition. It is calculated as follows:

$$E_r = \frac{1}{Zen_i} \frac{\partial p_i}{\partial r} - v_\theta B_\varphi + v_\varphi B_\theta, \qquad (8)$$

where $\frac{\partial p_i}{\partial r}$ the ion is pressure gradient, v_θ and v_ϕ are the poloidal and toroidal velocities, respectively, and n_i is the ion density, Z is the ion charge number and e is the elementary charge. Furthermore, the radial electric field depends on the gradient of temperature, density or pressure of the plasma species.

3. Simulation results

The simulations are carried out using the engineering parameters shown in table 1. For all simulations, an anomalous transport is calculated using MMM95 transport model and the auxiliary heating used the RF heating to induce the *H*-mode on the simulation as the ICRF and ECRH. The auxiliary heating profile is assumed to be a gaussian shape, centralized near the plasma center. The plasma parameters for simulation were chosen based on the *H*-mode plasma experiment on the HL-2A tokamak [14, 15].

Table 1. Plasma parameters for the simulation.

Parameter	Physical Description	Value	Unit
R	Major radius	1.65	M
A	Minor radius	0.40	M
I_p	Plasma current	0.48	MA
B_T	Toroidal field	2.8	T
K	Elongation	1.30	-
δ	Triangularity	0.30	-
Zeff	Effective charge	1.50	-
P_{aux}	Auxiliary heating	3	MW

Firstly, this simulation is carried out for the behaviours of plasma during the heating phase. Figure 1 shows the plasma stored energy (W_{TOT}) as a function of time. The plasma stored energy has similarity to the internal or total energy of the plasma tokamak. It is found that as the auxiliary heating is applied at 0.7 sec, the plasma stored energy is strongly increased, and then after the heating is turned off, it is decreased. The stored energy is strongly increased by around 50% when the auxiliary heating is included and is decreased by around 47 % as heating is turned off. When the transport barrier forms, the particle diffusivity is locally reduced near the edge regime. Therefore the electron temperature is increased on the heating phase (t=0.9 sec) as shown in figure 2. The figure shows electron temperature (T_e), electron diffusion transport (χ_e) and the temperature gradient (gradient) as a function of normalized radius at a time before (t=0.4 sec), during (t=0.9 sec) and after heating (t=1.2 sec).

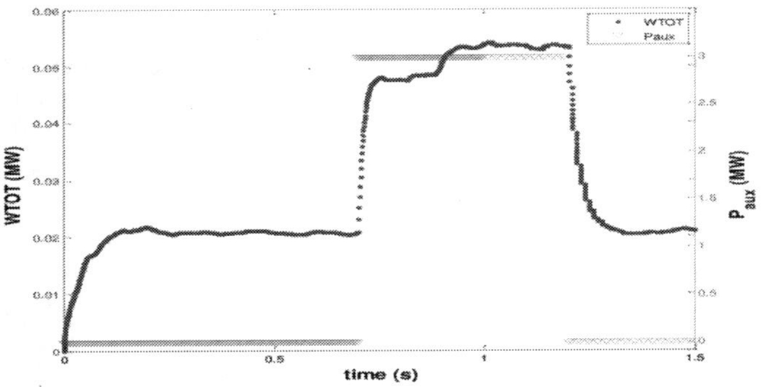

Figure 1. The evolution of plasma stored energy (W_{TOT}) and auxiliary heating power (P_{aux}) as a function of time, from BALDUR simulation.

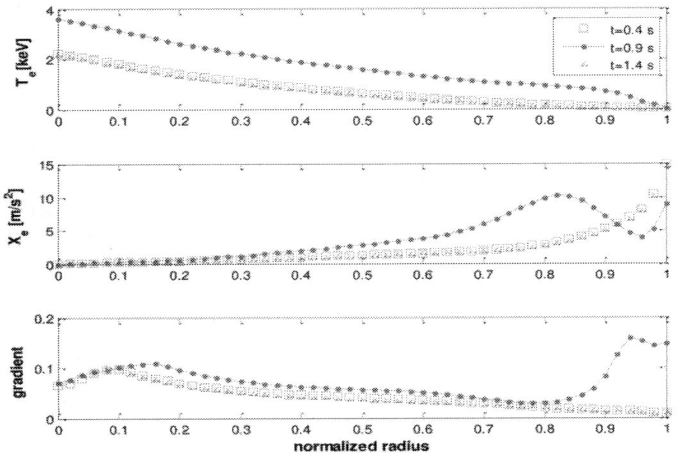

Figure 2. The electron temperature (top), electron thermal transport coefficient (center) and the electron temperature gradient (bottom) as a function of normalized minor radius at different times, from BALDUR simulation.

In the H-mode phase, the electron temperature increases with the strong temperature gradient near the plasma edge as shown in figure 2. Consequently, the H-mode plasma is reached, which causes the reduction of the diffusion transport near the edge plasma regime. The difference between L-mode and H-mode phase can be observed in the electron temperature and total electron transport profiles. During the H-mode phase, the turbulent transport is reduced near the edge regime, resulting in the formation of transport barrier.

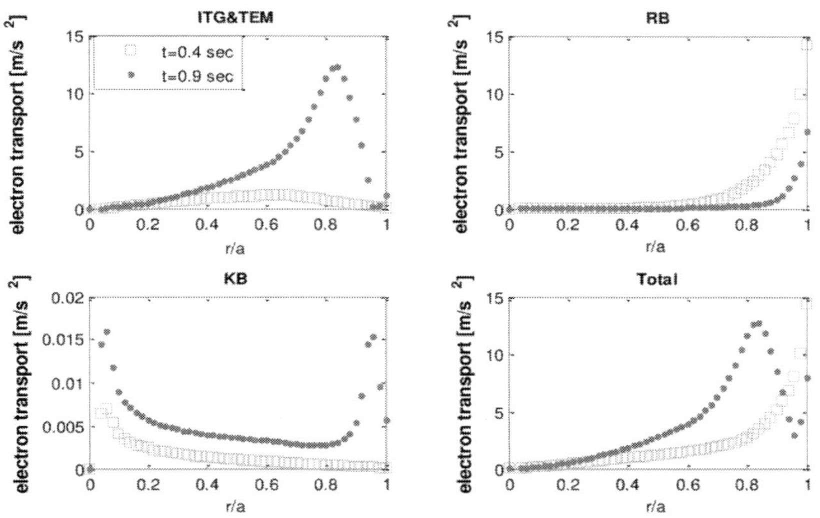

Figure 3. The radial profiles of the electron transport coefficient in *L*-mode (t=0.4 sec) and *H*-mode of plasma (t=0.9 sec), from BALDUR simulation.

In figure 3, the comparison between the *L*-mode (t=0.4 sec) and *H*- mode (t=0.9 sec) shows that the total transport decreases near the edge regime on model for the calculation. In the *H*-mode phase, the Weiland model has the ITG and TEM contribution which dominates near the edge regime. The resistive ballooning mode has the dominant contribution at the edge regime cause by the temperature gradient. The kinetic ballooning mode has smaller contribution than the other models. The total transport is calculated from the summation of the Weiland model and Guzdar-Drake resistive ballooning model. The result shows that the transport decreases near the edge plasma when the plasma undergoes *L-H* transition.

Next, the results on *H*-mode formation are compared between the theoretical simulations and the experiments. The *H*-mode threshold is used as the empirical scaling in the simulation and experimental data. It indicates the power transport across the separatrix (P_{sep}) must exceed the *H*-mode threshold value (P_{thr}) and then the *L-H* transition access. The *L-H* transition is confirmed with the *H*-mode threshold scaling during the heating phase. The power transport across the separatrix reaches than the *H*-mode threshold power for the transition.

Figure 4 shows the power transport across the separatrix in each of auxiliary heating power (0.5 and 3 MW) and the *H*-mode threshold power for the *L-H* transition. Before and after heating phase, P_{sep} of both heating is lower than the P_{thr} transition and the transport barrier does not occur in the electron temperature profile. During the heating phase, it is found that P_{sep} (P_{aux} = 3 MW) is larger than the threshold power that the electron temperature shows the ETB formation in figure 2 (T_e), which indicates the *H*-mode plasma phase and the turbulent transport.

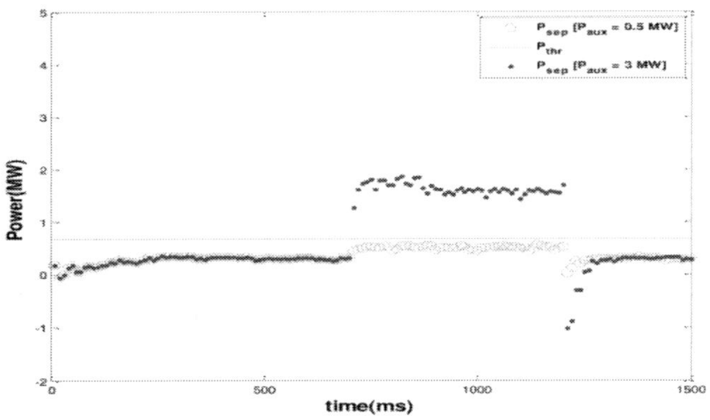

Figure 4. The power transport across the seperatrix (P_{sep}) and threshold power for transition (line) as functions of time, from BALDUR simulation.

Figure 5 illustrates the radial electric field and the flow shear stability as functions of normalized radius. When the *H*-mode occurs (t=0.9 sec), the radial electric field become more negative, resulting in increases the flow shear rate.

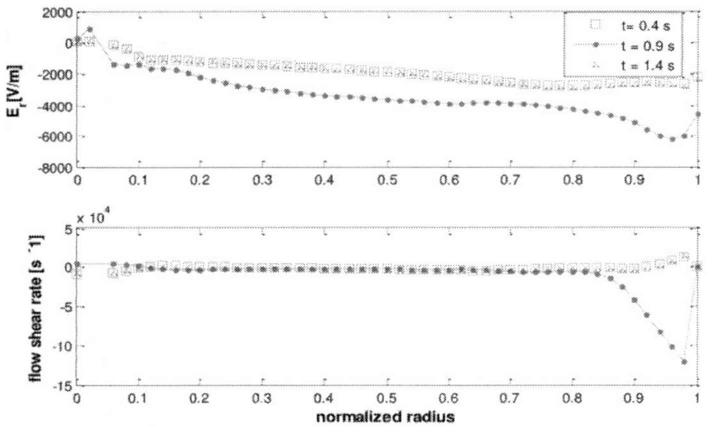

Figure 5. The evolution of (top) radial electric field (E_r) and (bottom) the flow shear rate as a function of normalized radius, from BALDUR simulation.

4. Conclusion

The BALDUR integrative predictive modelling code is used to simulate an evolution of plasma profiles during *L-H* transition. It is found that a self-transition from *L*-mode to *H*-mode can be achieved in the BALDUR simulations by using the MMM95 anomalous core transport. This transition can be achieved with the condition that an auxiliary heating power surpasses a certain threshold. The *H*-mode plasma on this simulation can be achieved with the flow shear suppression that causes the reduction of the transition and radial electric field. In the transport analysis, it is found that the resistive ballooning mode is the dominant mode near the plasma edge. When this mode is suppressed during the H-mode formation, it results in the transport barrier formation. The reduction of transport near the edge mainly occurs due to the resistive ballooning mode.

Acknowledgments
This work is supported by the Commission on Higher Education (CHE) and the Thailand Research Fund (TRF) under contract No.RSA5580041 and the National Research University Project of Thailand Office of Higher Education Commission.

References

[1] Glenn B, Onjun T and Amold H K 2003 *Plasma Physics and Controlled Fusion* **45** 1939
[2] Glenn B, Miguel A B, Onjun T, Arnold H K and a.A. Pankin 2003 *Physics of Plasmas* **10** 4358
[3] Glenn B, Arnold H K, Jon E. K, Aaron J. R, Jan W 1998 *Physics of Plasmas* **5** 1793
[4] David H, Glenn B, Jon K, Arnold H K, Onjun T and Pankin A 2001 *Physics of Plasmas* **8** 965
[5] Leonard W 2008 *Journal of Physics: Conference Series* **123** 012001
[6] Onjun T, Glenn B, Amold H K and Hammett G 2002 *Physics of Plasmas* **5** 5013
[7] Schneider P A *et al* 2012 *Plasma Physics and Controlled Fusion* **54** 105009
[8] Gohil P 2006 *Physique* **7** 606
[9] Singer C E *et al* 1998 *Phys. Commun* **49** 275
[10] Onjun T, Glenn B, Arnold H K and David H 2001 *Physics of Plasmas* **8** 975
[11] Rafiq T, Amold H K, Jan W, Pankin Y A and Luo L 2013 *Physics of Plasmas* **20** 1063
[12] Nguyen C N, Glenn B, Amold H K, Byrom C and Sykes A 2002 *Physics of Plasmas* **9** 3930
[13] Cohen B I *et al* 2013 *Physics of Plasmas* **20** 05596
[14] Duan X R *et al* 2010 *Nuclear Fusion* **50** 095011
[15] Liu C H *et al* 2012 *Acta Phys. Sin* **61** 205201

Preliminary validation of computational model for neutron flux prediction of Thai Research Reactor (TRR-1/M1)

S Sabaibang[1,3], S Lekchaum[1] and C Tipayakul[2]
[1] Department of Applied Physics, Faculty of Science , King Mongkut's Institute of Technology Ladkrabang, Latkrabang, Bangkok, Thailand,
[2] Thailand Institute of Nuclear Technology (Public Organization), Ongkharak, Nakhon Nayok, Thailand

E-mail: nemoshow2531@gmail.com

Abstract. This study is a part of an on-going work to develop a computational model of Thai Research Reactor (TRR-1/M1) which is capable of accurately predicting the neutron flux level and spectrum. The computational model was created by MCNPX program and the CT (Central Thimble) in-core irradiation facility was selected as the location for validation. The comparison was performed with the typical flux measurement method routinely practiced at TRR-1/M1, that is, the foil activation technique. In this technique, gold foil is irradiated for a certain period of time and the activity of the irradiated target is measured to derive the thermal neutron flux. Additionally, the flux measurement with SPND (self-powered neutron detector) was also performed for comparison. The thermal neutron flux from the MCNPX simulation was found to be 1.79×10^{13} neutron/cm^{2}s while that from the foil activation measurement was 4.68×10^{13} neutron/cm^{2}s. On the other hand, the thermal neutron flux from the measurement using SPND was 2.47×10^{13} neutron/cm^{2}s. An assessment of the differences among the three methods was done. The difference of the MCNPX with the foil activation technique was found to be 67.8% and the difference of the MCNPX with the SPND was found to be 27.8%.

1. Introduction

To utilize nuclear research reactors efficiently, neutron flux is generally an important parameter for the reactor users. In Thailand, there is a research reactor named "Thai Research Reactor–1/Modification 1 (TRR-1/M1)" which is a TRIGA Mark III swimming pool type. The operational power of the reactor power is 1.2 MW. It is used for training, research, isotope production and neutron activation analysis (NAA) [1]. The thermal neutron flux measurement at TRR-1/M1 adopts the gold foil activation technique. The measurement is conducted, at minimum, annually to monitor the flux changes within the irradiation facility. During the recent years, there have been efforts to develop and improve the computational model of Thai Research Reactor (TRR-1/M1). This model is expected to be capable of accurately predicting the neutron flux level and spectrum. In this research, the preliminary validation of the computational model was performed. The computational method of an in-core irradiation facility called CT (Central Thimble) was created. figure 1 shows the configuration of the TRR-1/M1 core including the location of CT irradiation facility within the core. The neutron flux result obtained by the computation model was compared with the results from two actual measurement methods to assess the

[3] To whom all correspondence should be addressed.

accuracy of the model. The first measurement method used the typical gold foil activation technique. The second measurement method employed SPND (self-powered neutron detector).

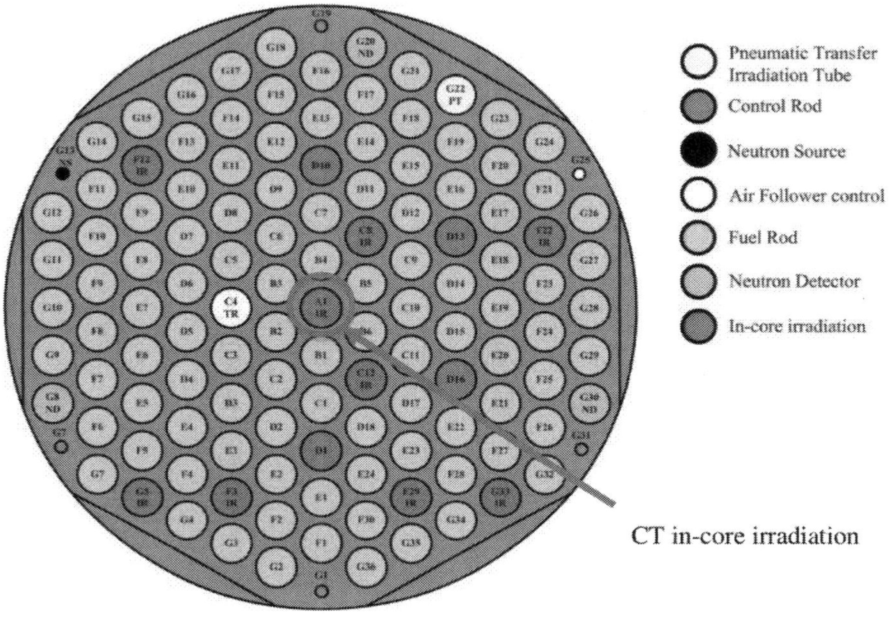

Figure 1. Configuration of TRR-1/M1 Core

2. Method

2.1. Computational Model

The computational model was created using MCNPX (Monte Carlo N-Particle Extended) computer code version 2.6 [2]. This computer code uses Monte Carlo method to simulate and calculate radiation transport. The model was expanded from the computational prototype [3], that is, the model of the CT irradiation facility was set up in two stages in order to save computational time. First, the model of the whole TRR-1/M1 core was created and the source distribution is set to be $5x10^8$ nps (number of particles) around the CT irradiation facility was recorded by SSR card of MCNPX [3]. In the second stage, the CT in-core irradiation facility was modeled in detail as shown in figure 2. This model uses the source distribution recorded from the previous stage as the source definition by SSW card of MCNPX [3]. By setting up the model in this two-stage scheme, the second stage requires much less computational time and therefore is easier to detect errors or make modeling adjustments. In this model, the size of the gold foil in the model was 0.25 cm in radius and 422.2 μm in thickness which is to mimic the size of the gold foil used in the foil activation measurement. The tally F4 card, the e0 and FM4 card [4] were used together in the simulation model in order to obtain thermal neutron flux spectrum and the reaction rate in different energy groups.

2.2. The gold foil activation method

In this experiment, the thermal neutron flux of the CT irradiation facility was measured by gold foil activation method. The gold foil measurement was performed according to the typical measurement procedure at TRR-1/M1. The dimensions of gold foil are shown in figure 3. To determine thermal neutron flux, two foils were irradiated. The first was a bare foil which was to measure the neutron flux

Journal of Physics: Conference Series **611** (2015) 012006 doi:10.1088/1742-6596/611/1/012006

in thermal and epithermal ranges while the second was covered with cadmium to measure the neutron flux only in the epithermal range. With these two foils, the thermal neutron flux can be calculated. The gold foil was loaded in the CT in-core irradiation facility for 20-minute irradiation. It was then left to decay for approximately 5 days. Finally, the radioactivity of the gold foil was measured to estimate the thermal neutron flux by a gamma spectrometer.

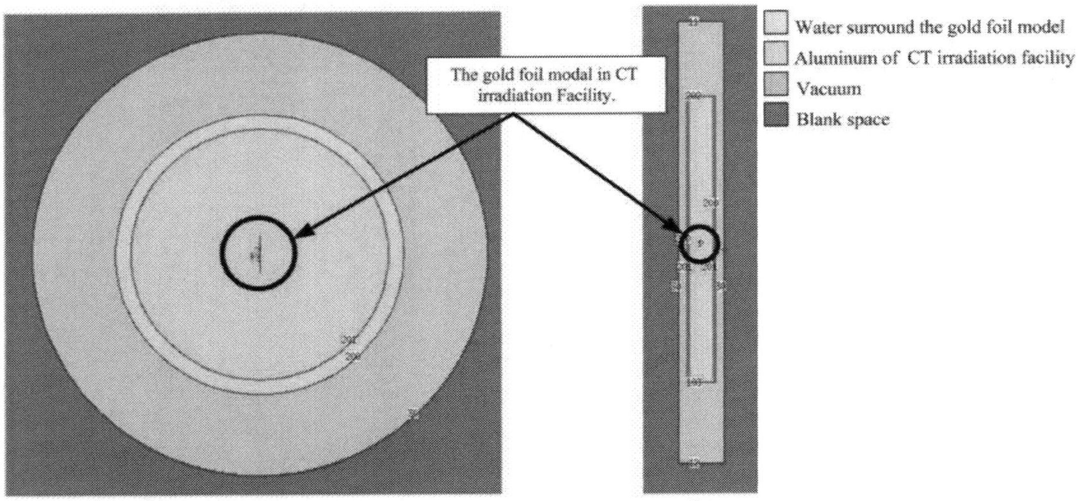

(a) (b)

Figure 2. The MCNPX model of CT irradiation model
(a) XY plane view.
(b) YZ plane view.

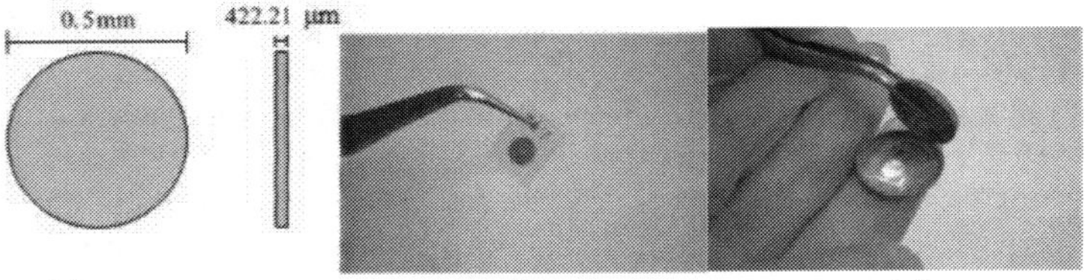

(a) (b) (c)

Figure 3. The preparation of gold foils for foil activation measurement
(a) The dimensions of the gold foil.
(b) The bare gold foil for thermal neutron measurement.
(c) The gold foil covered with cadmium for epithermal neutron measurement.

2.3. The SPND method

The SPND (self-powered neutron detector) was used for the measurement of the thermal neutron flux at the CT in-core irradiation facility for this research as well. The SPND is basically Rh based material which emits beta particles while being irradiated by thermal neutrons. The measurement of emitted beta particles is done in term of electrical current (A) which is proportional to the thermal neutron flux. As shown in figure 4, the SPND was attached to an aluminum stick and loaded into the CT irradiation position. The experiment of SPND was operated at the reactor powers of 10 kW, 100 kW, 500 kW,

1,000 kW and 1,300 kW respectively. The electrical current measured from the SPND was recorded and used for the estimation of the thermal neutron flux.

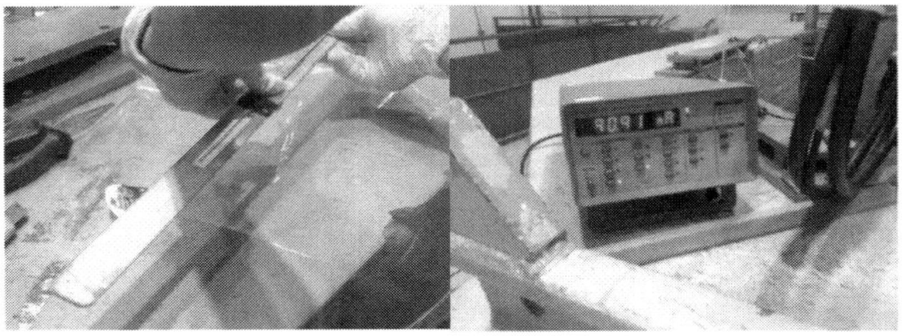

Figure 4. The preparation of SPND and the picoammeter setup.

3. Results

3.1. Result of MCNPX method

The thermal neutron flux in the gold foil model for the energy less than 0.025 eV (assuming this is the energy cutoff for thermal neutron flux) obtained by tally F4 of MCNPX was $4.29270\times10^{-4}\,(1/cm^2)$ per source. Then, the absolute thermal neutron flux (ϕ_{th}) at 1.2 MW was calculated from equation (1). The results of the thermal neutron flux at different reactor powers are shown in table 1.

$$\phi_{th(MCNPX)} = \text{Reactor Power} \times \text{fission constant} \times \phi_{TallyF4} \tag{1}$$

where *Reactor Power* is in the unit of MW,

fission constant = 3.467×10^{10} fissions/watt – sec [4, p5-102],

$\phi_{TallyF4}$ = Neutron flux as calculated from MCNPX, unit $1/cm^2$.

3.2. Result of the gold foil activation experiment

The thermal neutron flux was calculated for the gold foil activation by equation (2). The calculation of the thermal neutron flux at reactor power 1,200 kW is shown in table 1.

$$\phi_{th} = \frac{1}{Eff * f * \sigma_{th}}\left[\frac{CPS_{(bare)}}{N_{Au1}\left(1-e^{-\lambda_{tr1}}\right)e^{-\lambda_{d1}}} - \frac{CPS_{(Cd)}}{N_{Au2}\left(1-e^{-\lambda_{tr2}}\right)e^{-\lambda_{d2}}}\right] \tag{2}$$

where

ϕ_{th}	Thermal neutron flux (neutron/cm^{2}s)
Eff	The efficiency of gamma spectrometer
f	Decay fraction
σ_{th}	Microscopic activation cross section of $^{197}Au\,(n,\gamma)^{198}Au$
N_{Au1}	Number of atoms in the gold foil
N_{Au2}	Number of atoms in the cadmium-covered gold foil
CPS	Count per seconds. (for bare and cadmium-covered foils)

λ	Decay constant
tr	Irradiation time
td	Decay time

3.3. Result of the SPND

The constant of the SPND used in this experiment was 9.42×10^{-21} A/(neutron/cm^{2}s) and the thermal neutron flux was converted from the measured current by equation (3). The results of the thermal neutron flux measured by SPND at the reactor powers of 10, 100, 500, 1,000 and 1,200 kW respectively are shown in table 1.

$$\phi_{th(SPND)} = current\,(nA)\,/\,constant\;detector \tag{3}$$

where constant detector = 9.42×10^{-21} A/(neutron/cm^{2}s) (from SPND calibration)

Table 1. The result of the thermal neutron flux.

Reactor power (kW)	ϕ (neutron/cm^{2}s)		
	MCNPX	Gold foil activation	SPND
10	1.488×10^{11}	-	2.200×10^{11}
100	1.488×10^{12}	-	2.266×10^{12}
500	7.441×10^{12}	-	1.089×10^{13}
1,000	1.488×10^{13}	-	2.200×10^{13}
1,200	1.786×10^{13}	4.684×10^{13}	2.472×10^{13}

At 1,200 kW which is the typical operating power level, the thermal neutron flux from the MCNPX was found to be 1.79x10^{13} neutron/cm^{2}s. The thermal neutron flux from the gold foil activation measurement was 4.68x10^{13} neutron/cm^{2}s and the thermal neutron flux of SPND was 2.47x10^{13} neutron/cm^{2}s. The comparison of the thermal neutron fluxes between the MCNPX with the gold foil activation technique were found to be 67.8% and the difference of the MCNPX with the SPND was found to be 27.8% calculated from equation (4).

$$Percentage\,difference = \frac{|E-S|}{S}\times100\% \tag{4}$$

where E = the thermal neutron from the gold foil activation or the thermal neutron flux of SPND
S = the thermal neutron from the MCNPX

The MCNPX model gave a result closer to the SPND than the gold foil activation technique. The causes of discrepancies among the three methods are to be verified. It is believed that the major difference between the results from MCNPX model and those from other methods is due to the selected energy cutoff which could not be a good representation to the experiment. Moreover, the two experiment methods should be validated against each other as well. It is expected that the foil activation method should have higher measurement uncertainty because there are more steps requiring different types of measuring activities and hence each can contribute to the overall uncertainty. This foil activation experiment shall be repeated in the future with the uncertainty estimation. The SPND is believed to be the best reference method since it is more straightforward and measures the quantity which is directly proportional to the neutron flux.

4. Conclusion

Three methods, which are a computational model, foil activation technique and SPND, to obtain the thermal neutron fluxes were conducted at TRR-1/M1. The results from these methods were compared in this work. It was found that the computational model results were closer to the SPND method than the foil activation technique. It still cannot be concluded at the moment that the computational model is adequately accurate because the difference between the foil activation technique and SPND was not negligible. The foil activation and the SPND measurements may be subject to various statistical and systematic uncertainties which would affect the accuracy of the thermal neutron flux measurement. It is believed, however, the energy cutoff in the computational model could be the main cause of the large difference. More work is needed to make the thermal neutron fluxes measured by the foil activation technique and the SPND agree before making the final assessment of the accuracy of the MCNPX model.

Acknowledgements
The authors would like to thank the Department of Physics, King Mongkut's Institute of Technology Ladkrabang and the Reactor Operation Group, Thailand Institute of Nuclear Technology.

References

[1] Thailand Institute of Nuclear Technology 2010 *Safety Analysis Report for Thai Research Reactor - 1/Modification 1 (TRR-1/M1)* Revision 03 Chapter 5
[2] Pelowitz D B 2008 *MCNPX User's Manual, Version 2.6.0*, Los Alamos National Laboratory report LA-CP-07-1473
[3] Sabaibang S, Tipayakul C and Lekchaum S 2013 Analysis of neutron flux measurement in Thai Research Reactor-1/Modification 1 using the Monte Carlo Method *International Science, Social Science, Engineering and Energy Conference* **5** 402-405
[4] x-5 Monte Carlo Team 2008 *MCNP-A general Monte Carlo N-particle Transport code* Version 5 LA-UR-03-1987

International Nuclear Science and Technology Conference 2014 (INST2014) IOP Publishing
Journal of Physics: Conference Series **611** (2015) 012007 doi:10.1088/1742-6596/611/1/012007

Validation of the MCNP computational model for neutron flux distribution with the neutron activation analysis measurement

K Tiyapun[1], M Chimtin, S Munsorn and S Somchit
Reactor Center, Thailand Institute of Nuclear Technology, 16 Vibhavadi Rangsit St., Bangkok, 10900, Thailand

E-mail: kanokrat@tint.or.th

Abstract. The objective of this work is to demonstrate the method for validating the predication of the calculation methods for neutron flux distribution in the irradiation tubes of TRIGA research reactor (TRR-1/M1) using the MCNP computer code model. The reaction rate using in the experiment includes $^{27}Al(n, \alpha)^{24}Na$ and $^{197}Au(n, \gamma)^{198}Au$ reactions. Aluminium (99.9 wt%) and gold (0.1 wt%) foils and the gold foils covered with cadmium were irradiated in 9 locations in the core referred to as CT, C8, C12, F3, F12, F22, F29, G5, and G33. The experimental results were compared to the calculations performed using MCNP which consisted of the detailed geometrical model of the reactor core. The results from the experimental and calculated normalized reaction rates in the reactor core are in good agreement for both reactions showing that the material and geometrical properties of the reactor core are modelled very well. The results indicated that the difference between the experimental measurements and the calculation of the reactor core using the MCNP geometrical model was below 10%. In conclusion the MCNP computational model which was used to calculate the neutron flux and reaction rate distribution in the reactor core can be used for others reactor core parameters including neutron spectra calculation, dose rate calculation, power peaking factors calculation and optimization of research reactor utilization in the future with the confidence in the accuracy and reliability of the calculation.

1. Introduction

The research reactor (TRIGA- Mark III) at Thailand Institute of Nuclear Technology (TINT) referred to as TRIGA Research Reactor Modified I (TRR-1/M1) is most commonly used as neutrons sources for gem irradiation, isotope production, neutron radiography and scientific investigation using nuclear analytical techniques. Knowledge of the reactor physical parameters including neutron flux, gamma flux, neutron profiles, power profile and temperature profiles are very important to ensure safe operation, to manage for fuel loading pattern, to optimize for the fuel shuffling and to support experiment at the reactor. For the nuclear analytical techniques and neutron activation analysis (NAA), the accurate neutron spectral characteristics and neutron flux profile are crucial for the assessment of the uncertainty of the experiment and the accuracy of the method. Monte Carlo calculation with the MCNP code [1] has been validated against experimental measurements for the neutron flux and reaction rate. The computational methods, particularly geometry model of the reactor core using the

[1] To whom any correspondence should be addressed.

Monte Carlo is very powerful in providing the results for the neutron flux spectrum and the reaction rate distribution. The main objective of this study is to compare between the experimental and the calculated values from a computational model. Validation is conducted to test computational model capability in predicting the measured neutron flux distribution in the reactor core and irradiation tubes of TRR-1/M1. The results from this study can be used to support for the consistency of the MCNP results with the experimental results and to improve accuracy and reliability of the NAA results.

2. Experiments and Measurements

The neutron activation analysis was used to verify the calculated neutron flux distribution in different location of the TRIGA reactor. The experiment used aluminum 99.9%wt and gold foil (disk of 0.25 cm radius and 0.0005 cm thick) which were irradiated in 9 locations in the reactor core. The positions of the foils in the reactor core during irradiation are demonstrated in figure 1. Gold foils and gold foil covered with cadmium samples were irradiated with reactor power 10 kW for 20 minutes and 40 minutes, respectively. Aluminum samples were irradiated at 500 kW for 20 minutes. After the irradiation, the gamma ray count rates were measured with high purity germanium detector (HPGe) detector for calculating the specific activities of irradiated samples.

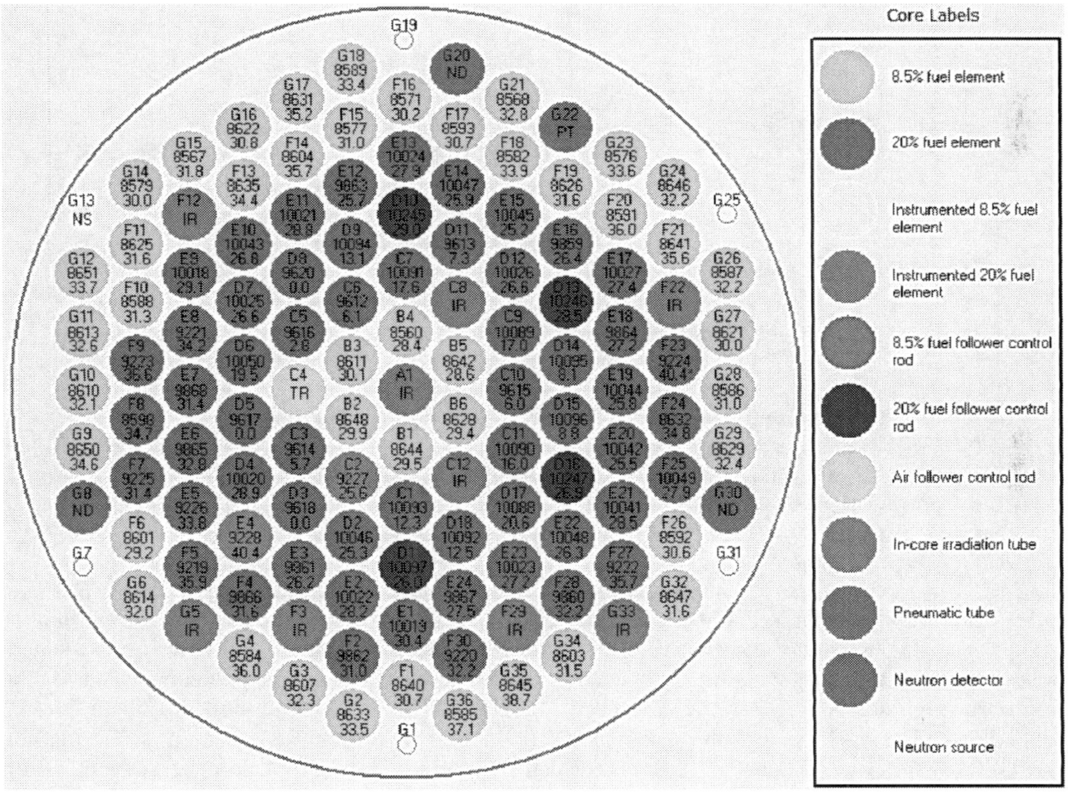

Figure 1. Schematic top view of TRIGA. The location of aluminium and gold foils used in the experiment are in In-core irradiation tubes (pink).

The activation reactions which were used in the experiments are as follows:

$$^{27}Al(n,\alpha)^{24}Na \xrightarrow{\gamma} {}^{24}Na \xrightarrow{t_{\frac{1}{2}}=15\,hr} {}^{24}Mg + e^- + \nu + \gamma\ (E = 1368.6\ keV) \tag{1}$$

$$^{197}Au(n,\gamma)^{198}Au \xrightarrow{\gamma} {}^{198}Au \xrightarrow{t_{\frac{1}{2}}=2.7d} {}^{198}Hg + e^- + \nu + \gamma\ (E = 411.8\ keV) \tag{2}$$

E is the energy of the most prominent gamma ray. The threshold for the first reaction which is illustrated in equation (1) has energy threshold at 5 MeV which is used for the fast neutron flux measurement. The cross section for the second reaction ^{197}Au(n, γ) is the highest for the energy below 10 keV, therefore, this reaction is useful for the investigation of the thermal neutron flux (E < 0.625 eV) and epithermal neutron flux (0.625 eV < E < 100 keV). Figure 2 shows the cross section for ^{197}Au(n, γ)^{198}Au and ^{27}Al(n, α)^{24}Na reactions from ENDF/B-VII nuclear data library [2].

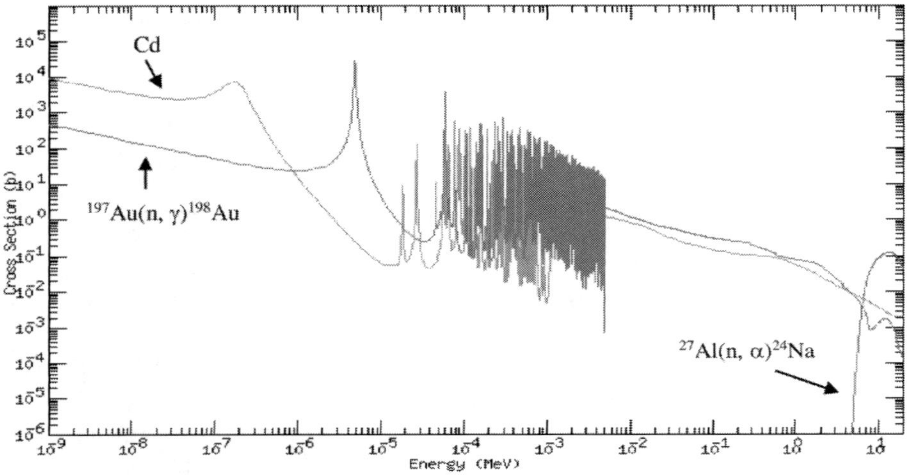

Figure 2. Cross sections for ^{197}Au(n, γ)^{198}Au and ^{27}Al(n, α)^{24}Na reaction and Cd from ENDF/B-VII nuclear data library [2].

3. MCNP Computational Method

3.1. MCNP Computer Code

The MCNP computer code was used for reaction rate and neutron flux calculation [1]. MCNP is a general-purpose Monte Carlo N-Particle code that can be used for neutron, photon, electron, or coupled neutron/photon/electron transport. Pointwise cross-section data typically are used, although group-wise data also are available. For neutrons, all reactions given in a particular cross-section evaluation (such as ENDF/B-VI) are accounted for. Thermal neutrons are described by both the free gas and S(α, β) models. For photons, the code accounts for incoherent and coherent scattering, the possibility of fluorescent emission after photoelectric absorption, absorption in pair production with local emission of annihilation radiation, and bremsstrahlung.

3.2. TRR-1/M1 Computational Model

The geometrical model of TRIGA-Mark III (TRR-1/M1) reactor was modelled for accurately calculated physical parameters of the TRIGA reactor. The model is compared to the criticality benchmark model, which is described and published in the International Handbook of Evaluated Criticality Safety Experiments [3]. The simplified model can be used in calculations of the effective multiplication factor k_{eff} [4], power peaking factor [5] and reactor kinetic parameters [6]. The model of the TRR-1/M1 simplification of the geometry is modelled by neglecting the surrounding detailed of the core to an extent that does not affect neutron flux and k_{eff} significantly. MCNP calculations were performed in kcode mode for criticality calculations using 3×10^5 neutrons per cycle and 1000 active cycles. The standard deviations were used as uncertainties for the calculations. A criticality source (ksrc card) was utilized in this study. The initial fission neutron source distribution used the source point as the input for ksrc card. Reaction rates of $^{197}Au(n, \gamma)^{198}Au$ and $^{27}Al(n, \alpha)^{24}Na$ reactions in individual irradiation tubes are calculated by multiplying neutron flux in the irradiation tubes with corresponding cross section (figure 2). The Monte Carlo results of neutron flux are normalized per one fission neutron. Therefore the results have to be properly scaled in order to compare the neutron flux values with the measurement quantities. The absolute values of neutron flux are directly proportional to the reactor power level. The calculation results have to be scaled according to the known power level. The scaling factor is defined in the following equation:

$$\Phi = \Phi_{\text{MCNP}} \frac{P\bar{\nu}}{wk_{eff}} \quad (3)$$

where P is the power level, $\bar{\nu}$ is the average number of neutrons per fission, w is the average energy per fission and k_{eff} is the effective multiplication factor determined in the MCNP run. The neutron flux parameters are referred to as thermal neutron flux, epithermal neutron flux and thermal to epithermal neutron flux ratio were calculated [7, 8].

4. Results

The MCNP calculated $^{197}Au(n, \gamma)^{198}Au$, $^{197}Au(n, \gamma)^{198}Au$ covered with cadmium and $^{27}Al(n, \alpha)^{24}Na$ reaction rates which normalized to 1200 kW power of all In-core irradiation tubes are plotted compared with the measurement of bare gold, gold covered with cadmium and aluminium. Since the Monte Carlo results of neutron flux are normalized per one fission neutron, the absolute value of neutron flux or reaction rate is directly proportional to the reactor thermal power level. The measurements and the MCNP calculations for thermal, epithermal and fast neutron flux are illustrated in figure 3, 4 and 5, respectively.

4.1. Comparison between the MCNP model and the measurements

In the MCNP simulation, the energy interval for thermal neutron is $E < 0.625$ eV and epithermal neutron is 0.625 eV $< E <$ 100 keV and fast neutron is $E >$ 100 keV. The average values for thermal flux in ring C was $(2.315 \pm 0.001)\times10^{13}$, ring F was $(1.157 \pm 0.001)\times10^{13}$ and ring G was $(8.492 \pm 0.001)\times10^{12}$ $n\cdot cm^{-2}s^{-1}$, respectively. The average values for epithermal flux in ring C was $(2.239 \pm 0.006)\times10^{12}$, ring F was $(8.614 \pm 0.005)\times10^{11}$ and ring G was $(4.691 \pm 0.004)\times10^{11}$ $n\cdot cm^{-2}s^{-1}$, respectively. The average values for fast flux in ring C was $(1.069 \pm 0.002)\times10^{13}$, ring F was $(6.520 \pm 0.002)\times10^{12}$ and ring G was $(3.194 \pm 0.001)\times10^{12}$ $n\cdot cm^{-2}s^{-1}$, respectively. In experimental results, the average values for thermal flux in ring C was 2.494×10^{13}, ring F was 1.070×10^{13} and ring G was 7.661×10^{12} $n\cdot cm^{-2}s^{-1}$, respectively. The average values for epithermal flux in ring C was 1.964×10^{12}, ring F was 8.411×10^{11} and ring G was 5.878×10^{11} $n\cdot cm^{-2}s^{-1}$, respectively. The average values for fast flux in ring C was 1.253×10^{13}, ring F was 7.140×10^{12} and ring G was 4.188×10^{12} $n\cdot cm^{-2}s^{-1}$, respectively. Figure 3 to figure 5 show comparisons of the MCNP calculations with the measurement values which show acceptable results.

4.2. The uncertainties and the discrepancies

However, it is expected that most variations in the measurement results were due to the position of the measuring point in the irradiation tubes. The discrepancies may results from the effect of fuel element burnup, the temperature effects, and the relative large uncertainty in the power calibration, which may directly affects the calculation values. The uncertainties of the calculation model may influence from the geometrical, dimension, material properties including composition and densities data, control rod positioning and the irradiation tubes positioning in the reactor core.

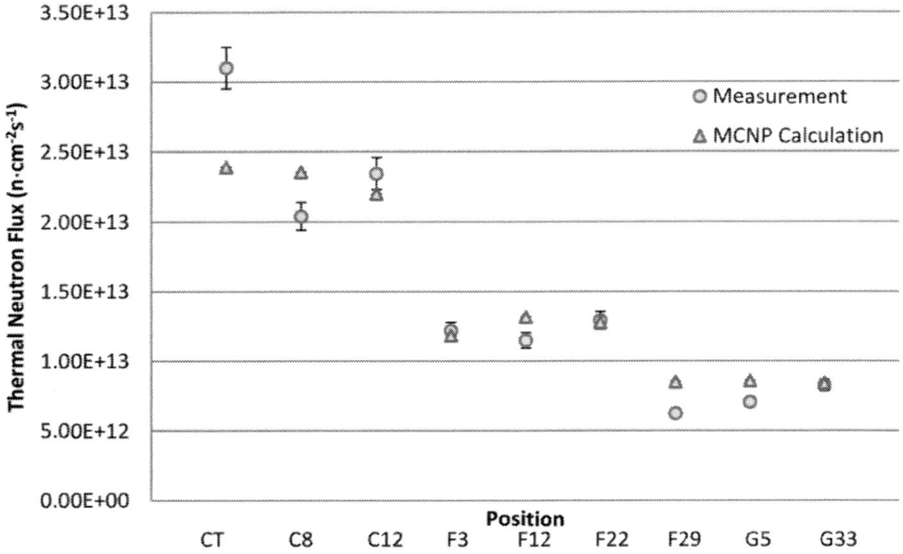

Figure 3. Thermal neutron flux results in 9 in-core irradiation tubes by the MCNP code and measurements.

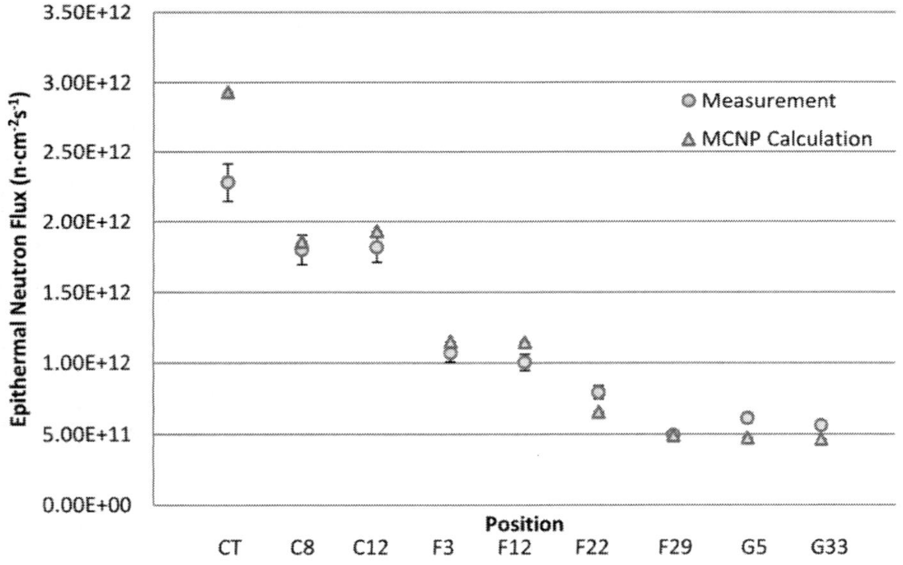

Figure 4. Epithermal neutron flux results in 9 in-core irradiation tubes by the MCNP code and measurements.

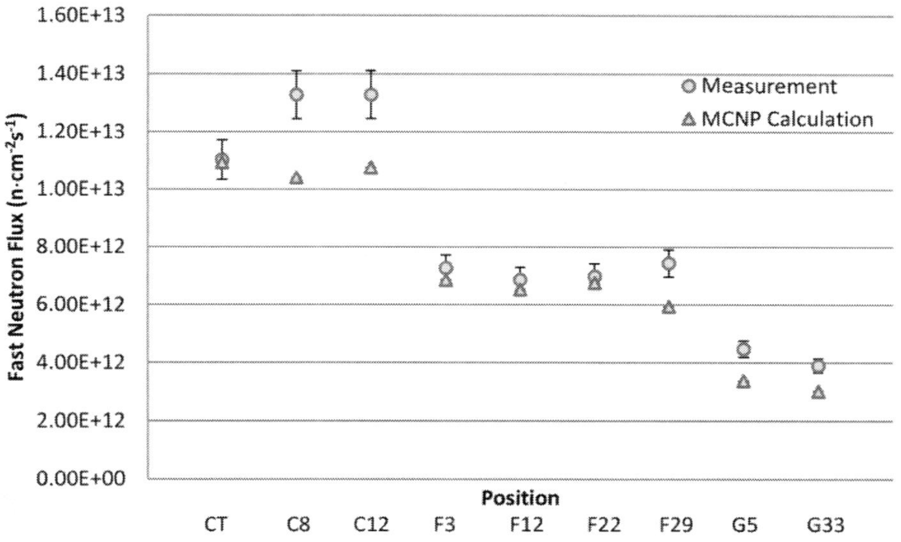

Figure 5. Fast neutron flux results in 9 in-core irradiation tubes by the MCNP code and measurements.

5. Discussion and Conclusion

Verification of the neutron flux distribution in the incore irradiation tubes was performed by comparing the MCNP calculation and the measurement of the neutron flux was determined from the $^{27}Al(n, \alpha)^{24}Na$ and $^{197}Au(n, \gamma)^{198}Au$ reaction rates. In this study the agreement between measured and calculated values are very good within the measurement and calculation uncertainties. The results indicated that the material and the geometrical properties used in the MCNP code are modelled well. The calculated neutron flux distribution will be more accurate with refinement in the future study. For future study, in order to observe the neutron flux influence from individual structures, the detailed or refinement model should be constructed such as graphite of the thermalizing and thermal column, triangular irradiation tubes, radial and tangential beam ports. The effect of fuel element burnup model should be modelled and radioisotopes which are significantly contributed to k_{eff} should be taken into account. The modification of the computational model for burnup and temperature effects should be conducted to validate the burnup and temperature effect uncertainties. The detailed sensitivity study of each parameter including xenon buildup, nitrogen dissolved, geometrical and material data (densities, compositions), the control rod position and cross section libraries which are used in the MCNP model and the effect from neighbouring irradiation should be performed to explain the systematic discrepancies between the neutron flux measurement and the MCNP calculation model. From this study, the neutron flux distribution can be utilized for spectra calculations, dose rate calculations, reaction rate calculations, power peaking factors, and kinetic parameters of the reactor core.

6. References

[1] X-5 Monte Carlo Team 2003 *MCNP - A General Monte Carlo N-Particle Transport Code, Version 5* (Oak Ridge: Los Alamos National Laboratory) LAUR-03-1987 LA-CP-03-0245

[2] Chadwick M B 2006 ENDF/B-VII *Next generation evaluated nuclear data library for nuclear science and technology, nuclear data sheets, special issue on evaluated nuclear data file ENDF/B-VII* (Brookhaven: National Nuclear Data Center)

[3] Jeraj R and Ravnik M 1999 *TRIGA Mark II Reactor: U(20)-zirconium hydride fuel rods in water with graphite reflector. IEU-COMP-THERM-003, International Handbook of Evaluated Critical Safety Benchmark Experiments* (Paris: Nuclear Energy Agency) NEA/NSC/DOC(95)03

[4] Ravnik M and Jeraj R 2003 Research reactor benchmarks *Nucl. Sci. Eng.* **145** 145-152

[5] Snoj L and Ravnik M 2008 Power peaking in mixed TRIGA cores *Nucl. Eng. Des.* **238(9)** 2473-2479

[6] Snoj L, Kavcic A, Zerovnik G and Ravnik M 2010 Calculation of kinetic parameter for mixed TRIGA cores with Monte Carlo *Nucl. Ene.* **37(2)** 223-229

[7] De Corte F 2001 The standardization of standardless NAA *J Radioanal. Nucl. Chem.* **248(1)** 13-20

[8] Simonits A, De Corte F, Moens L and Hoste J 1982 Status and recent developments in the k_o standardization method *J. Radioanal. Nucl. Chem.* **72(1)** 209-230

Journal of Physics: Conference Series **611** (2015) 012008 doi:10.1088/1742-6596/611/1/012008

Comparison of carbon density distribution in *L*-mode plasma discharges in TFTR tokamak using the Mixed Bohm/gyro-Bohm and Multi-Mode-95 transport models

W Buangam[1,4], T Onjun[1], N Poolyarat[2] and R Picha[3]
[1]Sirindhorn International Institute of Technology, Thammasat University, Pathumthani, Thailand
[2]Department of Physics, Thammasat University, Pathumthani, Thailand
[3]Thailand Institute of Nuclear Technology, Bangkok, Thailand

E-mail: w.buangam@gmail.com

Abstract. The behaviors of carbon density distribution in low confinement mode (*L*-mode) plasma in TFTR Tokamak are investigated using 1.5D BALDUR integrated predictive modeling code. In each simulation, either turbulent transport either Mixed Bohm/gyro-Bohm (Mixed B/gB) turbulent transport model or Multi-Mode (MMM95) turbulent transport model is used to describe thermal, particle, and impurity transports in the core region of the tokamak. It is found that, the simulations using both turbulent transport models agree equally well with experimental results, with an average RMS of 7-20% for the simulations with Mixed B/gB model and 9-16% for the simulations with MMM95 model.

1. Introduction

Although plasma in Low Confinement mode (*L*-mode) has lower performance than that of High Confinement mode (*H*-mode), it has better plasma stability and simplicity in operation [1]. If the understanding of *L*-mode plasma is improved, it could potentially enhance the *L*-mode performance. One of the important issues for an *L*-mode plasma experiment is the accumulation of impurity in tokamak. The presence of impurity can result in radiation losses and fuel dilution, which leads to a degradation of fusion performance [2]. As a result, it is important to investigate impurity density distribution in *L*-mode plasma.

In previous study by D.R. Baker *et al.* [3], there was an accumulation and peaking of impurities in the core in DIII–D discharges with centrally enhanced confinement and peaked density profiles, as was predicted by neoclassical theory. The work of G. Bateman *et al.* [4] used BALDUR code together with the Multi-mode (MMM95) core transport model to predict temperature and density profiles. It was found that those predicted profiles agreed with experimental data with an RMS deviation less than 15% for 41 *L*-mode and *H*-mode discharges from the Tokamak Fusion Test Reactor (TFTR), the Doublet III-D tokamak (DIII-D) and the Joint European Torus (JET). In Ref [5], T. Onjun *et al.* used BALDUR code with either the MMM95 or the Mixed B/gB core transport model to simulate plasma behaviours in tokamak for *L*-mode regime. Electron temperature and electron density profiles from these simulations were compared with experimental data from DIII-D and TFTR but they neglected to observe the impurity density profiles.

In this work, it aims to study the behaviors of carbon density distribution and impurity transport in *L*-mode tokamak in TFTR Tokamak by using BALDUR code. This work is organized as follows: brief descriptions of relevant components of the BALDUR code, including the MMM95 model and the Mixed B/gB model are presented in section 2; simulation results are described in section 3; and the conclusion is given in section 4.

[4] To whom any correspondence should be addressed.

2. BALDUR integrated predictive modeling code

The BALDUR code is an integrated predictive modeling code [6] developed for self-consistency simulating time evolution of various plasma properties, including electron and ion temperatures, electron and ion densities, helium and impurity densities. The BALDUR code combines various modules, in which each module is responsible for computing different physical phenomena, such as transport, heating and instability. It was shown in many reports that BALDUR code is capable to predict electron and ion temperatures, and electron density profiles within 10% relative RMS of experimental data [5, 7].

2.1. Multimode Core Transport Model (MMM95)

The Multi-Mode Model version 1995 (MMM95) [4] is a transport model based on a combination of transport due to different kinds of physics. It consists of the Weiland model for the ion temperature gradient (ITG) and trapped electron modes (TEM), the Guzdar–Drake model for drift-resistive ballooning (RB) modes, and modified kinetic ballooning (KB) modes. The thermal and particle transport coefficients can be expressed as:

$$\chi_i = 0.8\chi_{i,ITG\&TEM} + 0.65\chi_{i,KB}, \tag{1}$$

$$\chi_e = 0.8\chi_{e,ITG\&TEM} + \chi_{e,RB} + 0.65\chi_{e,KB}, \tag{2}$$

$$D_H = 0.8D_{H,ITG\&TEM} + D_{H,RB} + D_{H,KB}, \tag{3}$$

$$D_z = 0.8D_{z,ITG\&TEM} + D_{z,RB} + D_{z,KB}. \tag{4}$$

where χ_i is ion thermal diffusion coefficient in m^2/s, χ_e is electron thermal diffusion coefficient in m^2/s, D_H is hydrogenic particle diffusion coefficient in m^2/s and D_z is impurity particle diffusion coefficient in m^2/s.

2.2. Mixed B/gB model

The Mixed B/gB core transport model [9] is a semi-empirical transport model, which is a combination of Bohm (B) and gyro-Bohm (gB) terms. The transport diffusivity in Bohm model is proportional to the gyro-radius times the thermal velocity, while the transport diffusivity in gyro-Bohm model is proportional to the square of the gyro-radius times the thermal velocity divided by the plasma major radius. The expressions of transport coefficients in this model are:

$$\chi_i = 0.5\chi_{gB} + 4.0\chi_B, \tag{5}$$

$$\chi_e = 1.0\chi_{gB} + 2.0\chi_B, \tag{6}$$

$$D_H = D_z = \left(0.3 + 0.7\frac{r}{a}\right)\frac{\chi_e\chi_i}{\chi_e + \chi_i}, \tag{7}$$

$$\text{where } \chi_{gB} = 5\times10^{-6}\sqrt{T_e}\left|\frac{\vec{\nabla}(T_e)}{B_T^2}\right|, \tag{8}$$

$$\chi_B = 4\times10^{-5}R\left|\frac{\vec{\nabla}(n_eT_e)}{n_eB_T}\right|q^2\left(\frac{T_e\left(\frac{r}{a}=0.8\right) - T_e\left(\frac{r}{a}=1\right)}{T_e\left(\frac{r}{a}=1\right)}\right), \tag{9}$$

where r/a is the normalized minor radius of the plasma, χ_{gB} is the gyro-Bohm contribution in m^2/s, χ_B is the Bohm contribution in m^2/s, T_e is the local electron temperature in keV, B_T is the toroidal magnetic field in Tesla, and n_e is the local electron density in 10^{19} m^{-3}.

3. Simulation results

In this work, the BALDUR integrated predictive modeling code is used to simulate the radial profiles of impurity density in *L*-mode scenario from 4-different auxiliary heating power discharges of TFTR tokamaks. Some important engineering parameters of these discharges are listed in table 1.

For each simulation, either MMM95 anomalous transport or Mixed B/gB anomalous transport model is used to describe both thermal and particle transports. It is assumed that pedestal impurity density in the simulation is the same as that in the experiment.

Table 1. List of engineering parameters in TFTR discharges.

Tokamak Shot No.	Unit	45966	50911	50921	62270
Major radius (R)	m	2.45	2.45	2.45	2.45
Minor radius (a)	m	0.80	0.798	0.797	0.80
Plasma current (I_p)	MA	2.00	1.78	0.89	1.78
Magnetic field (B_T)	T	4.76	4.23	2.14	4.77
Elongation (κ_{95})	-	1.00	1.00	1.00	1.00
Line average electron density ($\bar{n}_e$)	10^{19} m^{-3}	3.44	4.37	1.77	3.23
Auxiliary heating power (P_{aux})	MW	11.30	17.72	4.66	19.20
Diagnostic time (t_{diag})	s	3.47	3.93	3.95	4.17

3.1. Profile comparison

Figure 1 shows the radial profiles of the ion thermal diffusion coefficient (χ_i) from the MMM95 model and the Mixed B/gB model for discharges no. 45966 and 50921. It is found that χ_i from Mixed B/gB is higher than that from MMM95 model. Figure 2 shows the radial profiles of the electron thermal diffusion coefficient (χ_e) from the MMM95 model and the Mixed B/gB model for discharges no. 45966 and 50921. χ_e of Mixed B/gB tends to be higher than that of MMM95 model in most of the region for both discharges, except at the edge of the plasma. It is also found that the ion temperature gradient term (ITG) provides the largest contribution in the center of the plasma as shown in figure 4. Note that, the

Journal of Physics: Conference Series **611** (2015) 012008 doi:10.1088/1742-6596/611/1/012008

peak in figure 3 of MMM95 come from the error of the model because this model consists of many models to predict the simulation. Figure 5 shows the radial profiles of the impurity (carbon) diffusion

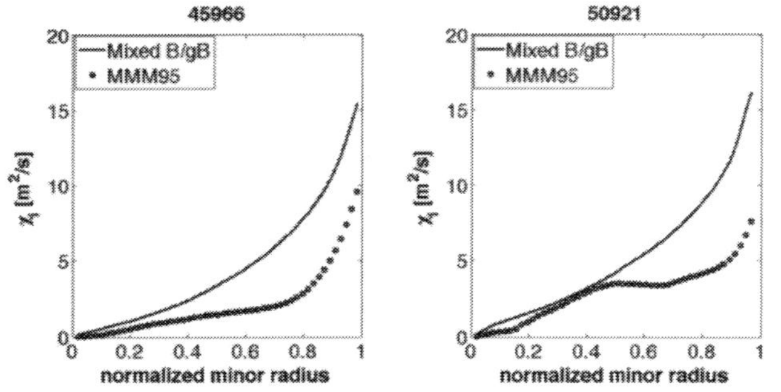

Figure 1. Ion thermal diffusion coefficient (χ_i) as a function of normalized minor radius for TFTR discharge 45966 (left) and TFTR discharge 50921 (right). The solid lines represent Mixed B/gB model while the dotted lines represent MMM95 model.

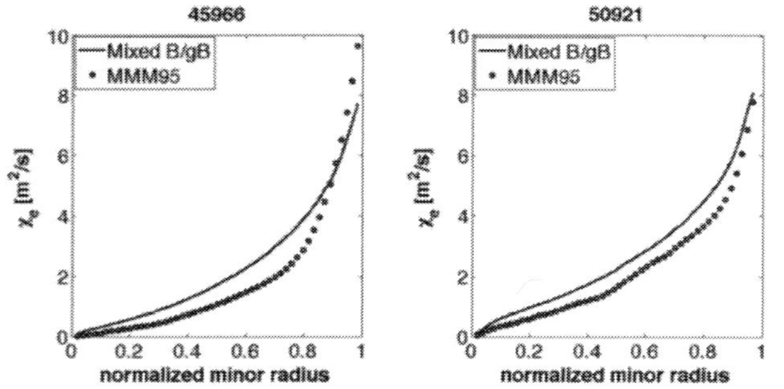

Figure 2. Electron thermal diffusion coefficient (χ_e) as a function of normalized minor radius for TFTR discharge 45966 (left) and TFTR discharge 50921 (right). The solid lines represent Mixed B/gB model while the dotted lines represent MMM95 model.

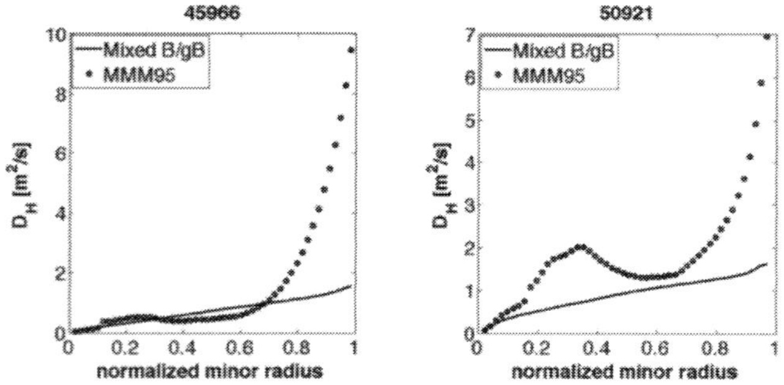

Figure 3. Hydrogenic particle diffusion coefficient (D_H) as a function of normalized minor radius for TFTR discharge 45966 (left) and TFTR discharge 50921 (right). The solid lines represent Mixed B/gB model while the dotted lines represent MMM95 model.

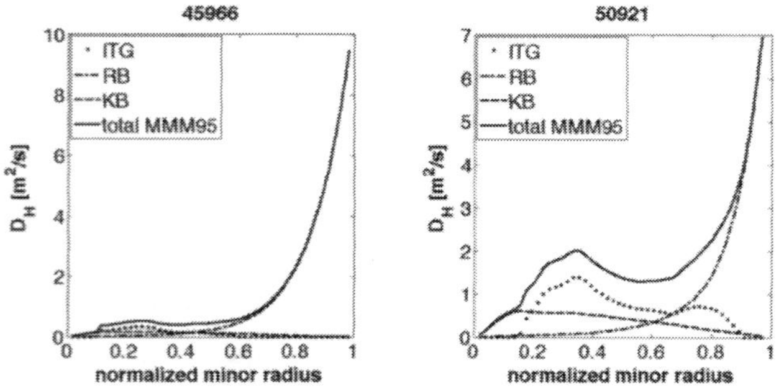

Figure 4. Contributions of hydrogenic particle diffusion coefficient (D_H) of MMM95 model as a function of normalized minor radius for TFTR discharge 45966 (left) and TFTR discharge 50921 (right). The dotted lines show the (ITG) contribution, the dashed-dotted lines show the resistive ballooning (RB) contribution, the dashed lines show the kinetic ballooning (KB) contribution, and the solid lines show the total effective diffusivity (tot MMM95).

coefficient (D_z) from either the MMM95 model or the Mixed B/gB model for discharges no. 45966 and 50921. D_z of Mixed B/gB model is higher than MMM95 model in the core of plasma but kinetic ballooning term (KB) provides the largest contribution in the center of the plasma while the resistive ballooning contribution term (RB) provides the largest contribution near the edge of the plasma. Figure 7 shows impurity (carbon) density profiles for discharges no. 45966 and 50921 from TFTR.

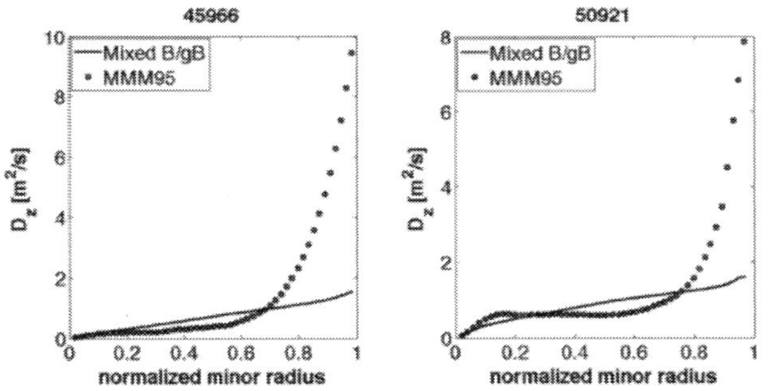

Figure 5. Impurity diffusion coefficient (D_z) as a function of normalized minor radius for TFTR discharge 45966 (left) and TFTR discharge 50921 (right). The solid lines represent Mixed B/gB model while the dotted lines represent MMM95 model.

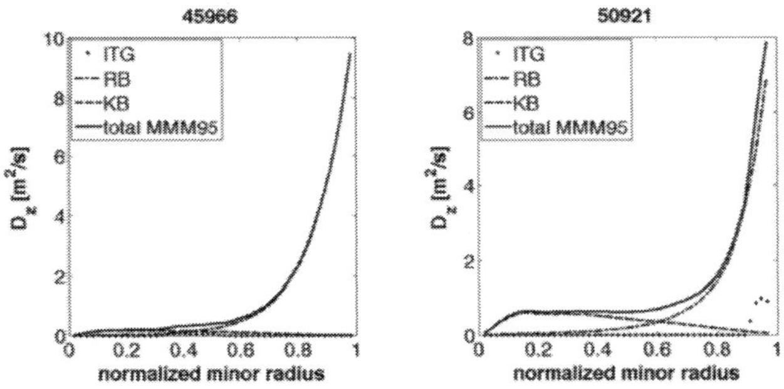

Figure 6. Contributions of Impurity diffusion coefficient (D_z) of MMM95 model as a function of normalized minor radius for TFTR discharge 45966 (left) and TFTR discharge 50921 (right). The dotted lines show the (ITG) contribution, the dashed-dotted lines show the resistive ballooning (RB) contribution, the dashed lines show the kinetic ballooning (KB) contribution, and the solid lines show the total effective diffusivity (total MMM95).

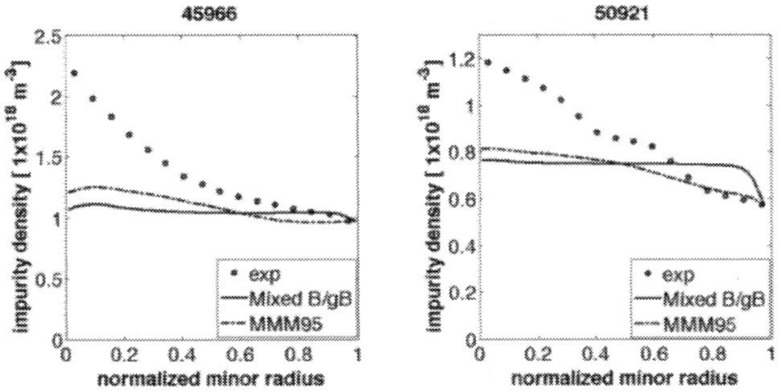

Figure 7. Impurity (carbon) density as a function of normalized minor radius for TFTR discharge 45966 (left) and TFTR discharge 50921 (right). In each panel, dotted lines represent results of experimental data, solid lines represent Mixed B/gB model and dot-dashed lines represent results of MMM95 model.

It is found that the impurity density profiles simulated using MMM95 model and Mixed B/gB model are lower than the impurity density profiles from the experimental data, which can be explained by the hydrogenic particle diffusion (D_H) and the impurity diffusion coefficient (D_z) of the simulation being higher than that of the experiment. If the transport is higher, the lower the impurity density profiles will be lower. In addition, the impurity density profiles for the simulations using MMM95 model are higher than those using Mixed B/gB model for both discharges.

3.2. Statistical analysis

Two statistical quantities: the relative root-mean-square deviation (RMSD) and the relative offset are used to quantify the comparison between the simulations and the experiments. Both are computed based on the difference between simulation profiles and experimental data. The RMS deviation for impurity (carbon) density is defined as

$$\sigma_X = \sqrt{\frac{1}{N}\sum_{j=1}^{N}\left(\frac{X_j^{\exp} - X_j^{sim}}{X_{\max}^{\exp}}\right)^2} \quad (10)$$

where $X_j^{\exp}$ and $X_j^{\rm sim}$ are the $j^{\rm th}$ data point of the simulation and experimental profiles, respectively, while $X_{\max}^{\exp}$ is the maximum data point of the experimental profile of X as a function of radius which has N points in total. The RMS deviation of impurity (carbon) density profile is designated by σ_i ,
then the distribution of the RMS deviation over all the discharges can be characterized by the average RMS deviation as

$$\bar{\sigma} = \frac{1}{N_S}\sum_{i=1}^{N_X}\sigma_i \, , \quad (11)$$

and the RMS deviation is defined as

$$\sigma_\sigma = \sqrt{\frac{1}{N_S - 1}\sum_{i=1}^{N_S}(\sigma_i - \bar{\sigma})^2} \, , \quad (12)$$

where N_s is the number of all discharges. The relative offset of impurity (carbon) density is defined as

$$f = \frac{1}{N}\sum_{j=1}^{N}\left(\frac{X_j^{exp} - X_j^{sim}}{X_{max}^{exp}}\right). \quad (13)$$

This statistical analysis is used for comparing the simulation profiles and experimental profiles. The results of the statistical analyses for simulations are presented in figures 8 and 9. The RMS deviation for each discharge using the Mixed B/gB model varies from 7.76% to 20.35% while that using the MMM95 model varies from 9.73% to 16.94%. The offsets shown in figure 9 vary from about -1.77% to 13.93% by using the Mixed B/gB model and vary from about 4.52% to 12.25% by using the MMM95 model. The main reason that makes the results of RMS deviation and offset significantly different between some discharges is auxiliary heating power. The higher auxiliary heating power discharge yields the lower of RMS deviation and offset except for discharge no. 50921. The lowest auxiliary heating power for discharge no. 50921 gives the lowest of RMS deviation and offset; this may be explain by the others engineering parameters. The average RMS deviation ($\bar{\sigma}$) and the RMS deviation of the RMS deviation (σ_σ) for impurity (carbon) density profiles are summarized in table 2. It can be seen that the magnitude of the average RMS deviations of the MMM95 model is smaller than the Mixed B/gB model. However, the two models match experiment data equally well except for one case (discharge 50911).

Table 2. The average RMS deviation ($\bar{\sigma}$) and the RMS deviation of the RMS deviation (σ_σ) for impurity (carbon) density profiles produced by simulations using the Mixed B/gB model and MMM95 model.

Profile	Mixed B/gB		MMM95		$\lvert\bar{\sigma}_{Mixed} - \bar{\sigma}_{MMM95}\rvert$	$\sigma_{\sigma\,Mixed} + \sigma_{\sigma\,MMM95}$
	$\bar{\sigma}$	σ_σ	$\bar{\sigma}$	σ_σ		
C density	15.53	5.54	13.66	3.37	1.87	8.91

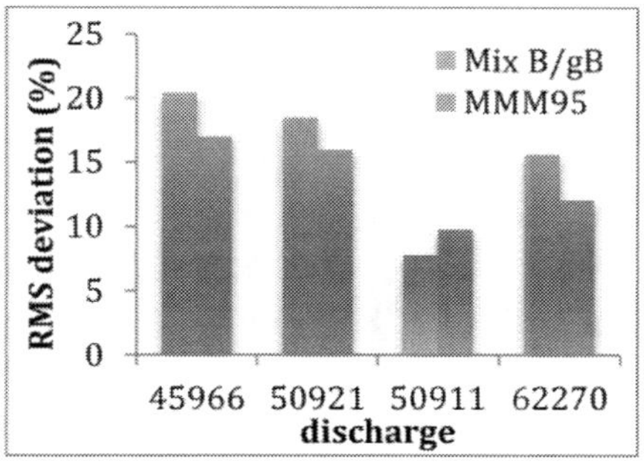

Figure 8. RMS deviation for impurity (carbon) density profiles produced by simulations using the Mixed B/gB model and MMM95 model compared with experimental data for TFTR discharge 45966, 50921, 50911 and 62270.

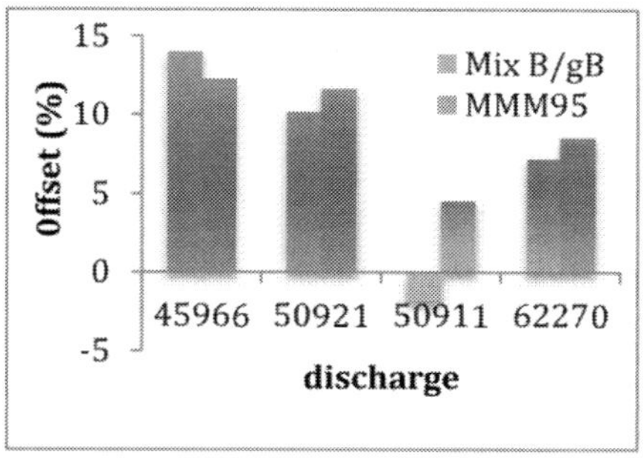

Figure 9. Offset from impurity (carbon) density profiles produced by simulations using the Mixed B/gB model and MMM95 model compared with experimental data for TFTR discharge 45966, 50921, 50911 and 62270.

4. Conclusion

The carbon density distribution in the *L*-mode of TFTR tokamak are carried out using BALDUR code with two choices of turbulent transport models, Mixed B/gB and MMM95 models. It is found that the both turbulent transport models match with the experiment data for carbon density profiles equally well.

Acknowledgement

This project is supported by the National Research University Project of Thailand Office of Higher Education Commission. W. Buangam thanks the Development and Promotion of Science and Technology Talents Project for DPST grant support.

References

[1] Wesson J 1997 *Tokamak* (New York: Oxford University Press) p 175
[2] Stangeby P C 2000 *The plasma boundary of magnetic fusion devices* (UK: Institute of Physics Publishing) p 277
[3] Baker D R, Wade M R, Baylor L R, DeBoo J C , Greenfield C M, Houlberg W A, and Stallard B W 2000 *Transport of Particles and Impurities in DIII–D Discharges with Internal Regions of Enhanced Confinement* (Work supported by U.S. Department of Energy under Contracts DE-AC03-99ER54463 and DE-AC05-96OR22462)
[4] Bateman G 1998 *Phys. Plasma* **5** 1793
[5] Onjun T 2001 *Phys. Plasma* **8** 975-985
[6] Singer C E 1988 Baldur: A one-dimensional plasma transport code *Comput. Phys. Commun.* **49** 275-398
[7] Hannum D, Bateman G, Kinsey J 2001 *Phys. Plasmas* **8** 964-974
[8] Tala T J J, Heikkinen J A and Parail V V 2001 *Plasma Phys. Control. Fusion* **43** 507-524

Determination of the total inductance of TPF-I

T Kunamaspakorn[1, 4], N Poolyarat[1], R Picha[2], J Promping[2] and T Onjun[3]
[1] Department of Physics, Faculty of Science and Technology, Thammasat University, Pathum Thani, Thailand.
[2] Thailand Institute of Nuclear Technology, Bangkok, Thailand
[3] School of Manufacturing Systems and Mechanical Engineering, Sirindhorn International Institute of Technology, Thammasat University, Pathum Thani, Thailand.

E-mail: bank.thanet@gmail.com

Abstract. Thailand Plasma Focus I (TPF-I) is a dense plasma focus device which was developed as a collaborative project among Thailand Institute of Nuclear Technology, Sirindhorn International Institute of Technology and Thammasat university. It aims to be a radiation source for academic research. This device has a power of 6 kJ. In this work, a Rogowski coils was used for measuring high speed current pulse to capture electric signal during operation TPF-I. By determining period of the discharge, it was found that the total inductance of TPF-I is about 216±12 nH, which is in agreement with Lee model.

1. Introduction

Dense Plasma Focus (DPF) is a device, which was invented in 1964 by J.W. Mather [1] and N.V. Filipov [2]. DPF can produce a short-lived plasma that is hot and dense by electromagnetic acceleration and compression enough to produce many products, for example, neutron [3], electron [4], proton [5], x-ray [6], etc. The electromagnetic compression of the plasma is called a *pinch*. The signals from DPF can be obtained by using Rogowski coil and we can compare with the signal from Lee model code [7-8]. The electrodes are enclosed in a vacuum chamber with a low pressure gas (the fuel for the reaction) filling the space between them.

Thailand Plasma Focus I (TPF-I) was developed as a collaborative project among Thailand Institute of Nuclear Technology, Sirindhorn International Institute of Technology and Thammasat university. It aims to be a radiation source for academic research. A diagram of this device's assembly is shown in figure 1.

[4] To whom any correspondence should be addressed.

Journal of Physics: Conference Series **611** (2015) 012009 doi:10.1088/1742-6596/611/1/012009

Figure 1. TPF-I assembly. The chamber is on the top of the mobile platform, vacuum system and a capacitor bank are under the chamber, the trigger system is on a capacitor bank.

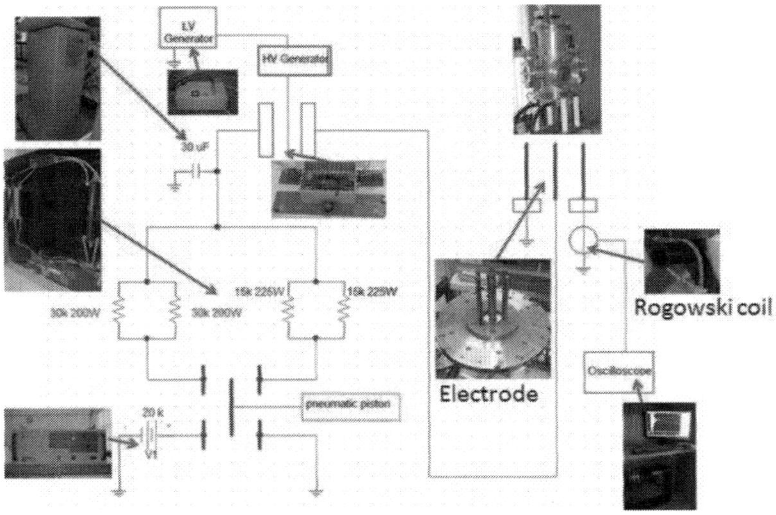

Figure 2. Circuit diagrams of TPF-I.

2. Components of TPF-I and design parameters

The TPF-I consists of 7 subsystems; 1.) 20 kV dc power supply, 2.) a 30 μF 20kV capacitor bank, 3.) trigger system, 4.) chamber, 5.) measurement system, 6.) vacuum system and 7.) gas system. The design parameters are given in table 1.

Table 1. Parameters of TPF-I.

Parameter	Description	Value
a	Anode radius	0.95 cm.
b	Cathode radius	1.85 cm.
z_0	Anode length	13.5 cm.
c_0	Capacitance	30 μF
V_0	Voltage	Up to 20 kV

3. Determining of total inductance of TPF-I

A discharge of DPF can be considered as a discharge of RLC circuit (figure 3). So, RLC discharge equations was used to find inductance form this device which shown in equation (1) and equation (2). So, we can determine the inductance the inductance from the period of discharge.

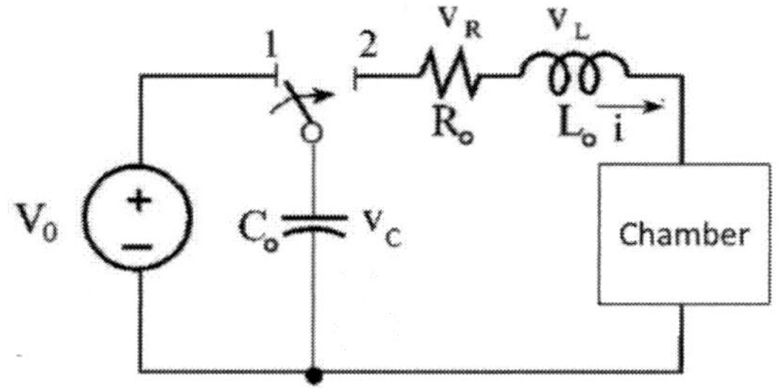

Figure 3. The equivalent circuit of dense plasma focus device.

$$f = \frac{1}{2\pi\sqrt{LC}} \quad (1)$$

$$L = \frac{T^2}{4\pi^2 C} \quad (2)$$

The Lee model code [9] is a well-known code and can be used as a benchmark for simulating a current waveform of any conventional Mather-type plasma focus. Lee model code requires input parameters such as inductance, capacitance, electrode radii and length, and operating parameters; charging voltage and fill gas pressure. In this work, after we determine total inductance, we will use it to simulate in Lee model code and verify the finding.

3.1 Finding the total inductance from the period

To find the total inductance we use the Rogowski coil capture the discharge signal from one of the six cables connected to cathode inside the chamber. The experiment with two different operating voltages, at 12kV and 10kV, are applied. Half of the periods is determined from several points of these signals as indicated the figure 4. The values of half period are shown in table 2, also their corresponding the total inductance. The average inductance is found to be 216±12 nH

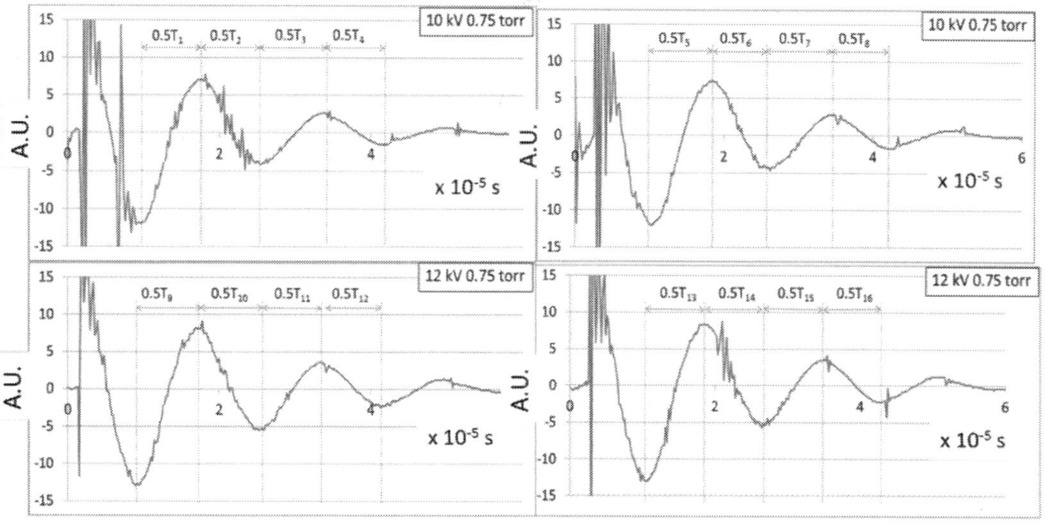

Figure 4. This figure shows four typical discharge signals from Rogowski coil. Vertical lines indicate points used for determining half of period.

Table 2. The computation of the inductance from signal period. In this table, T = period, f = frequency and L = inductance.

First condition (10 kV 0.75 torr)				Second condition (12 kV 0.75 torr)			
No. (i)	$0.5T_i$(μs)	T_i (μs)	L_i (nH)	No. (i)	$0.5T_i$(μs)	T_i (μs)	L_i (nH)
1	7.82	15.64	207	9	8.00	16.00	216
2	7.80	15.60	205	10	8.20	16.40	227
3	8.40	16.80	238	11	8.00	16.00	216
4	8.00	16.00	216	12	8.00	16.00	216
5	7.60	15.20	195	13	7.80	15.60	205
6	7.80	15.60	205	14	8.20	16.40	227
7	8.40	16.80	238	15	8.20	16.40	227
8	7.80	15.60	205	16	8.00	16.00	216
Average Inductance							216
Standard deviation							12
Range of Inductance = Average ± Standard deviation							204-228

3.2. Confirming inductance with Lee model code

The Lee model is used to predict the plasma behaviors in TPF-I. In this part, we use the obtained lower and upper limit of inductance, which are 204 nH and 228 nH respectively, from part 3.1 to simulate the current waveform and compare with experimental waveform. The other parameters used in simulation are given in table 2. The results of the experimental signal and the simulation signal are shown in figure 5. We simulate at two operating voltage; 10 kV and 12 kV. Note that the peak experimental signal is normalized to the peak of simulation signal using average inductance from the graph, it can be seen that the experimental signal are generally within the upper and lower limits of simulation signals. Some of discrepancies could be due to parameters, massf currf massfr and currfr, which are assumed to be similar to those in UNU/ICTP [10].

Table 2. Model parameters of Lee model code.

Lee Model parameter	Description	Value
Lo	Inductance	Vary
Co	Capacitance	30 μF
b	Cathode radius	1.85 cm.
a	Anode radius	0.95 cm.
ro mohm	System resistance	9
zo	Anode length	13.5 cm.
Vo	Voltage	12 kV, 10 kV
Po	Pressure	0.75 torr
MW	Mass weight of gas (Argon)	40
A	Atomic number of gas (Argon)	18
At-1 mol-	Operational parameter	1
massf	Axial mass swept-up factor	0.0635
currf	Axial Current flow factor	0.7
massfr	Radial mass swept-up factor	0.16
currfr	Radial Current flow factor	0.7

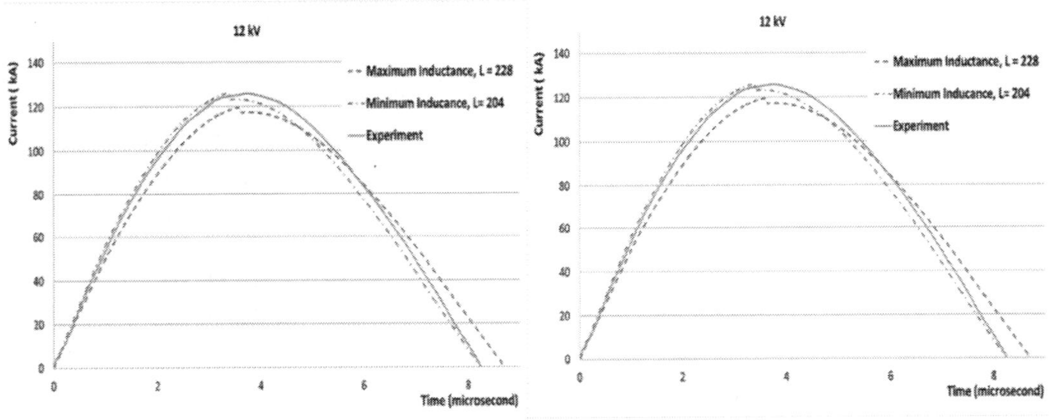

Figure 5. Comparison the result between simulation(dotted line) and experiment(solid line). (Voltage: Left 10 kV, Right 12 kV).

4. Conclusion

TPF-I or Thailand Plasma Focus I device has been completely constructed. We captured the discharge signal of the TPF-I using Rogowski coil. The signals were then used to determine the total inductance of TPF-I, which is found to be 216±12 nH. This result of total inductance generally agrees with Lee model simulation. The disagreement could be because the parameterization of the massf, currf, massfr, and currfr variables has not been investigated. The study of these variables is planned for the future work.

Acknowledgments

The authors thank Assoc. Prof. Rajdeep Sing Rawat and his team for helpful discussion and suggestions on my experiment, Prof. Sing Lee for Lee model code and Mr. Arlee Tamman for his help to build a pneumatic switch. This research work was supported by memorandum of understanding between Thailand Institute of Nuclear Technology, Sirindhorn International Institute of Technology and Thammasat University.

References

[1] Mather J W 1964 *Phys. Fluids Supply* **5** 5
[2] Petrov D P, Filippov N V, Filippova T I, Khrabrov V A 1958 *Plasma physics and controlled thermonuclear reactions* **4** 170
[3] Asai H and Ueno I 1989 *Fusion Eng. Des.* **7** 335
[4] Harries W L, Lee J H, Mcfarland D R 1978 *Plasma Phys.* **20** 95
[5] Smith Jr, Luo C M, Rhee MJ, Schneider RF 1985 *Phys. Fluids* **28** 2305
[6] Kato Y, Ochiai I, Watanabe Y, Murayama SJ 1988 *J. Vac. Sci. Technol. B* **6** 195
[7] Lee S and Saw S H 2008 *Appl. Phys. Lett* **92** 021503
[8] Lee S, Saw S H, Lee P C K, Rawat R S and Schmidt H 2008 *Appl Phys Letters* **92** 111501
[9] Lee S, Lee P, Saw S H and Rawat R S *Plasma Phys. Control. Fusion* **50** 065012
[10] Lee S, Moo S P, Wong C S and Chew A C 1998 *Proc. Satellite Meeting of the International Meeting on Frontiers of Physics* **5**

Comparison of discharge power in dielectric barrier discharge (DBD) with stainless steel and KCl liquid electrode

T Traikool[1, 3], **N Poolyarat**[1] **and T Onjun**[2]
[1]Department of Physics, Faculty of Science and Technology, Thammasat University. Pathum Thani, Thailand
[2] School of Manufacturing Systems and Mechanical Engineering, Sirindhorn International Institute of Technology, Thammasat University, Pathum Thani, Thailand

E-mail: tipwimol.traikool@gmail.com

Abstract. A Dielectric Barrier Discharge (DBD) device is developed to generate cold atmospheric plasma. Two different materials used as a top electrode: stainless steel and KCl liquid are considered. The bottom electrode is stainless steel for both cases. The gap of this DBD can be varied up to 3 mm. The DBD is powered by an AC high voltage in the range of ±10 kV with frequency in a range of 2-7 kHz. It is found that the discharge with KCl liquid electrode yield higher power than that with stainless steel electrode.

1. Introduction

Dielectric barrier discharge (DBD) device is a typical non-equilibrium plasma device that can be operated at atmospheric pressure [1]. A typical DBD consists of two electrodes and, at least one dielectric barrier in between. The dielectric barrier limits charge transported in the discharge. To generate plasma, an AC high voltage is applied such that the applied voltage is greater than breakdown voltage of the medium gas. There have been many studies about effects involved parameters such as electrode structure, frequency, applied voltage, gas gap, gas types etc. These parameters can affect electrical characterization of DBD. Okazaki et al. [2] worked with different gases including helium, neon, argon, nitrogen, oxygen and air at atmospheric pressure, in which different performances were observed. Fang et al. [3] studied DBD using metal wire mesh electrodes in nitrogen, Ar, air in which similar results were obtained. Rahel et al. [4] studied DBD using two water electrodes separated by two 1 mm thick Pyrex sheets of glass in air. Wang et al. [5] various studied the different electrode of DBD in atmospheric pressure. Titov et al. [6] studied physical characteristics of atmospheric pressure dc discharge with liquid electrode by using distilled water and aqueous solution (KCl, $CuCl_2$, $FeSO_4$). Fuangfung et al. [7] studied different types of electrode by using KCl liquid electrode in atmospheric pressure for waste water treatment. The DBD have applications for the industry. Xu [8] studied applications of DBD such as excimer radiation sources, UV/VUV lamps, ozone generators and their applications in pollution control.

In this paper, the comparison of discharge characteristics of DBD in atmospheric pressure air with two different structure electrodes: stainless steel electrode and KCl liquid electrodes is reported. The electrical measurements are used to characterize the discharges.

---

[3] To whom any correspondence should be addressed.

2. Experimental setup

A schematic diagram of the experimental setup in this work is shown in figure 1. This DBD system can be used for generating plasma in atmospheric pressure. The discharges are created between two parallel electrodes. Two different materials are used as the top electrode: stainless steel 304 electrodes and liquid electrodes in a Petri dish. The bottom electrode, used-for both cases, is stainless steel with a dimension of 110 mm diameter and 3 mm thickness. The liquid electrode in this work is potassium chloride (KCl) solution with a concentration of 0.2g/150ml. In other case, the top electrode is made of stainless steel 304 with the same dimension as the bottom electrode. All the electrodes are covered by glass plates with a thickness of 3 mm. The discharge gas gap can be adjusted in the range of 1-3 mm. The AC high voltage was applied in a range of ±10 kV with frequency in a range of 2-7 kHz. Note that, for a fixed gap, when the frequency changes, the applied voltage will change too. A capacitor of 470 nF, connected between bottom electrode and ground, are used for measuring the accumulated charge on the dielectric plate. The measured signals are recorded by an oscilloscope RIGAL DS1052E. Figure 2 shows typical plasma occurred during the discharge. Figure 2(a) shows the side view of the plasma and figure 2(b) shows the top view of the plasma.

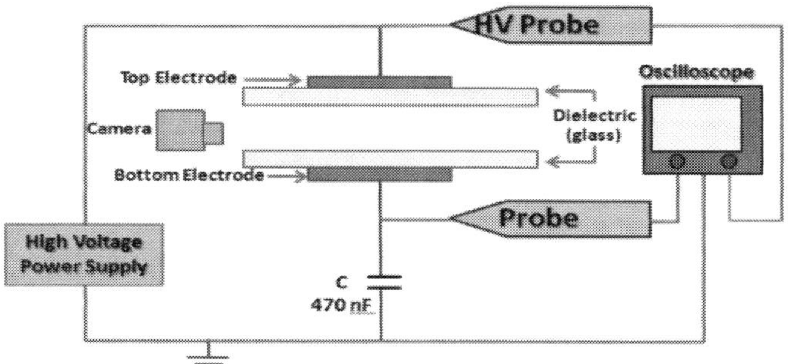

Figure 1. Schematic diagram of DBD.

(a)

(b)

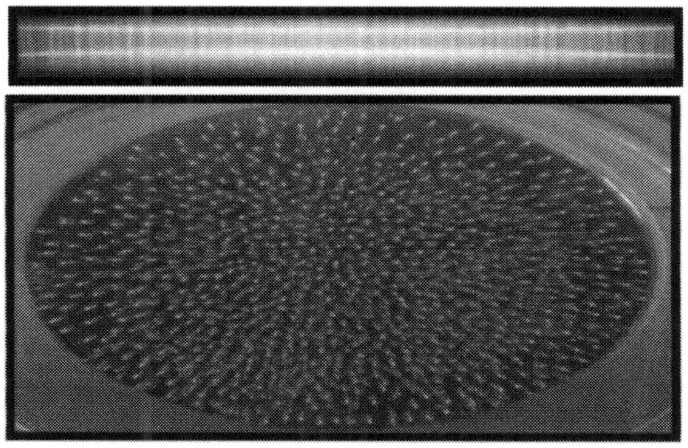

Figure 2. Plasma discharge (a) side view
(b) top view when using KCL electrode.

3. Results and discussions

The discharge power of the DBD can be determined by using a high voltage probe measuring voltage across the top and bottom electrodes and the charge transferred (Q) across them. The charge transferred can be calculated from the voltage across the 470 nF capacitor and its capacitance. The plots of Q-V diagram, called the Lissajous diagram, as shown in figure 3 can be used for determining the discharge power *(P)* of the DBD, which is the enclosed area in *Q-V* Lissajous diagram. However, in this experiment, there are streamers, micro filaments or micro arc discharges occurred in each of the discharges, resulting in a distortion of data. Streamers discharge is a type of electrical discharges which can form when there is a voltage difference. A computational method is used to process the experimental data by using sinusoidal wave function $y = A\sin(\omega t + \phi)$ to fit the experimental data. The fit curves are shown as the solid lines in figure 3.

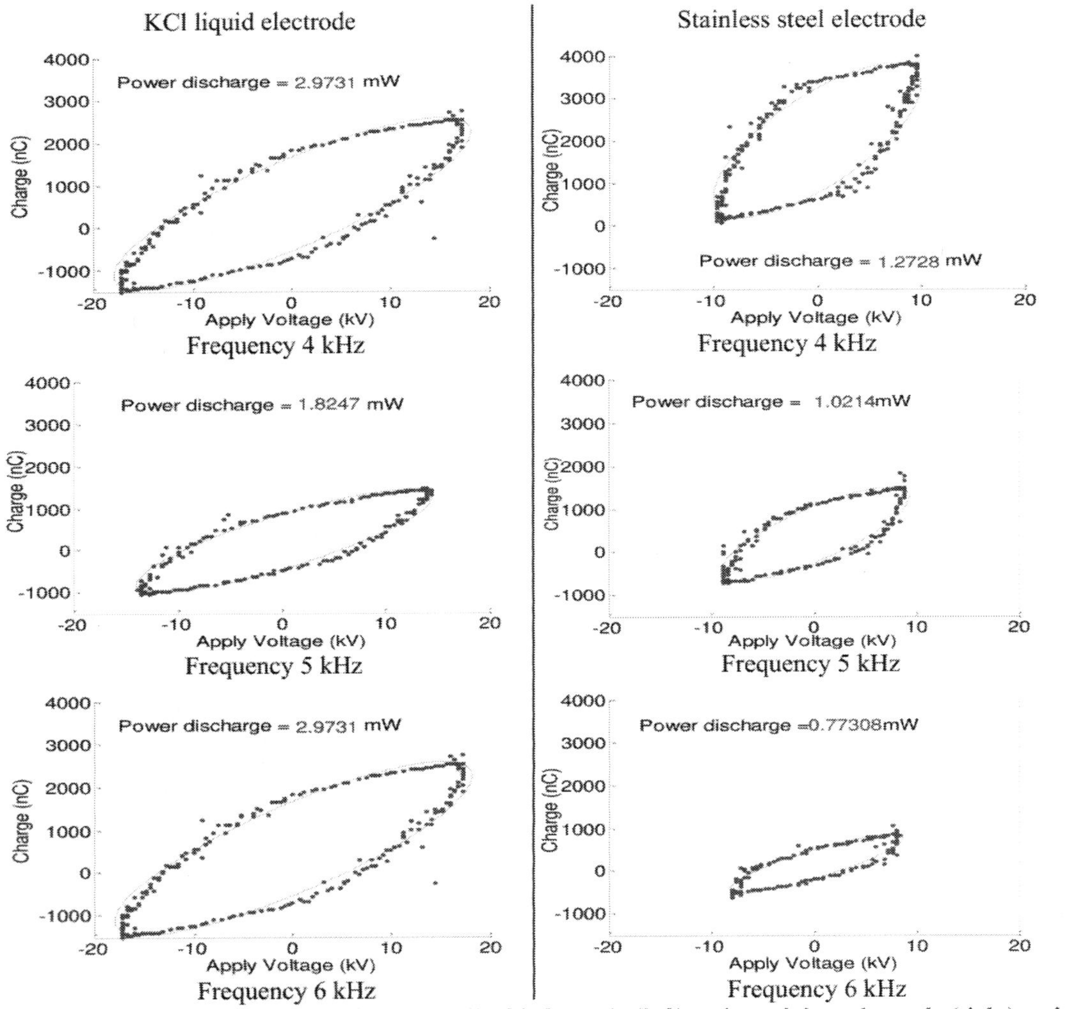

Figure 3. Q-V Lissajous of the DBD with a liquid electrode (left) and a stainless electrode (right) at the frequency of 4-6 kHz.

The discharge power in the DBD plasma can be calculated by dividing the energy discharge with the period. The energy discharge can be calculated from the *Q-V* area plot, which is the power from the discharge of one cycle. Discharge power can be calculated from equation (1).

$$P = \frac{W}{T}, \qquad (1)$$

where P is the discharge power , W is the energy from the discharge, and T is the period of one cycle of discharge energy.

The energy from the discharge can be calculated from

$$W = \int_{t_0-T/2}^{t_0+T/2} v(t)\bullet i(t)\,dt \ , \qquad (2)$$

where T is the period, $i(t)$ is the current caused by the discharge, and $v(t)$ is the voltage applied.

The current can be calculated from the voltage drop across the capacitor (v_c) and the current time changes as shown in equation (3).

$$i(t) = \frac{dq}{dt} = C\frac{dv_c}{dt}, \qquad (3)$$

when q is the amount of charge generated by the discharge, C is the capacitance of the connected capacitor.

Hence, the energy from the discharge of one cycle can be calculated from equation (4).

$$W = \int_{t_0-T/2}^{t_0+T/2} v(t).Cdv_c = \int_{t_0-T/2}^{t_0+T/2} v(t).dq(t) \qquad (4)$$

Moreover, in this work, the discharge power is measured at various gap distances, in the range of 1-3 mm. The AC high voltage is applied by varying the frequency in the range of 2-7 kHz to operate the plasma discharge. Note that, when the frequency changes, the applied voltage will change. In the case of KCl liquid electrode, the discharge power peaks at frequency 4 kHz for 1 mm and 2mm gap and at 5-6 kHz for 3 mm gap. In the case of stainless steel electrode, the discharge power peaks at 3 kHz for 1 mm and 2mm gap, and at 5 kHz for 3 mm gap. These findings could be due to different resonance frequency of each material. In addition, it is found that when expanding the gap, the discharge power is decreasing. However, the stainless steel electrode yields higher discharge power than the KCl liquid electrode in the frequency range of 2-3 kHz. In the frequency range of 4-7 kHz the discharge power from the KCl electrode are higher than those from the stainless steel electrode. The results are shown in figure 4.

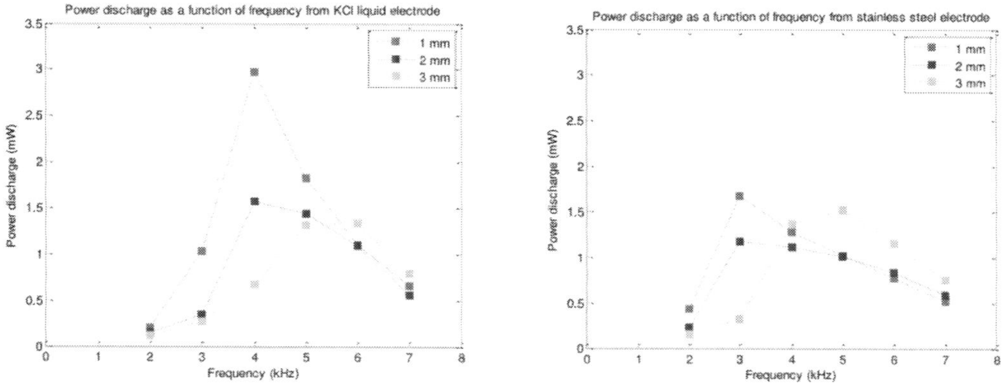

(a) KCl liquid electrode (b) Stainless steel electrode

Figure 4. Discharge power as a function of frequency for two types of electrodes.

4. Summary

A cold plasma generated by DBD at atmospheric pressure using high voltage in the range ±10 kV and frequency 2-7 kHz from two different electrode structures, stainless steel electrode and liquid electrode, is achieved. The discharge power of these two different electrodes are obtained and compared with each other. In most cases, it is found that the discharge power using a KCl liquid electrode with a stainless steel electrode are higher than those using stainless steel electrode.

Acknowledgments

This project is supported by the annual government research budget through Thammasat University. The authors would like to thank Assoc. Prof. Manu Fuangfung of Thammasat University and Mr. Arlee Tamman of Walailuk University for useful comments and helpful suggestions.

References

[1] Chirokov A, Gutsol A and Fridman A 2005 *Pure Appl. Chem.* **77** 487–495
[2] Okazaki S, Kogoma M, Uehara M and Kimura Y 1993 *J. Phys. D Appl. Phys.* **26** 889
[3] Fang Z, Qiu Y and Luo Y 2003 *J. Phys. D Appl. Phys.* **36**2980
[4] Rahel J and Sherman D M 2005 *J. Phys. D Appl. Phys.* **38** 547
[5] Changquan W, Guixin Z and Wang X 2012 *Vacuum* **86** 960-964
[6] Titov V 2009 *IPCS Proceedings.*
[7] Fuangfung M and Chimaoy T 2014 *PIERS Proceedings.* 2727-2730
[8] Xu X 2001 *Thin Solid Films* **390** 237-242

A nuclear method to authenticate Buddha images

S Khaweerat[1], W Ratanatongchai, J Channuie, S Wonglee, R Picha, J Promping, K Silva and T Liamsuwan
Nuclear Research and Development Division, Thailand Nuclear Institute of Technology (TINT), Chatuchak, Bangkok, Thailand

E-mail: sasiphank@tint.or.th, khaweerat@gmail.com

Abstract. The value of Buddha images in Thailand varies dramatically depending on authentication and provenance. In general, people use their individual skills to make the justification which frequently leads to obscurity, deception and illegal activities. Here, we propose two non-destructive techniques of neutron radiography (NR) and neutron activation autoradiography (NAAR) to reveal respectively structural and elemental profiles of small Buddha images. For NR, a thermal neutron flux of 10^5 n cm^{-2} s^{-1} was applied. NAAR needed a higher neutron flux of 10^{12} n cm^{-2} s^{-1} to activate the samples. Results from NR and NAAR revealed unique characteristic of the samples. Similarity of the profile played a key role in the classification of the samples. The results provided visual evidence to enhance the reliability of authenticity approval. The method can be further developed for routine practice which impact thousands of customers in Thailand.

1. Introduction

Thai people traditionally often make small soil images of Lord Buddha as well as the legendary monks. It has been a long-time belief that making or holding Buddha images can bring a great protective power and a good fortune in various aspects. The commercial value of these small images vary from a few hundred to several million Baht. The more famous the model is, the more expensive it becomes. For example, some special models of Luang-Pu-Thuad, similar to the ones in this report, may worth more than 1 million Baht. Due to the high demand of famous models, they are often copied and traded throughout the market. Differentiation between the replica and the genuine ones is normally done using individual skills and a simple magnifier. Since their appearances are often very similar, the authenticity approval is always ambiguous. Here, the proving evidence based on scientific fundamentals offers great benefits to make a confident judgement that will help to eliminate controversies and illegal activities.

In general, absolute dating provides an affirmative conclusion. However, it destroys more or less some amounts of the samples – the requirement that is unacceptable for the owners of Buddha images. The present study employs two complimentary techniques: neutron radiography (NR) and neutron activation autoradiography (NAAR) to reveal either similarities or differences of particular Buddha images without sample destruction. Comparison of the resulting profiles leads to sample classification and subsequent authenticity approval.

X-ray and gamma radiography are well known in non-destructive testing communities. They are commonly used to reveal the inner structure of many objects; however, some limitations exist. Both X-ray and gamma ray are electromagnetic radiation that is difficult to penetrate through materials with a

[1] To whom any correspondence should be addressed.

high atomic mass. The visualization of light elements that are buried within heavy metals is hardly achieved. In contrast, neutron attenuation coefficient is material specific and atomic mass independent, and neutron has a relatively high sensitivity for some light elements including hydrogen, boron, nitrogen and carbon. As a result, neutron radiography could be used in various applications, where X-ray and gamma radiography are incapable, such as the determination of moisture content, solvent concentration, resin structure and liquid flow pattern inside a metal container. Similar to other radiography techniques, neutron radiography offers a visual image of radiation transmitted through material on the detector such as film, an imaging plate, a micro channel plate or a CCD camera. The transmission of X-ray, gamma ray and neutrons through materials can be given by Beer-Lambert law [1]:

$$I = I_0 \cdot e^{-SK \cdot z} \tag{1}$$

where I and I_0 are the transmitted and initial intensity, respectively; z is the sample thickness, and SK is the total attenuation coefficient for the material, which can be described as follows:

$$SK = \sum_i \sigma_i \cdot N_i \tag{2}$$

where σ_i is the microscopic cross section, and N_i is the density of each contributing element i. Since the macroscopic cross section of neutrons is unique and different from X-ray and gamma ray, the radiographic images obtained from these techniques are different.

Similar to other techniques, NR has its strengths and limitations. NR can illustrate the inner structure profile but it is impossible to determine the elemental composition using NR. On the other hand, NAAR is a combination of two techniques: neutron activation analysis and autoradiography. Neutron activation analysis generally uses a nuclear research reactor to generate a sufficient thermal neutron flux at the activation position. Interaction between thermal neutrons and the elements can transform stable elemental isotopes into radioisotopes with an appreciable half-life and the characteristic beta or gamma rays emission [2]. The neutron-induced gamma and beta rays can be recorded using a conventional radiographic method [3]. The neutron activation parameters including neutron beam characteristics; brilliance, divergence and spatial distribution, have major effects on the pattern of the resulting visual image that indicates the localization of radioactive substances in the sample. The sensitivity of neutron activation analysis has been recognized in solving real analytical problems since the 1960s [4], and now it becomes a valuable analytical tool for cultural heritage study [3]. The study presented in this report utilized NAAR in complimentary to NR to investigate the nature of particular Buddha images and classify them based on similarities or differences of the image profiles.

2. Methodology

Three darkish soil Buddha images were measured to record their dimensions, photographed, and then attached onto an imaging plate BAS-ND 2040 (Fujifilm) with a cassette. The Buddha images were triangular in shape and their dimensions were approximately 1.5 x 2.5 cm^2. The average thickness was 0.4 mm. Two of them were delivered from the same area but they were different from one another. The cassette was put in an adjustable slot against the 8" south beam tube of TRR-1/M1 reactor at Thailand Institute of Nuclear Technology for neutron radiography (figure 1). The samples were aligned at the centre and placed close to each other to avoid inhomogeneity of neutron flux. At the exposure position, the L/D ratio which refers to the ratio of the collimator length (L) and the effective diameter of aperture (D) [5] was 45 and the frame of view (FOV) was 30 x 30 cm^2. The neutron flux was 1.08×10^6 n cm^{-2} s^{-1}, and the exposure time was varied from 60 - 120 seconds. After exposure, the imaging plate was scanned using Typhoon FLA 7000 scanner (GE Healthcare Life Sciences, USA).

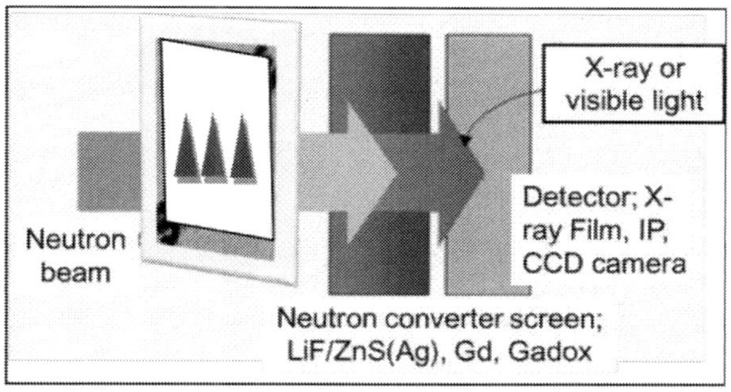

Figure 1. NR setup.

In NAAR experiment, all Buddha images were activated in an in-pool vertical tube for 60 seconds. The neutron flux at the activation position was ~10^{12} n cm^{-2} s^{-1}. Afterwards, they were left to decay for 5, 15, and 20 minutes. After each particular decay time, the samples were placed on a 20 x 40 cm^2 BAS-MS imaging plate (Fujifilm) for two minutes; at the same time, a $LaBr_3(Ce)$ scintillation detector was set next to the samples for gamma ray measurement (figure 2). The gamma spectrum provided qualitative information on the radioisotopes found in the samples at a particular decay time. Finally, the determination of long-lived isotopes was performed in similar practice but the activation time and the decay time were prolonged to 1 hour and 3 days, respectively. Image processing and data analysis was performed using Multi Gauge V3.0 software (Fujifilm).

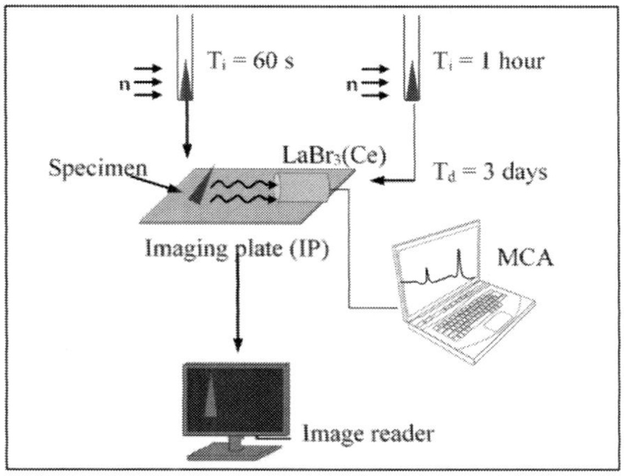

Figure 2. NAAR setup.

3. Results and discussion

If the geometry of NR set up, the neutron flux and the exposure time were fixed, the image intensity would correspond to the elemental content, the structure and the density of material compositions of the Buddha images. Neutrons transmitted through the object, which were recorded on the imaging plate, was originally shown in grey scale. The Multi Gauge V3.0 software had a provision to change a grey-scale image into a multi-colour image for viewing intensity variation in more details. The different intensity patterns of three Buddha images are shown in figure 3. The colour scheme represented in red meant the maximum number of neutrons recorded on the imaging plate, which was mainly observed in the background area; blue colour represented the most attenuated area. The colours orange, yellow and green represented the transmission levels between the red and the blue levels. Two out of three samples, N1 and N3, showed similarity in their colour profile implying similar material compositions. They were probably made using the same technology and raw materials. The appearance of sample N2 looked similar to others but its NR results showed significant differences. It appeared that N2 attenuated neutrons more than N1 and N3; as a result, N2 was probably made of materials with a higher attenuation ability than others.

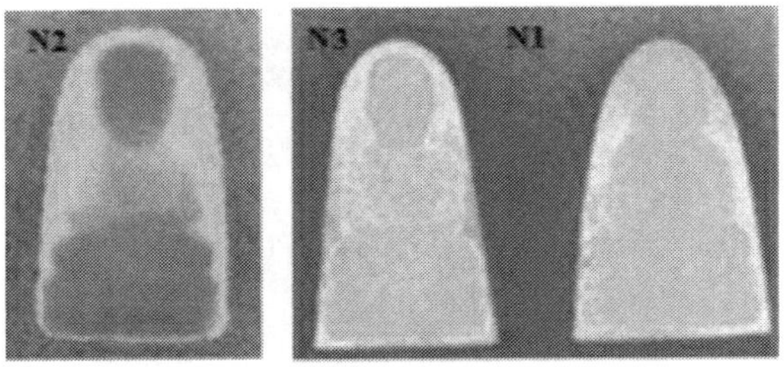

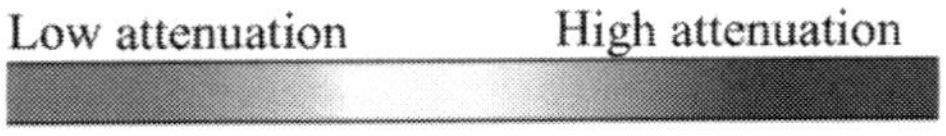

Figure 3. Neutron radiography revealed different intensity patterns of three Buddha images (from left to right; sample ID: N2, N3 and N1).

In addition, the intensity of neutrons passing through the Buddha images and recorded on the imaging plate was analysed quantitatively. Results of scanned images were presented in Quantum Level (QL) value which corresponded to the pixel data stored after scanning. QL is a logarithmic data generated specifically on the system [6]. The value provided numerical expression to explain typical features of the samples. For instance, the average QL value measured from left to right as demonstrated in figure 4 indicated mixing homogeneity. The QL value of N1 and N3 was identical but 1.4 times higher than the value of N2. Since N1 and N3 were from the same origin, the QL value confirmed the relationship of provenance. The distinct characteristics of N2 implied a different provenance. Even though the NR profile and the QL value were in agreement, sample investigation based on elemental analysis of particular Buddha images were conducted to strengthen the conclusion.

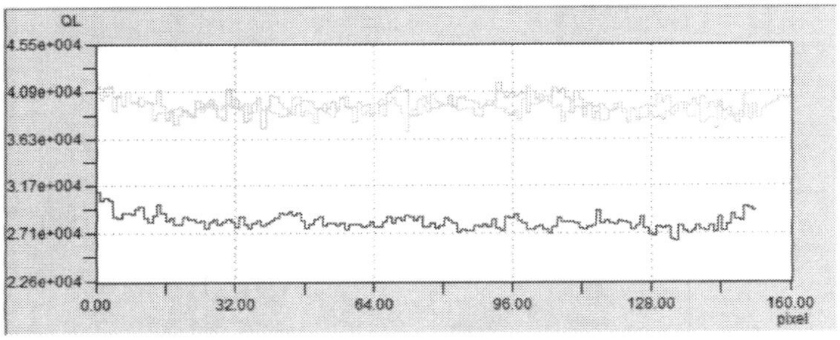

Figure 4. The comparison of QL values of three Buddha images: N1 (green), N2 (blue), and N3 (light blue).

Elemental analysis by neutron activation was performed in parallel to autoradiography. The position and the area under each photo peak of the gamma spectrum are related to the elemental species and their quantity present in sample. The analysis indicated that samples N1, N2 and N3 exhibited similar elemental profiles which included aluminium, manganese, arsenic, sodium, potassium and lanthanum. However, they were not entirely the same. The elemental concentrations were slightly different which could be identified by unequal peak areas and the background radioactivity level. The similar spectra of N1 and N3 supported the NR results. Meanwhile, the spectrum of N2 was different, especially in its background level and the area under each photo peak. The comparison of gamma spectra of the three samples obtained after being activated for 1 hour and left to decay for 3 days were illustrated in figure 5.

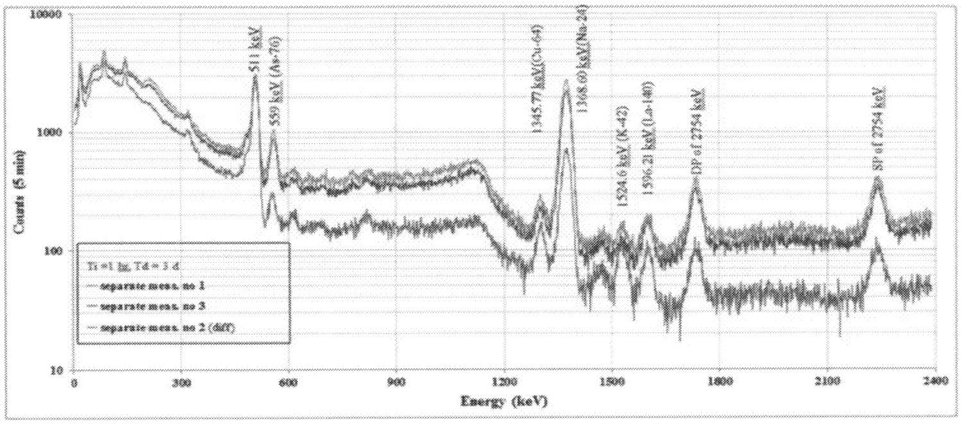

Figure 5. The gamma spectra of three Buddha images after being activated for 1 hour and left to decay for 3 days.

The images obtained from autoradiography suggested typical patterns at a time (figure 6). Even though the conditions for image construction were exactly the same, the intensity level and the pattern implied unique characteristics of individual samples. Autoradiography of N1 and N3 confirmed that they could be categorised into the same group. In addition, the results revealed the distribution of elements within the sample. The smooth pattern of N2's image suggested the homogeneity of raw materials in N2 which distinguished the sample from the others. The intensity differences corresponded well with the isotopic ratios (figure 7) in particular samples. Two Buddha images, N1 and N3, consisted of similar elemental ratios providing the similar intensity levels, while N2 showed differences in both the elemental ratio and the intensity level.

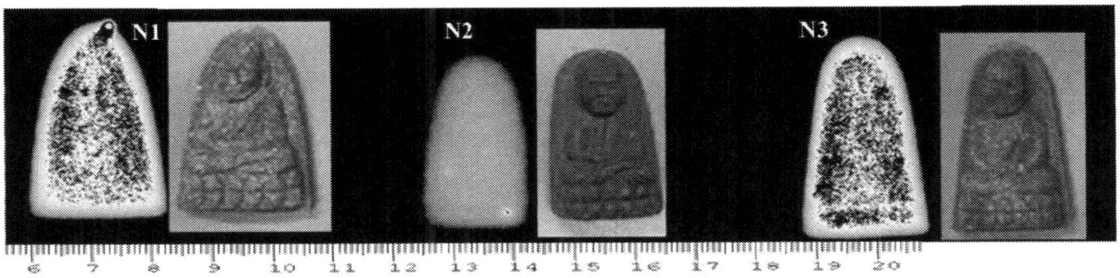

Figure 6. Autoradiographs of three Buddha images obtained under the same conditions: 1 minute activation, 50 minutes decay time and 15 minutes exposure on the imaging plate.

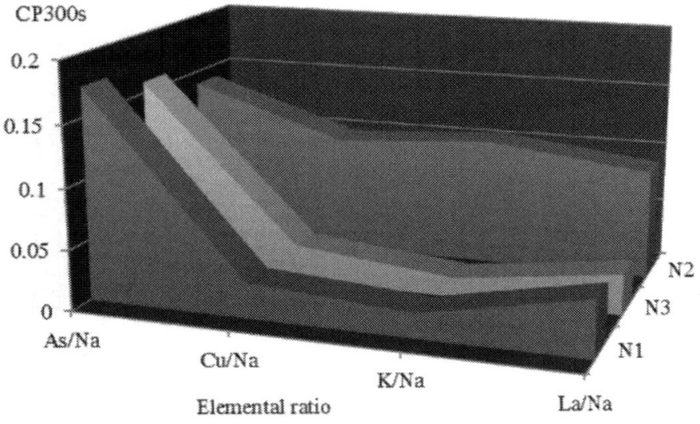

Figure 7. The comparison of elemental ratios of three Buddha images: N1, N2 and N3.

4. Conclusions

Five pieces of evidence from NR and NAAR are in agreement. The visual image from neutron radiography and autoradiography, the QL value, the gamma spectra, and the elemental ratios contributed to sample classification. This study used known samples to confirm that Buddha images from different provenances would result in different output information. The complimentary techniques of NR and NAAR are powerful non-destructive testing tools for completing the information of unseen structures as well as qualitative elemental compositions. Further development should be taken into account for establishing the standard approach for authenticity certifying of small Buddha images. A number of small Buddha images should be gathered and analysed in order to establish a reliable database. Due to the potential of retrieving historical data of valuable and significant objects without sample invasion, these techniques should be applicable in several fields of studies and could generate an impact on a number of customers in Thailand.

Acknowledgements

The authors truly appreciated the support of the International Atomic Energy Agency (IAEA) Coordinated Research Project fund under the CRP-F11018 "Application of Two and Three Dimensional Neutron Imaging with Focus on Cultural Heritage Research".

References

[1] Lehmann E H and Mannes D 2012 Wood investigations by means of radiation transmission techniques *J. Cultural Heritage* **13** S35-43

[2] Kalicki A *et al.* 2001 Neutron autoradiography: working-out method and application in investigations of test paintings *Radiation Measurements* **34** 567-9

[3] Andreani C, Gorini G and Materna T 2009 Novel Neutron Imaging Techniques for Cultural Heritage Objects *Neutron Imaging and Applications* ed H Z Bilheux, R McGreevy, and I S Anderson (New York: Springer) pp 229-52

[4] Michael D G and Hector N 2003 Neutron activation analysis and provenance research in archaeology *Measurement Science and Technology* **14** 1516

[5] MacGillivray G *Neutron Radiography Collimator Design* (Nray Services Inc.)

[6] Paterson I J *et al.* 2008 Image plate response for conditions relevant to laser–plasma interaction experiments *Measurement Science and Technology* **19** 095301.

Journal of Physics: Conference Series **611** (2015) 012012 doi:10.1088/1742-6596/611/1/012012

EPR study of free radicals in cotton fiber for its potential use as a fortuitous dosimeter in radiological accidents

W Sudprasert[1], P Insuan and S Khamkhrongmee

Department of Applied Radiation and Isotopes, Faculty of Science, Kasetsart University, Bangkok 10900, Thailand

E-mail: fsciwasu@ku.ac.th

Abstract. Electron paramagnetic resonance (EPR) spectroscopy was applied to characterize radiation-induced free radicals in the cotton fiber in order to determine the possibility for using cotton as a fortuitous dosimeter in accidental exposures to radiation. Cotton fabrics were irradiated at 0.1, 0.5, 1, 2, 10, 50 and 500 Gy using a ^{137}Cs gamma source. The irradiated samples were then stored in the dark under controlled environmental conditions for 1, 15, 35 and 60 days. The EPR spectra were observed in samples using a Bruker EMX X-band spectrometer equipped with a TE_{102} rectangular cavity. The EPR signal intensities of irradiated samples were determined from peak-to-peak amplitudes of EPR spectra and compared to those of unirradiated samples. The following optimum parameters were used: 100 kHz modulation frequency, 9.84 GHz microwave frequency, 1.8 mT modulation amplitude, 1.0 mW microwave power, 655 ms time constant, 41 ms conversion time and 41.98 s sweep time. The EPR spectra of unirradiated samples showed a singlet line with $g = 2.006$ due to pre-existing stable organic radicals in the cotton fibers, whereas those of irradiated samples showed the same pattern with different signal intensities according to the doses. Irradiation increased the signal intensity in a dose dependent manner. The signal intensity exhibited an exponential decay with storage time from 1 to 60 days. Obviously, the degree of fading of EPR intensity did not depend on the absorbed dose from 0.1 - 50 Gy. The maximum fading was about 60% at 60 days of storage for irradiated samples at all doses. However, this post-irradiation signal appeared to be detectable for up to 60 days. The EPR study results indicated the potential of using cotton as a fortuitous dosimeter in radiological accidents.

1. Introduction

Ionizing radiation is utilized and applied worldwide in many areas including medicine, industry, agriculture and research. The risk of radiological accidents might generally be increased and consequently the radiation emergency preparedness is of great concern in most countries. Following a radiological or nuclear incident, determination of the absorbed dose to individuals exposed to critically high radiation doses is of importance for initial medical treatment decisions. Therefore, the search for fortuitous dosimeters is currently needed especially when the accident happens in a local suburb area where people do not generally have a personal dosimeter, or the dosimeter is not properly worn at the appropriate position.

Electron paramagnetic resonance (EPR) spectroscopy is applicable to paramagnetic molecules with one or more unpaired electrons, i.e. radicals and transition metal ion complexes [1]. EPR spectral studies

[1] To whom any correspondence should be addressed.

on several materials having potential to be used as a fortuitous dosimeter have been reported, such as mineral glasses from watch [2-3] and mobile phone LCDs [4], wallboards [5], sugar [6], eggshells [7] and cotton from clothing fabrics [8]. Among these materials, clothing fabrics have been shown to be suitable as a fortuitous EPR dosimeter due to their advantages in availability and practicality to provide dose mapping throughout the body. Viscomi *et al* [8] reported the preliminary EPR study in untreated cotton yarns following a high dose (500 Gy) and intermediate doses (50 to 10 Gy) of irradiations. The results highlighted the presence of a specific post-irradiation signal distinguishable from the pre-irradiation signal at all doses studied. In this work, the EPR signals in cotton fabrics were investigated covering the dose ranging from 0.1-500 Gy to explore the possibility of using cotton existing in the clothing fabrics as a fortuitous dosimeter. The parameters of EPR spectrometer were initially optimized. The dose-response relationship was obtained by using the optimized parameters. The fading of free radicals was demonstrated by studying the stability of EPR signals during storage.

2. Materials and methods

2.1. Cotton samples
Cotton fabrics used in this study were obtained from a textile company (anonymous). They were cut into small pieces and put into plastic bags for irradiation.

2.2. Gamma irradiation
Irradiation of cotton fabrics was carried out using a ^{137}Cs gamma source (Mark I) at a dose rate of 3.7 Gy min^{-1} confirmed by a Fricke dosimeter, giving doses of 0.1, 0.5, 1.0, 2.0, 10, 50 and 500 Gy. The irradiated samples were sealed in plastic bags and maintained in the dark at room temperature until EPR readings.

2.3. EPR measurements
EPR spectra were measured using a Bruker EMX X-band EPR spectrometer equipped with a TE_{102} rectangular cavity. The samples were cut into small pieces with suitable size to be put into the quartz tube. The ESR tube containing 100 mg of sample was adjusted to a fixed position to maintain reproducibility in the cavity. Initial EPR measurements were carried out using a 500 Gy-irradiated cotton fabric to optimize the parameters including microwave power, modulation amplitude and time constant. The optimal parameters used for the dose dependent study were: 100 kHz modulation frequency, 9.84 GHz microwave frequency, 1.0 mW microwave power, 1.8 mT modulation amplitude, 655 ms time constant, 41 ms conversion time and 41.98 s sweep time. The intensity of the EPR signal was measured by determining the peak-to-peak height from the minimum to the maximum of the spectrum. The fading characteristics of EPR signal for the storage time of 1, 15, 35 and 60 days were examined. All reported intensities represented the mean of three individual experiments.

3. Results and discussion

3.1. EPR spectrum optimization
As previously described, the magnitude of the microwave power affected the saturation of EPR signal intensity [9]. The initial value for microwave power was set at 0.01 mW, while the modulation amplitude was fixed at 1.8 mT. The EPR signal intensity was observed when the microwave power was increased from 0.01 to 5.0 mW. The relationship between EPR signal intensity and the square root of microwave power is shown in figure 1a. The signal amplitude increased with microwave power from 0.01 to 1.0 then gradually decreased. A microwave power of 1 mW was selected for all experiments to avoid saturation effects.

Next, the modulation amplitude was optimized by acquiring the spectrum at a set modulation varied from 0.1 to 2.5 mT. It was found that the intensity increased while the pattern of EPR spectra was not changed (figure 1b). The modulation amplitude at 1.8 mT was set since the best spectrum was obtained.

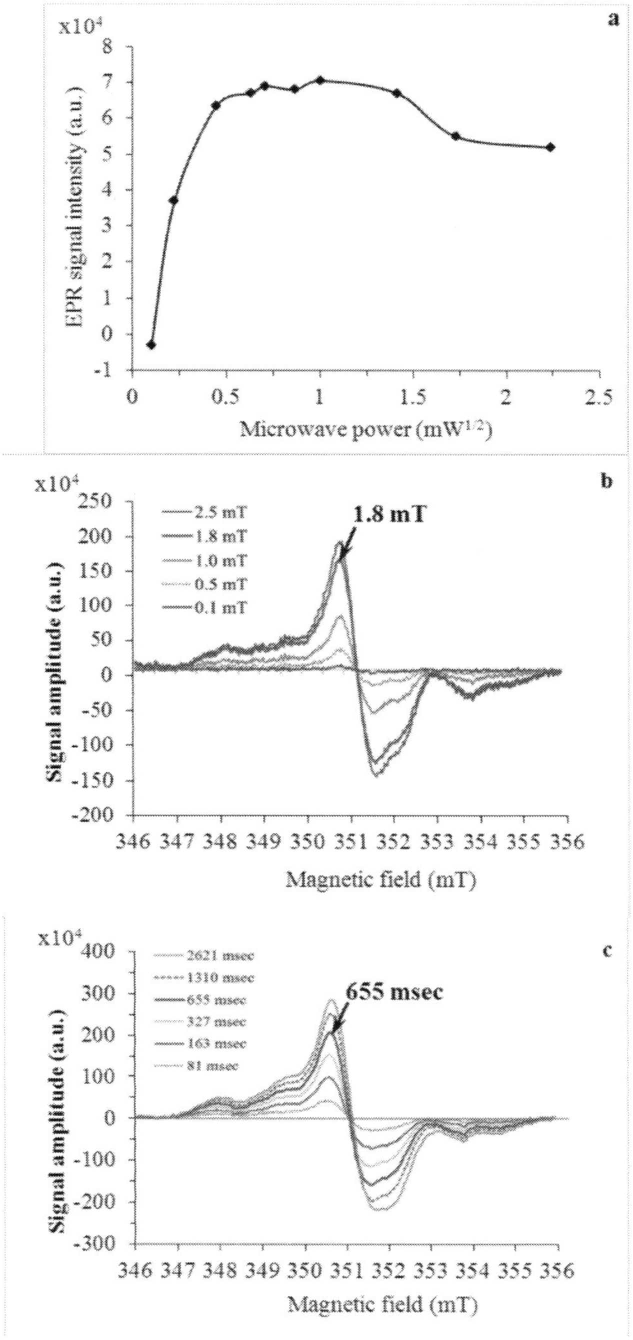

Figure 1. EPR spectrum optimization. (a) Variation of EPR signal intensity with microwave power for irradiated cotton (500 Gy). (b) EPR spectra with varied modulation amplitude. (c) EPR spectra with varied time constant.

Finally, the time constant was varied to maximize the signal-to-noise ratio. It appeared that the intensity increased as the time constant was increased from 81 to 2,621 ms; however, distortion was not obviously present (figure 1c). Since the time constant is related to the scan time, the moderate time constant at 655 ms was selected to obtain the appropriate scan time to collect the spectra under this experiment.

3.2. EPR spectra of unirradiated and irradiated samples

EPR spectral investigations were carried out in unirradiated and irradiated samples. As shown in figure 2, the EPR spectrum of an unirradiated sample exhibited a singlet lines centered at $g = 2.006$ probably arising from stable free radicals originated from the processing of the cotton fabrics. This might be due to the effect of several parameters on cotton fibers during its fabrication such as chemical and physical processes. The results were in agreement with the previous work reported by Viscomi *et al* [8].

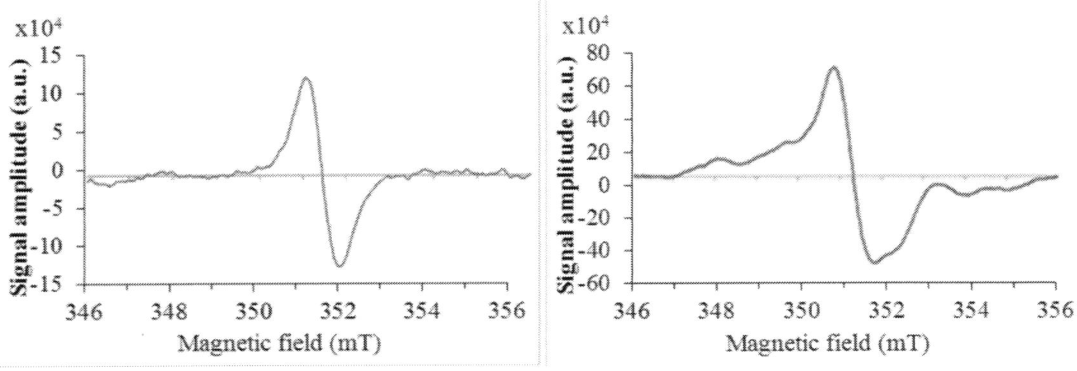

Figure 2. EPR spectra of unirradiated (left) and 500 Gy-irradiated (right) cotton samples.

After irradiation of the cotton samples, there was no difference observed except that the signal intensity of EPR spectrum was changed. It is generally known that cotton is nearly pure cellulose consisting of polysaccharide chains. This EPR signal originated from the carbon-centered radicals produced in the cotton cellulose by gamma irradiation and stabilized at room temperature. It was previously expected that 5 different types of radicals would be produced by H atom elimination from a C-H bond in the molecule of saccharides [10]. The yield of free radicals varied depending on the dose received. Even though these free radicals were detected in unirradiated cotton fibers, their EPR spectrum showed a weak signal compared to samples irradiated at the lowest dose of 0.1 Gy under this study. The EPR signal of 0.1 Gy-irradiated samples exhibited an increase of 40% relative to that of unirradiated samples.

The EPR spectra in relation to the doses of irradiated cotton are shown in figure 3. The pattern of EPR spectra did not change as the dose increased from 0.1 - 50 Gy. It was observed that the EPR signal intensity of irradiated cotton linearly increased with the absorbed dose (data not shown). The construction of the dose-response curve might be further developed in order to be used to estimate the dose received in unknown samples.

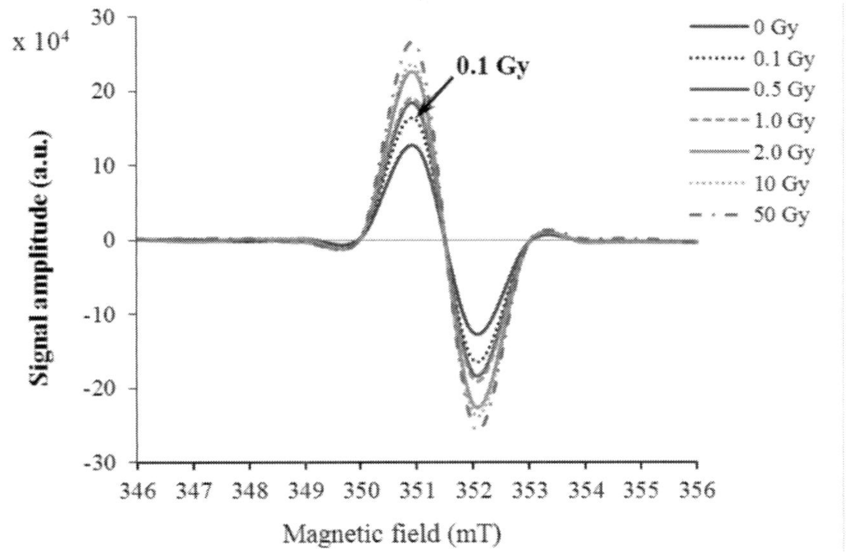

Figure 3. EPR spectra of unirradiated and irradiated cotton samples.

3.3. The stability of EPR signals

When irradiated samples were stored at room temperature in the dark, their EPR spectrum exhibited a broadened signal with a slight shift in the resonance signal and centred at $g = 2.234$. The signal intensity exhibited an exponential decay with storage time from 1 to 60 days (figure 4). However, the degree of fading of EPR intensity was not related to the absorbed dose from 0.1-50 Gy. The intensity of signal decreased between 55 and 60% after 60 days of storage of the irradiated samples at all doses. However, the post-irradiation signal was still detectable for up to 60 days. Further investigation should be performed to explore its persistence for periods longer than 60 days to assess the possibility of using the EPR signal as a marker of radiation exposure.

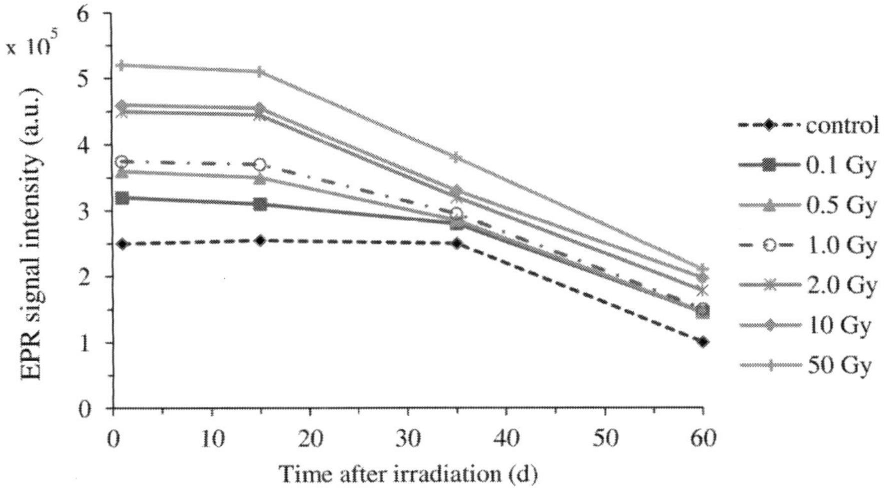

Figure 4. The stability of the EPR signals of cotton samples during the storage period.

4. Conclusions

The measurement of free-radicals in the cotton fabrics using EPR spectroscopy is possible for the determination of the radiation dose received by affected individuals in a radiological accident. The dose distribution obtained will provide useful information for medical treatments. As cotton is used to make a number of textile products and can be blended with other materials such as fibers, including rayon, and synthetic fibers such as polyester, further efforts are required to develop it a fortuitous dosimeter.

Acknowledgements

The authors gratefully acknowledged the support of an EPR instrument by the Office of Atoms for Peace, Thailand. This study was funded by Kasetsart University Research and Development Institute.

References

[1] Hagen W R 2013 EPR Spectroscopy *Practical Approaches to Biological Inorganic Chemistry* ed R Louro and R Crichton (Amsterdam: Elsevier) chapter 3 pp 53-75
[2] Teixeira M I, Da Costa Z M, Da Costa C R Pontuschka W M and Caldas L V E 2008 *Radiat. Meas.* **43** 480-2
[3] Longo A, Basile S, Brai M, Marrale M and Tranchina L 2010 *Nucl. Instrum. Meth. Phys. Res.* B **268** 2712-18
[4] Trompier F, Monaca, S D, Fattibene P and Clairand I 2011 *Radiat. Meas.* **46** 827-31
[5] Thompson J W, Atiya I A, Rink, W J and Boreham D 2009 *Radiat. Meas.* **44** 243-8
[6] Da Costa Z M , Pontuschka W M and Campos L L 2005 *Appl. Radiat. Isot.* **62** 331-6
[7] Da Costa Z M, Pontuschka W M, Ludwig V, Giehl J M, Da Costa C R and Duarte E L 2007 *Radiat. Meas.* **42** 1233-6
[8] Viscomi D, De Angelis C and Fattibene P 2011 *Radiat. Meas.* **46** 978-83
[9] Sudprasert W, Monthonwattana S and Vitittheeranon A 2012 *Radiat. Meas.* **47** 640-3
[10] Takács E, Wojnárovits L, Borsa J, Földváry Cs, Hargittai P and Zöld O 1999 *Radiat. Phys. Chem.* **55** 663-6

Estimation of the indoor radon and the annual effective dose from granite samples

P Sola[1], W Srinuttrakul and P Kewsuwan
Thailand Institute of Nuclear Technology, Ongkharak, Nakhon Nayok, Thailand

E-mail: ph_solar@hotmail.com

Abstract. Inhalation of radon and thoron daughters increases the risk of lung cancer. The main sources of indoor radon are building materials. The aim of this research is to estimate the indoor radon and the annual effective dose from the building materials. Eighteen granite samples bought from the markets in Thailand were measured using an ionization chamber (ATMOS 12 DPX) for the radon concentration in air. Radon exhalation rates were calculated from the radon concentration in chamber. The indoor radon from the granite samples ranged from 10.04 to 55.32 $Bq{\bullet}m^{-2}{\bullet}h^{-1}$ with an average value of 20.30 $Bq{\bullet}m^{-2}{\bullet}h^{-1}$ and the annual effective dose ranged from 0.25 to 1.39 $mSv{\bullet}y^{-1}$ with an average value of 0.48 $mSv{\bullet}y^{-1}$. The results showed that the annual effective doses of three granite samples were higher than the annual exposure limit for the general public (1 $mSv{\bullet}y^{-1}$) recommended by the International Commission on Radiological Protection (ICRP). In addition, the relationship between the colours and radon exhalation rates of granite samples was also explained.

1. Introduction

Building materials are enriched with radioactive elements and thus are considered as NORMs (naturally occurring radioactive materials). Radioactive elements in NORMs, especially radon (Rn-222) may have an effect on human health. Radon emits high energy alpha particles (5.5 MeV) which can cause damage to the DNA (deoxyribonucleic acid) of human. Therefore, long-term exposure to high levels of radon concentration increased the risk of lung cancer [1-2]. Radon is a radioactive gas and can be released from the building materials into the air by leaking through a crack of the room floor and accumulated in the houses [3-4]. Radon exposure can occur from all building materials made from substances containing significant amount of radium. In some cases, radon levels are very low, but in other instances, building materials may contribute highly to a significant radon exposure. These building materials sometimes contain comparatively high amounts of uranium, radium and thorium, of which radon is a decayed product [5-7]. Reports from many countries as well as Thailand revealed that granite has higher levels of the radon exhalation rate than other building materials [8]. Thus, in this research, eighteen granite samples were bought from several markets in Thailand. Their radon exhalation rates were determined using an ionization chamber (ATMOS 12 DPX) by the closed-chamber method [9]. The rates were calculated based on the measured radon concentration in the chamber. After that, the radon exhalation rates were used for the estimation of indoor radon and the effective dose from the building materials. The resulting indoor radon concentrations were compared with the average of radon background value of 20 $Bq{\bullet}m^{-3}$ in Thailand surveyed by Moontep *et al* and Boonyaprapa *et al* [10-11] and the guidelines

[1] To whom any correspondence should be addressed.

level for indoor radon of 148 Bq•m^{-3} recommended by the United States Environmental Protection Agency (USEPA) [12]. The resulting annual effective doses were also compared with the corresponding results of different sources and the dose limit for the general public of 1 mSv•y^{-1} recommended by the International Commission on Radiological Protection (ICRP) [13]. In addition, we have explained the relationship between the colour index of each granite sample and its radon exhalation rate.

2. Materials and methods

Granite samples (figure 1a) were bought from a building material store in Thailand. These building materials may be imported from aboard or locally produced. Radon in the pore space of any materials will migrate to the atmosphere in a process called exhalation. In this research, building materials with known surface area were estimated for their radon exhalation rate (production rate) using an ionization chamber (ATMOS 12 DPX) by the closed-chamber method. Each sample was put in a closed exhalation chamber made of stainless steel with a volume of 0.05 m^3 (figure 2b-c). The closed chamber was tested for the leakage rate by evacuating the container to zero on the vacuum gauge and left for 6 h. Any drop of the vacuum-gauge needle would indicate air leakage. The radon-free air was flowed through an activated charcoal canister into the closed chamber for about 2 min (or until it reached the normal pressure). The closed chamber was kept until the radioactive equilibrium was reached (which normally took more than 30 days).

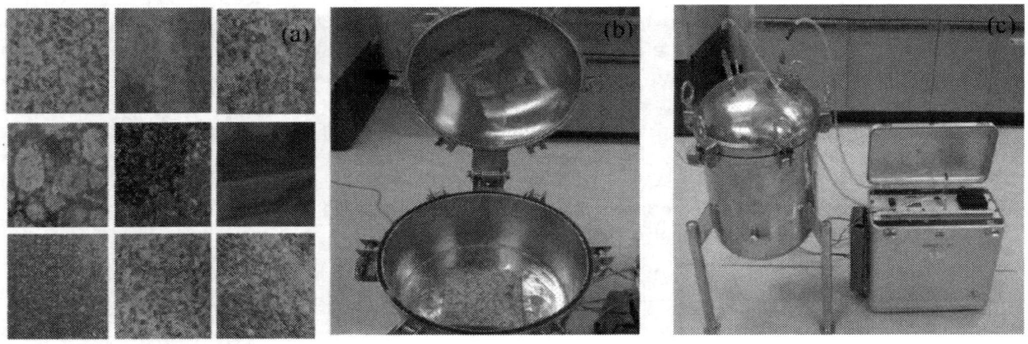

Figure 1. (a) Colour of the granite samples used; (b) the exhalation chamber; (c) the experimental set up used in this study.

Radon concentration in the closed exhalation chamber was measured by an ionization chamber (ATMOS 12 DPX) (a continuous radon monitor) [14]. An extrapolation of the radon concentration to an infinite time point could be calculated by equation 1. The calculation of radon exhalation rate from building materials using radon concentrations measured in the closed chamber under the radioactive equilibrium state was described by Kovler *et al* [14] as shown in equation 2:

$$C_t = C_\infty \left(1 - e^{-\lambda_{Rn} t}\right) \quad (1)$$

$$E = \frac{C_\infty \lambda_{Rn} V}{S} \quad (2)$$

where C_t was the concentration of radon measured at time t (Bq•m^{-3}); C_∞ was the maximum radon concentration in equilibrium (Bq•m^{-3}); t was the accumulation time of the sample (h); λ_{Rn} was the decay constant of radon (h^{-1}); E was the radon exhalation rate (Bq•m^{-2}•h^{-1}); V was the volume of the chamber (m^3), and S was the surface area of the source (sample) (m^2).

Radon exhalation rates of the building materials showed a good correlation with indoor radon concentrations under the determining parameters such as the surface area of total wall, the room volume, the outdoor radon concentration and the ventilation rate in the house etc. The estimation of indoor radon concentration in a room due to radon exhalation rates from building materials could be calculated by equation 3 [15-16]:

$$C = \frac{ES}{V(\lambda_{Rn} + \lambda_v)} + \frac{C_o \lambda_v}{\lambda_{Rn} + \lambda_v} \quad (3)$$

where C is radon concentration (in a room) (Bq•m^{-3}); E is radon exhalation rate (Bq•m^{-2}•h^{-1}); C_o is outdoor radon concentration which is about 5 - 10 (Bq•m^{-3}); V is the room volume (m^3); S is the surface area of total wall (m^2); λ_{Rn} is the radon decay constant (h^{-1}), and λ_v is the ventilation rate at 0.5 (h^{-1}).

The room volume and the surface area of total wall in this research were 56 m^3 and 90.4 m^2 respectively, by referring to the radiation protection report RP-112 [17]. In this research, for the highest safety, the ventilation rate was set at 0.5 h^{-1} [18]. Generally, ventilation is the key factor that affects the indoor radon concentration, and on the average, an air-conditioned room has a higher radon concentration than a non-air-conditioned room in the same category of building. At the same time, the use of the building materials only slightly influences the indoor radon concentration [19].

The indoor radon concentration values estimated were used for the calculation of the annual effective dose due to living in a dwelling. Estimation of the annual effective dose was calculated using equation 4 [20]:

$$H = C \times F \times T \times D \quad (4)$$

where H is the annual effective dose rate (mSv•y^{-1}); C is the radon concentration (Bq•m^{-3}); F is the radon-equilibrium factor (0.4); T is the indoor occupancy factor (7000 h•y^{-1}); D is the dose conversion factor (9.0×10^{-6} mSv•h^{-1} per Bq•m^{-3}).

3. Results and discussion

Radon concentrations from granite samples in a closed chamber under a radioactive equilibrium state, radon exhalation rates, indoor radon concentrations and estimates of the annual effective dose rate are shown in table 1.

The radon concentrations of the granite samples in closed chamber ranged from 14.36 ± 3.31 to 3,400.61 ± 4.39 Bq•m^{-3}. The radon exhalation rates ranged from 0.06 to 14.30 Bq•m^{-2}•h^{-1}. The estimated indoor radon concentrations ranged from 10.04 to 55.32 Bq•m^{-3}. The estimated annual effective doses varied from 0.25 to 1.39 mSv•y^{-1}. From table 2, the highest radon exhalation rate was found in both pink and orange granite samples, such as GS13, GS 16 and GS18, while the lowest radon exhalation rate was found in both black and white granite samples, such as GS1, GS3, GS4, GS7, GS12 and GS15.

Estimated indoor radon concentration and the annual effective dose rate, in comparisons to the background, the guideline level and the dose limit, were shown in figure 2.

Table 1. The origins of the samples, their colour index, the equilibrium radon concentrations, radon exhalation rates, estimated indoor radon concentrations and the annual effective dose rates.

No.	Origin of sample	Colour index	Radon concentration	Radon exhalation rate	Estimated indoor radon concentration	Estimated annual effective dose rate
			($Bq \cdot m^{-3}$)	($Bq \cdot m^{-2} \cdot h^{-1}$)	($Bq \cdot m^{-3}$)	($mSv \cdot y^{-1}$)
GS1	Thailand	Black [a], White	14.36±3.31	0.06±0.01	10.04	0.25
GS2	Thailand	Grey [a], Black, Pink	1061.38±8.84	4.46±0.04	24.04	0.61
GS3	Thailand	Black [a], White	28.65±4.52	0.12±0.02	10.23	0.26
GS4	Thailand	Black [a], White	263.81±9.77	0.80±0.03	12.39	0.31
GS5	Thailand	White [a], Grey, Black	479.33±16.57	2.02±0.02	16.26	0.41
GS6	Thailand	Grey [a], White, Black	744.40±20.98	3.13±0.04	19.80	0.50
GS7	Thailand	Black [a], White	52.33±5.81	0.22±0.02	10.55	0.27
GS8	Brazil	Red [a], Black	1343.01±11.04	5.65±0.05	27.81	0.70
GS9	India	Black [a], Grey	80.97±2.61	0.34±0.06	10.93	0.28
GS10	China	Black [a], Grey	60.07±7.84	0.25±0.03	10.65	0.27
GS11	Brazil	Golden [a], Red, Black	67.08±6.71	0.25±0.03	10.66	0.27
GS12	China	Black [a]	43.60±16.77	0.18±0.07	10.43	0.26
GS13	Vietnam	Orange [a], Grey, Pink	3400.61±4.39	14.30±0.02	55.32	1.39
GS14	Finland	Orange [a], Pink, Black	521.30±7.73	2.19±0.03	16.82	0.42
GS15	South Africa	Black [a]	109.70±9.14	0.46±0.04	11.32	0.29
GS16	Brazil	Pink [a], Golden, Black	2672.52±3.88	11.24±0.02	45.59	1.15
GS17	India	Red [a], Black	980.63±23.26	4.12±0.10	22.96	0.58
GS18	Brazil	Pink [a], Golden, Black	2229.00±4.18	9.37±0.065	39.65	1.00

[a] The first colour was the colour index.

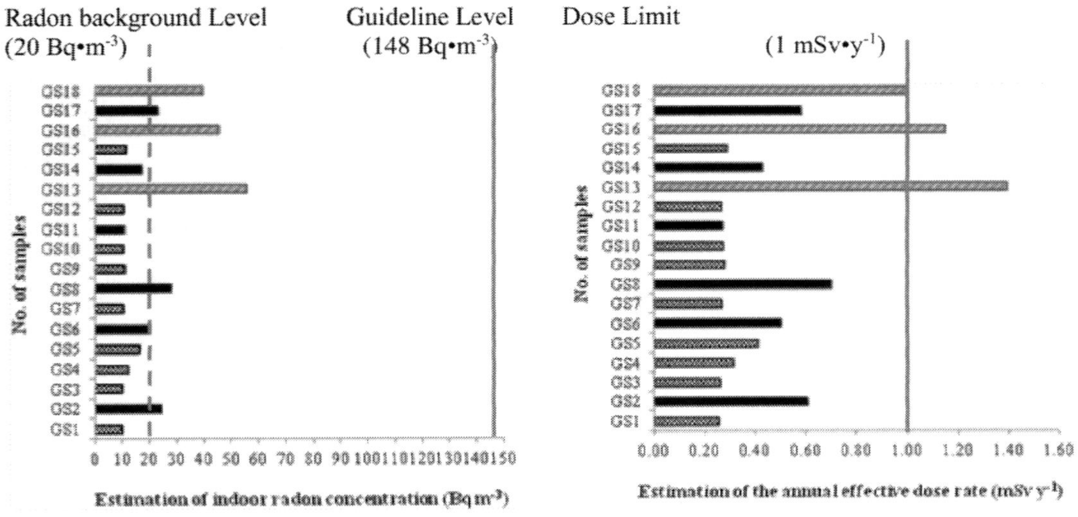

Figure 2. Comparison of the estimations of indoor radon concentration (a) and the annual effective dose rate (b) in comparison to the background level of radon, the guideline level of Thailand and the dose limit. ▩ represented black and white granite samples; ▨ represented pink and orange garnite samples; ■ represented granite samples of other colours.

Figure 2 showed that the estimations of indoor radon concentration of three granite samples (GS13, GS16 and GS18) were higher than the dose limit for the public. Interestingly, these three granite samples appeared pink and orange in colour. While black and white granite samples produced the lowest radon exhalation rate. Different colours of granites appeared to exhale different amounts of radon. However, additional experiments should be conducted as radon exhalation rates also depend on other factors. Radon exhalation rates also depend on the composition of each granite sample, which contains naturally-occurring uranium and radium. These radioactive elements can lead to a rather high radon exhalation rate. However, from this experiment, differences in the colour of granite samples may be used to assess the amount of the radon exhalation rate from granite. This finding may be useful in the selection of building materials for home owners.

4. Conclusion

In this research, granite samples of multiple origins were bought from a material store in Thailand. The radon concentrations of the granite samples in closed chamber under radioactive equilibrium were monitored by using an ionization chamber (ATMOS 12 DPX). The values of radon concentrations were used for the calculation the radon exhalation rate from granite samples. Estimations of the indoor radon and annual effective dose from the building materials were calculated. The estimations of indoor radon concentration were compared with the background level of Thailand (20 Bq•m^{-3}) surveyed by Moontep *et al* and Boonyaprapa *et al* [10-11] and the guideline of indoor radon concentration (148 Bq•m^{-3}) recommended by USEPA. Estimations of the annual effective dose were compared with the annual effective dose limit for the public (1 mSv•y^{-1}) recommend by ICRP. Radon concentrations from granite samples ranged from 14.36 ± 3.3 to 3,400.61 ± 4.39 Bq•m^{-3}. Radon exhalation rates from granite samples ranged from 0.06 to14.30 Bq•m^{-2}•h^{-1}. The estimations of the indoor radon concentration ranged from 10.04 to 55.32 Bq•m^{-3}. The estimations of indoor radon concentration of six granite samples were higher than the average radon background level of Thailand. However, they were below the guideline level for indoor radon concentrations. Estimations of annual effective dose ranged from 0.25 to 1.39 mSv•y^{-1}. The annual effective dose of three granite samples were higher than the recommended annual effective dose limit for the public. Granite samples with different colours exhaled different amounts of radon. However, additional experiments should be carried out to investigate the relationship between granite colours and radon exhalation rates.

Acknowledgments

The authors would like to thank the staffs of Nuclear Engineering Division, Thailand Institute of Nuclear Technology, Mr. Kosol Moombansao, Mr. Sanee Thaveesuk and Mr. Charoen Kheawngarm for their helpful comments and supports in making the radon chamber.

References

[1] National Council on Radiation Protection and Measurements 1984 *Evaluation of Occupational and Environmental Exposures to Radon and Radon Decay products in the United States NCRP Report No. 78* (Bethesda Maryland, MD: National Council on Radiation and Measurements)
[2] World Health Organization 2009 *WHO Handbook on Indoor Radon, A Public Health Perspective* (Geneva: WHO Press)
[3] Nowak Mo Song B H 1990 *Int. Symp. on Radon and Radon Reduction Technology Vol. V: VIII* (Atlanta, GA: EPA)
[4] Cothern C R and Smith J E 1987 *Environ. Sci. Res.* **35** 218
[5] Allen J G, Minegishi T, Myatt T A, Stewart J H, McCarthy J F and Macintosh D L 2010 *J. Expo. Sci. Environ. Epidemiol.* **20** 263
[6] Righi S and Bruzzi L 2006 *J. Environ. Radioactivity* **88** 158
[7] Ai-Jarallah M 2001 *J. Environ. Radioactivity* **53** 91
[8] Sola P, Srinuttrakul W, Laoharojanphand S and Suwankot N 2014 *DAe-BRNS 5th Symp. On Nuclear Analytical Chemistry (NAC-V)* (Mumbai: BARC) pp 82-3

[9] Kovler K 2006 *Proc. Int. Radon Symp. (*Kansas City, MO) pp 31-7
[10] Moontep C, Suparom K, Sola B, Chantarachote W, Wanabongse P and Bovornkitti S 2008 *Thammasat Medical J* **8** 455
[11] Boonyaprapa S, Wanabongse P, Cheepsattayakorn A, Saeung S, Sola B and Bovornkitti S 2008 *J. Health Systems Res.* **2** 460
[12] US EPA 1992 National emission standards for hazardous air pollutants; Emission standards for radon emissions from phosphogypsum stacks; Final rule (40CFR61) *Federal Register* **57** 23305
[13] ICRP 1991 *Annals of the ICRP; 1990 Recommendation of the Int. Commission on Radiation Protection* (*ICRP Publication 60)* (Oxford: Pergamon Press)
[14] Kovler K, Perevalov A, Steiner V and Metzger L A 2005 *J. Environ. Radioactivity* **82** 321
[15] Krisuk E M 1980 *Health Phys* **38** 199
[16] Quindos L S, Newton G J and Wilkening M H 1987 *Radiat. Prot. Dosim.* **19** 125
[17] European Commission 1999 Radiological protection principles concerning the natural radioactivity of building materials *Radiation Protection Report RP-112* (Luxembourg: European Commission)
[18] Krisiuk E M 1980 *Health Phys.* **38** 199
[19] Leung J K, Tso M Y W and Ho C W 1999 *Health Phys.* **76** 537
[20] United Nations Scientific Committee on the Effects of Atomic Radiation 2000 Sources and effects of ionizing radiation *2000 Report to the General Assembly with Annex B: Exposure from Natural Source of Radiation* (New York: UNSCEAR)

Journal of Physics: Conference Series **611** (2015) 012014 doi:10.1088/1742-6596/611/1/012014

Safety and security management of disused sealed radioactive sources in Thailand

N Ya-anant[1,2], P Nuanjan[1], A Phattanasub[1], T Akharawutchayanon[1], A O-manee, N Prasertchiewchan[1] and J C Benitez-Navarro[2]
[1] Head, Radioactive Waste Management Section,
Radioactive Waste Management Center, Thailand Institute of Nuclear Technology, Chatuchak, Bangkok, Thailand
[2] Nuclear Engineer, Waste Technology Section, Division of Nuclear Fuel Cycle and Technology, Atomic Department of Nuclear Energy, International Energy Agency, Vienna, Austria

E-mail: nanthavan@tint.or.th

Abstract. When sealed radioactive sources are no longer in use, they should be returned back to the country of origin. However, most of them could not be returned to the origin; therefore, disused sealed radioactive sources (DSRS) have to be managed locally to ensure the safety and the security for long term storage before final disposal. The Radioactive Waste Management Center, Thailand Institute of Nuclear Technology, is authorized to operate the treatment, conditioning and storage of DSRS in Thailand. This paper will describe the operational procedures on characterization technique, identifation of unknown sources, volume reduction technique, re-packaging, registration and record keeping of DSRS. The successful results included that the record keeping of DSRS has been developed, and the national inventory of stored DSRS has been made up to date. The results confirmed that the quality control at the DSRS storage facility at Thailand Institute of Nuclear Technology was established and well implemented to ensure safe and secure management.

1. Introduction

A sealed radioactive source is a container of encapsulated radioactive material, which usually has the appearance of a small, harmless piece of metal. The capsule or material of a sealed source is strong enough to maintain leak tightness under the conditions of use for which the source was designed, and also under foreseeable mishaps. In more technical terms, it is radioactive material that is permanently sealed in a capsule or closely bound in a solid form. The source is designed to contain the radioactive material under normal operating conditions and usually has high concentration of radioactive material in a small volume. The earliest uses of radioactive sources typically involved radium-226 particularly in medical applications in Thailand since 1950 [1].

Up to now sealed radioactive sources are used in various applications in medicine, agriculture, industry, transportation, construction, geology, mining, research, etc. In most of these areas the use of sealed radioactive sources either cannot be replaced by other methods or provides results that are unmatched by other methods. The first sealed radioactive sources were used for medical purposes from

[2] To whom any correspondence should be addressed.

Journal of Physics: Conference Series **611** (2015) 012014 doi:10.1088/1742-6596/611/1/012014

the beginning of the twentieth century. Until the 1940s sources were only using naturally occurring isotopes (radium and polonium). Today sealed radioactive sources are made of numerous artificial radioisotopes. The gamma rays emitted by the source were used to treat cancer. The radioactive source material, over 1,000 curies of cobalt-60, was contained in a doubly-encapsulated stainless-steel cylinder approximately one inch in diameter. The rest of the capsule (approximately two inches in diameter) consisted of a tungsten alloy housing threaded to fit into the teletherapy unit. The top of the housing contained an opening through which the stainless-steel source container could be seen [2].

When sealed radioactive sources are no longer in use, and there is no intention of using them again, they are usually called "disused sealed radioactive source" or "DSRS". Worldwide, the number of sources that are considered disused is very large [2] and warrants dedicated efforts for their management in a safe and secure manner. If lost or not properly controlled, disused sealed sources can be a threat to human health and the environment. Exposure to large doses of radiation from an unshielded high activity source can be lethal or cause severe radiation injury. If the source capsule is damaged the radioactive material can be released and dispersed, resulting in contamination to the environment, social and economic impacts.

2. Why we need the safety and security management on DSRS

The serious radiological accident in Thailand occurred in Samut Prakan province in 2000, when cobalt-60 head of a disused teletherapy unit was partially dismantled, and taken from that storage to be sold as scrap metals [3]. Three victims died and 10 people received high doses from the radioactive source. It was an expensive lesson to be learned in Thailand [1]. Therefore, we do really need the safety and security management of DSRS. The picture of the Samut Prakan radiological accident is shown in figure 1.

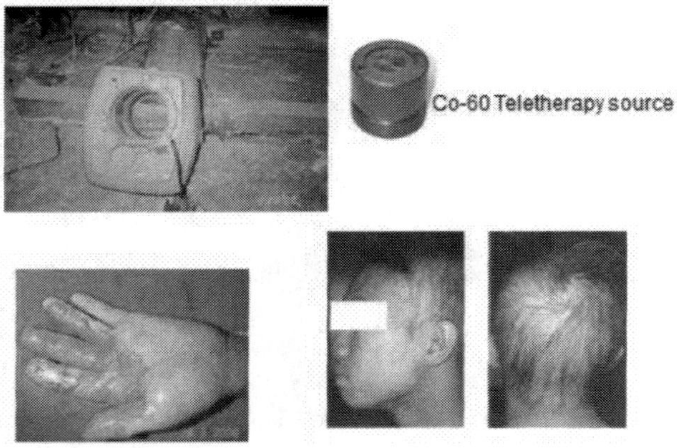

Figure 1. Photographs from the Samut Prakan Radiological Accident in 2000 [2] (Permission to publish was granted by IAEA).

The safety and security concerning sealed radioactive sources became a highlighted topic after the attack on the United States on 11 September 2001. The US itself, including the International Atomic Energy Agency (IAEA) and other countries put an increased awareness of the need for safety and security measures to protect radioactive sources against terrorism [4]. The planning of security measures of radioactive source must take greater account of the potential for deliberate acts to attack or use radioactive sources to expose people and cause contamination. These include fatal and injurious radiation exposure, contamination of the environment, the serious economic and psychosocial costs; the total effect of which is mass disruption. As a consequence, the management of DSRS is highly justified, and the need for special security measures to protect against terrorism should then become part of the safety assessment.

3. Management of DSRS

In Thailand, when a radioactive source is no longer in use, it should be returned back to the country of origin [5]. However, most of DSRS could not be returned to the origin; therefore, it became the responsibility of the Radioactive Waste Management Center (RWMC), Thailand Institute of Nuclear Technology (TINT), to manage these DSRS under the regulatory controlled of the Office Atoms for Peace (OAP), the Thai nuclear and radiation regulatory body. The procedures of DSRS management in Thailand are as follows:

- History of source / record keeping
- Identification
- Characterization
- Volume reduction (if possible)
- Conditioning / re-packaging
- Interim storage and final disposal (in the future)
- Inventory of DSRS / registration of DSRS

4. Procedures

The reorganization operation of DSRS has been conducted at TINT radioactive waste storage facility no.1 located in Bangkok, Thailand. The operation included revision and development of proper techniques on identification, characterization, and record-keeping. An inventory of DSRS has also been developed.

The technical procedures were as follows:

1) The operating area was prepared by using plastic sheets to cover the designated area;

2) Each DSRS was checked for radioactive contaminations before it was moved to the operating area as shown in figure 2;

3) Each individual source was taken to the operating area and checked for dose rate (figure 3);

4) The identification label was checked, and its photo was taken (figure 3);

Figure 2. Checking for contamination and dose-rate of DSRS.

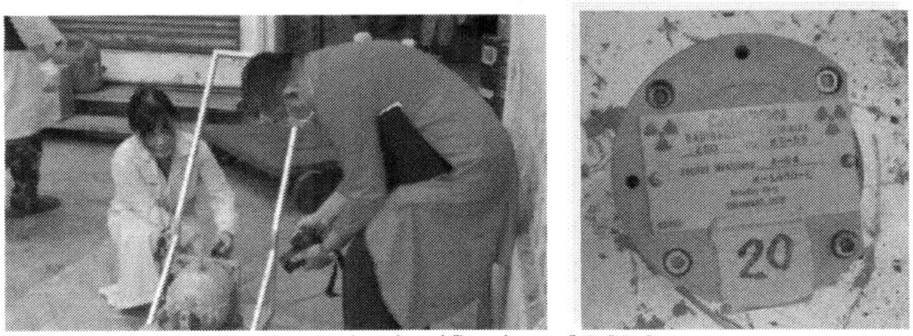

Figure 3. Identification of DSRS.

5) The sources were placed in an empty room prepared for DSRS storage, organized by the type of radioactive sources and radionuclides (segregation) as in figure 4;

Figure 4. Segregation and grouping by type of radioactive sources and radionuclides.

6) Volume reduction of DSRS was achieved by removing DSRS from their large containments and re-packing them in 200 liter drums as shown in figures 5 and 6;

Figure 5. Volume reduction by removing DSRS from their containment and re-packing them.

Figure 6. Re-packing of DSRS in 200-liter drums.

7) DSRS were inventoried;
8) The storage conditions were improved by cleaning the floor and the cellar;
9) The sources were arranged in baskets and put on shelves, grouped by radionuclides and type of radiation emitted;

10) A record was kept for each operation and DSRS. For record keeping, the following information was recorded:
- Former owner who possessed the source and associated device, including the contact information,
- Unique identification of the source (manufacture, model number, serial number, and date of manufacture),
- Unique identification of the associated device (manufacture, model number, serial number, and date of manufacture),
- Radionuclide, the sources activity and the date on which the activity was measured,
- Category of the source,
- Form of the radioactive material (physical and chemical) including its special form status,
- Record of where the source was received from or transferred to,
- Date on which the source and associated device was entered into the register,
- Planned disposition of the sources such as the planned date of its transfer to a waste storage or a disposal facility.

Figure 7. Before and after the reorganization of DSRS at TINT storage facility no.1.

5. Results and Discussion

The total number of DSRS which were characterized and re-organized at TINT storage facility no.1 was 418 pieces. These DSRS included Am-241, Am-241/Be, Co-60, Cs-137, Kr-85, Sr-90, Ni-63, Pm-147 and Ra-226. Most of DSRS were removed from the old containers and then re-allocated to new pallets and new metal boxes. The number of pallets for re-allocated DSRS was 53; they were stacked into 3 shelves. Four unknown sources were identified to the known sources by identifier tool and the international source information catalogue [6]. Also, old records in TINT waste storage facilities must be checked and updated regularly. Sometimes workers who have been employed for a long time and are near retirement age will need to provide ascertained information about unknown sources. It is especially important to tap into this information resource before such individuals disappear. From this mission, the Radioactive Waste Management Center has to revise the working instruction (WI) of DSRS storage and to add the characterization procedure into the WI of DSRS storage.

The new registration of all DSRS in storage facility no.1 was included in the process as recommended by the OAP. From this mission, the security and safety process of DSRS at storage facility no. 1 has been put in place and well controlled by the operator (TINT) and the regulator (OAP). The access control, key cards, fire extinguisher, and the emergency preparedness and response plan at the storage facility of DSRS were revised and implemented during this mission by the advice and supports of TINT Safety Unit.

The outputs achieved as the results of activities implemented were as follows:
- The record keeping of 418 sources was developed and updated;
- The reorganization of DSRS was well done within the storage facilities as shown in figure 7

- Unknown DSRS was re-categorized as known DSRS based on characterization results.
- The national inventory of stored DSRS was made up-to-date.
- Proper operational procedures and quality control at DSRS storage facility was established and implemented, such as characterization procedures, the emergency preparedness and response plan at the storage facility.

6. Conclusion

In general, the users and owners of radioactive sources would have some information about their sources, and may also have documents or records of sources that they possessed or used in the past. In fact, some sources may have been transferred to others without regulatory control. This sometimes happens; therefore, the Radioactive Waste Management Center, TINT, has to characterize and identify the stored DSRS to get the correct information for record keeping. The re-organization of DSRS was very useful and really needed. The characterization and categorization of disused radioactive sources, and identification of each DSRS with a label to complete the records are very important steps to organize the DSRS according to the radionuclide and the type of sources. The outcome of this work was the updated inventory of DSRS control. Additional advantage of this work was the volume reduction of DSRS at the TINT storage facilities.

Acknowledgments

The authors wish to express their deep gratitude to the IAEA, especially Dr. Juan Carlos Benitez Navarro, who has been giving all his support, valuable advice, and guidance throughout this work. Many suggestions and technical support provided by Ms. Mercedes Salgado and Mr. Juan Miguel, the external experts from Cuba, were deeply appreciated. The authors were very grateful to Ms. Pattama Na Nakorn, Mr.Pisit Suntarapai, and staff from the Bureau of Radiation Safety Regulation, the Office of Atoms for Peace, Thailand, for their kind co-operation. Finally, the authors wish to extend their sincere appreciation to the manager of the RWMC, and the team from RWMC and also TINT Safety Unit staff, for their assistance and equipment as well as the facilities throughout this work.

References

[1] Ya-anant N,Tiyapun K and Saiyut K 2011 *Radiation Protection Dosimetry* **146** 111–4
[2] International Atomic Energy Agency (IAEA) (n.d.) *What is a Sealed Radioactive Source* Retrieved March 13, 2015, from http://www.iaea.org/OurWork/ST/NE/NEFW/Technical-Areas/WTS/sealedsources-sealedsources.html
[3] International Atomic Energy Agency 2002 *The Radiological Accident in Samut Prakan*, (Vienna: IAEA)
[4] United States General Accounting Office 2003 *GAO-03-804: Nuclear Security Federal and State Action Needed to Improve Security of Sealed Radioactive Sources* (Washington D.C.: GAO) pp 5-6
[5] Ministry of Science and Technology 2003 *Regulation on Procedure on Radioactive Waste Management B.E.2546* Bangkok, Thailand
[6] International Atomic Energy Agency (n.d.) *International Catalogue of Sealed Radioactive Sources and Devices (ICSRS)* Retrieved March 13, 2015, from http://www.iaea.org/OurWork/ST/NE/NEFW/Technical-Areas/WTS/information-SOURCE.html

Journal of Physics: Conference Series **611** (2015) 012015 doi:10.1088/1742-6596/611/1/012015

Carbohydrate based materials for gamma radiation shielding

F Tabbakh[1], V Babaee and Z Naghsh-Nezhad
Institute of Nuclear Science and Technology, Plasma and Nuclear Fusion Research School, Tehran, 14399-51113, Iran

E-mail: ftabbakh2000@yahoo.com

Abstract. Due to the limitation in using lead as a shielding material for its toxic properties and limitation in abundance, price or non-flexibility of other commonly used materials, finding new shielding materials and compounds is strongly required. In this conceptual study carbohydrate based compounds were considered as new shielding materials. The simulation of radiation attenuation is performed using MCNP and Geant4 with a good agreement in the results. It is found that, the thickness of 2 mm of the proposed compound may reduce up to 5 % and 50 % of 1 MeV and 35 keV gamma-rays respectively in comparison with 15% and 100% for the same thickness of lead.

1. Introduction

There are many reasons for searching for new materials for radiation shielding. For example lead is one of the most commonly used materials in gamma-ray shielding which known as a toxic substance with high risk for workers and it is recommended to search for the lead-free shields [1]. The most widely used materials for gamma-ray shielding are made of heavy elements or concrete. The conventional gamma-ray shielding materials due to their weight, price or mechanical properties my not be suitable for some applications such as astronomic applications [2, 3], portable sources or plasma focus systems (PF) but, polymeric compound [1-7] may be promising candidate for such applications.

In this work, we conceptually studied the attenuation of gamma-ray (or X-ray) by a polymeric compound doped with gadolinium. Two typical energies of 35 keV and 1 MeV for photons were considered corresponding to the average energy of photons emitted from a small plasma focus and two tokamaks (Alvand and Damavand) at Plasma and Nuclear Fusion Research School in Iran respectively. Furthermore, the attenuation of thermal neutrons using the proposed compounds was investigated too.

The calculations were conducted using MCNPX2.4.0 [8] and GEANT4.9.4.p02 [9] Monte Carlo simulation tools which play important role in radiation shielding calculations especially for the problems with the complicated geometries or compound materials. Also, a comparison between the mass attenuation coefficients of the proposed compounds and some common shielding materials are presented in the results and discussion section.

2. Materials and methods

The Monte Carlo techniques give the results based on the statistical calculations and cross-sections. This will cause some discrepancies between the simulation results and the experimental data [10, 11]. In the standpoint of nuclear safety, to decrease the uncertainty in calculations we have used MCNPX [4, 12-15] and GEANT4 [15-18] tools and compared the results for more reliable prediction of experimental

[1] To whom any correspondence should be addressed.

data. A comparison between the mass attenuation coefficients of the proposed compounds and widely used shielding materials are presented later in this work.

At the stage of conceptual design, two typical photon energies of 35 keV and 1 MeV corresponding to the average energy of photons emitted from a small plasma focus and two tokamaks (Alvand and Damavand) at Plasma and Nuclear Fusion Research School in Iran respectively. Later, the attenuation of thermal neutron was studied to determine the ability of these materials for attenuating the both gamma and thermal neutron radiation.

Regarding to gadolinium large cross section for thermal neutron [15] the calculations were performed for different atomic percentages of gadolinium and lead for comparison. Table 1 presents the atomic fraction of hydrogen, carbon, oxygen, silicon and gadolinium or lead with the total densities from ~2 g/cm^3 to ~8 g/cm^3 defining the input materials in simulation. Conventionally, we called the compounds with 2% and 6% dopant as CHGd2 and CHGd6 with ~2 g/cm^3 to ~8 g/cm^3 densities respectively.

Table 1. The elemental atomic fraction.

Element	Atomic fractionin the compound
H	0.32 –0.37
C	0.15 –0.17
O	0.25 –0.27
Si	0.13 –0.16
Gd/Pb	0.02 -0.06

In both MCNPX and GEANT4 we used one directional point source and the slab shield with surface dimension of 10cm×10cm. MCNPX was tallied by f1 as surface current and both f6 and *f4, average energy deposited over a cell. Since for photon/electron problems with thin shield electron may not deposit their energies locally, we used both f6 and *f4 tally to reduce the results uncertainties [8].

In GEANT4 calculation we used EMLOW6.19 and G4ENDL3.14 for photon neutron libraries. Neutron interactions were considered by means of G4NeutronHP data set in the physics lists.

3. Results and discussion

Figure 1 and figure 2 show the attenuation of 1 MeV photon versus target thickness for CHPb6 and CHGd6 respectively. The circular symbol depicts GEANT4 results and square show MCNPX results. As it was explained, CHPb6 corresponds to 6 % Pb and CHGd6 corresponds to 6 % Gd. The compound density is 8 g/cm^3. Less than 2 % discrepancies between two results for larger thicknesses are due to the small differences between the two libraries used by each tools. The increase in the thickness means the more interaction of photons with material and more different results consequently. This could be the good reason for using more than one method when there are no experimental references.

As one can see, 2 mm thick CHPb6 attenuates 5 % of 1 MeV photons, while the same attenuation is achievable at 2.6 mm thickness when using CHGd6. The discrepancy between the results for the two materials in this case is less than 4 %.

Figure 3 illustrates the attenuation of 35 keV photons versus the thickness for CHGd2. One can see, up to 75 % attenuation for the thickness of 2 mm. The discrepancy of less than 2 % shows the agreement between MCNPX and GEANT4 and the reliability of the results consequently. The same reason as what was discussed for Figures 1 and 2 can explain the small discrepancies between MCNPX and GEANT4 in this figure.

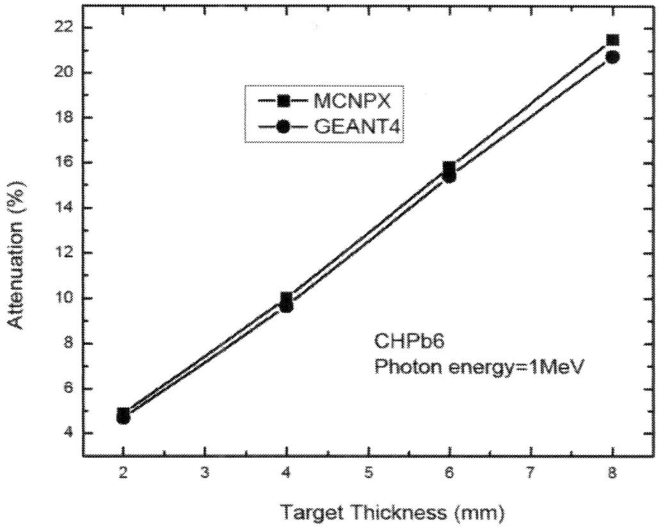

Figure 1. Attenuation of1MeVgamma-rayin CHPb6.

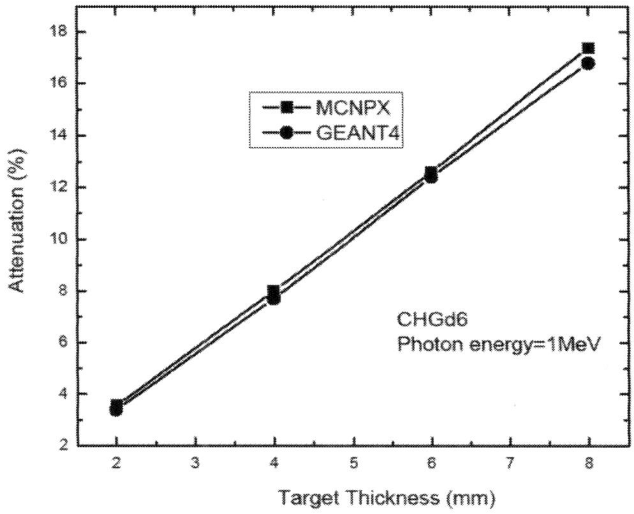

Figure 2. Attenuation of1MeVgamma-rayinCHGd6.

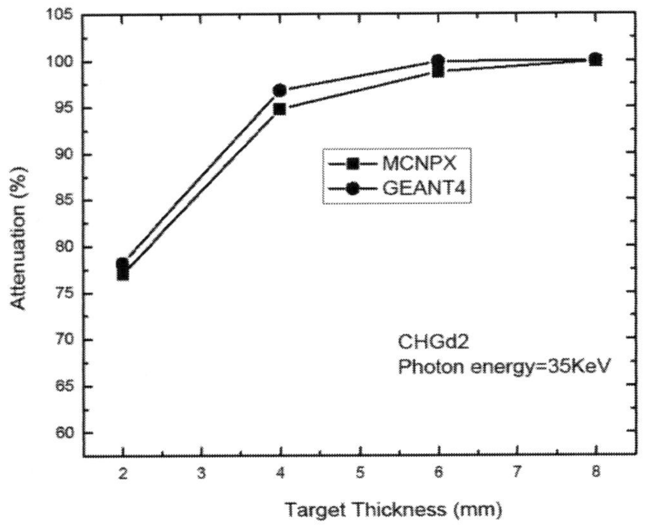

Figure 3. Attenuation of 35keVgamma-ray in CHGd2.

Table 2 shows the mass attenuation coefficients (μ/ρ) of CHGd6 and CHPb6 as comparison with standard data [19, 20] for lead, tungsten and concrete. The energies of 0.035 MeV, 0.05 MeV, 0.5 MeV and 1 MeV were selected typically. Accordingly, the two proposed compounds have relatively smaller attenuation coefficients than lead and tungsten but comparable with concrete. The presented values for attenuation coefficient were obtained from the gradient of the line fitted to the curve of logarithmic intensity versus the thickness. It can be expressed as below;

$$\mu = -\frac{Ln(\frac{I}{I_0})}{x} \tag{1}$$

where, I and I_o are the intensity of the incident and outgoing photons respectively, x is the thickness and μ is the linear attenuation coefficient.

Table 2. CHPb6, CHGd6, lead, tungsten and concrete mass attenuation coefficients (μ/ρ).

Photon Energy (MeV)	mass attenuation coefficients (cm^2/g)				
	CHPb6 (ρ=8 g/cm^3)	CHGd6 (ρ=8 g/cm^3)	lead (ρ=11.3 g/cm^3)	tungsten (ρ=19.3 g/cm^3)	concrete (ρ=2.4 g/cm^3)
0.035	0.94	0.9	20.1	10.67	0.9
0.05	2.05	2.1	8.041	5.949	0.341
0.5	0.0805	0.069	0.1614	0.1378	0.089
1	0.03	0.02	0.07	0.06618	0.065

Figure 4 illustrates the ratio of the deposited energy to the incident photon energy in terms of the incident photon energy for CHGd6 compound. The absorption edge was calculated and shown as a peak at 90 keV which was explained by the both simulation results as an enhancement in photoelectric effect in the material at this energy.

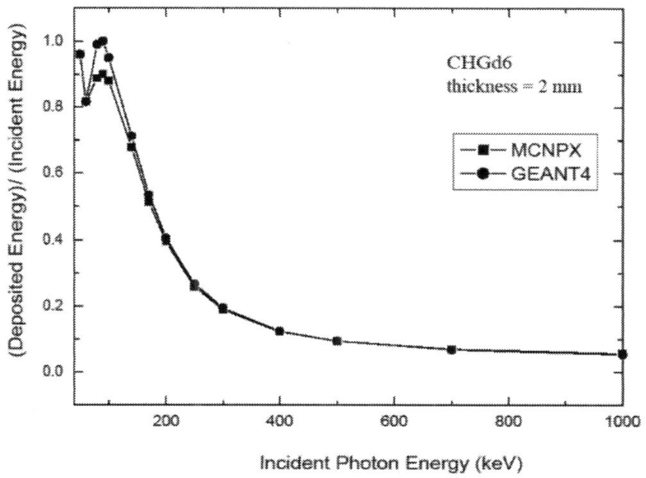

Figure 4. The ratio of the deposited energy to the incident energy versus the incident photon energy for CHGd6.

Shielding the fast neutrons needs relatively thick hydric materials while the thicknesses of the discussed materials in this work are less than 1 cm and hence, we only concentrated on the thermal neutrons at this stage. Figure 5 shows the calculation results of thermal neutron attenuation for CHGd2. As mentioned before, due to gadolinium large cross-section for thermal neutron, CHGd2 with 2 % gadolinium atomic fraction has been studied as thermal neutron shield. According to the results which show a good agreement between the two tools, the thicknesses of 2 mm and 16 mm of the compound attenuate 5 % and 50 % of thermal neutrons respectively.

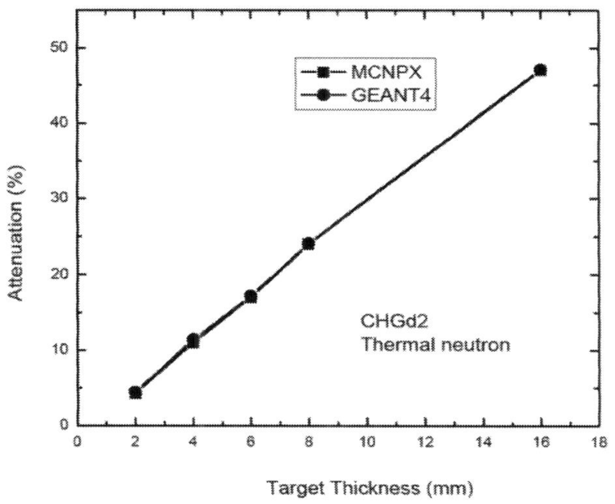

Figure 5. Thermal neutron attenuation in CHGd6.

4. Conclusion

In present work, we studied the attenuation of gamma-ray and thermal neutron using some typical carbohydrate compounds. The depicted results may be useful for designing the new shields for several kinds of applications such as, shielding the radioactive sources, plasma focus systems or low energy medical electron accelerators. Hopefully, the conducted calculations and studies can be developed for more energetic sources to extend the applications of these new shielding materials.

In this conceptual study, we used Monte Carlo tools but there are more considerations in the detail studies of polymeric shields. For example, the produced secondary particles like alpha particle in fast neutron shielding, thermal properties for thermodynamical calculations, physical damage and changes in properties of these materials due to continuous irradiation that it will require more precise studies.

References

[1] Scuderi G J *et al* 2006 Evaluation of non-lead-based protective radiological material in spinal surgery *Spine Journal* **6** (5) 577-582

[2] Atxaga A *et al* 2012 Radiation Shielding of Composite Space Enclosures *Proc. of 63th International Astronautical Congr.* (Italy: Naples/ ORBi) pp 1-10

[3] Bhowmik S and Benedictus R 2007 Performance of Durable polymeric Nano Composite Under Electromagnetic Radiation at Low Earth Orbit *Proc. Applied Electromagnetic Conf.* (Kolkata: IEEE) pp 1-4

[4] Odano N, Konnai A and Asami M 2014 Development of High-performance Gel-type Radiation Shielding Material Using Polymer Resin *Progress in Nuclear Science and Technology* **4** 639-642

[5] Hu H 2011 *Composite Material for Shielding Mixed Radiation, Advances in Composite Materials for Medicine and Nanotechnology* ed Brahim Attaf (Shanghai: InTech) pp 565-592

[6] Haruvy Y 1990 Radiation Durability and Functional Reliability of Polymeric Materials in Space Systems *Int. J. Rad. Appl. Instrum. C* **35** 204-212

[7] Amato E and Lizio D 2009 Plastic Materials as a Radiation Shield for Beta Sources: a Comparative Study Through Monte Carlo Calculation *J. Radiol. Port.* **29** 239-250

[8] Waters L S 2003 MCNPX user's manual version 2.4.0 (Los Alamos: Los Alamos National Laboratory)

[9] Geant4 Collaboration 2008 *Introduction to GEANT4 user documentation version 4.9.4*

[10] Harmon C D and Busch R D 2006 *Criticality Calculation with MCNP5* (Los Alamos: Los Alamos National Laboratory)

[11] Tabbakh F and Ahmadinyar A 2008 Analytical and Statistical Calculation of GammaDose Rate for the Accident of Losing the Shield for Tehran Reasearch Reactor *Chinese Phys. C* **32** 1-5

[12] Hughes H G *et al* 2002 Monte Carlo N-particle Code System for Multiparticle and High Energy Application (New Mexico: Los Alamos National laboratory)

[13] Osei-Mensah W, Fletcher J J and Danso K A 2012 Assessment of Radiation Shielding Properties of Polyester Steel Composite Using MCNP5 *Int. J. Sci. and Tech.* **2** (7)455-461

[14] Sharapov E I *et al* The Upscattering of Ultracold Neutrons from the Polymer $[C_6H_{12}]_n$ *Phys. Rev. C* **88** 064605 1-7

[15] Enger S A, Rosenschold P M, Rezaei A and Lundqvist H 2006 Monte Carlo Calculation of Thermal Neutron Capture in Gadolinium: A Comparision of GEANT4 and MCNP with Measurements *Med. Phys.* **33** 337-341

[16] Geant4 Collaboration 2003 Geant4-a Simulation Toolkit *Nucl. Instr. Meth. A* **506** 250-303

[17] Allison J *et al* 2006 Geant4 Developments and Application *IEEE Trans. Nucl. Sci.* **53** 270-278

[18] Aguayo E, *et al* 2013 *Monte Carlo Simulation Tool Installation and Operation Guide* (Oak Ridge: U.S. Department of Energy)

[19] ANSI/ANS-6.4.3 1991 *American National Standard for Gamma-Ray Attenuation Coefficient and Buildup factors for engineering materials* (La Grange Park: American Nuclear Society)

[20] Hubbell J H and Seltzer S M 1993 Table of X-Ray Mass Attenuation Coefficients *Radiation Research* **136** 147-170

Recent developments in GEM-based neutron detectors

K Saenboonruang[1]
Kasetsart University, Chatuchak, Bangkok, 10900, Thailand

E-mail: fscikssa@ku.ac.th

Abstract. The gas electron multiplier (GEM) detector is a relatively new gaseous detector that has been utilized for less than 20 years. Since the discovery in 1997 by F. Sauli, the GEM detector has shown excellent properties including high rate capability, excellent resolution, low discharge probability, and excellent radiation hardness. These promising properties have led the GEM detector to gain popularity and attention amongst physicists and researchers. In particular, the GEM detector can also be modified to be used as a neutron detector by adding appropriate neutron converters. With properties stated above and the need to replace the previous expensive ^{3}He-based neutron detectors, the GEM-based neutron detector could be one of the most powerful and affordable neutron detectors. Applications of the GEM-based neutron detectors vary from researches in nuclear and particle physics, neutron imaging, and national security. Although several promising progresses and results have been shown and published in the past few years, further improvement is still needed in order to improve the low neutron detection efficiency (only a few percent) and to widen the possibilities for other uses.

1. Introduction

On the evening of 8 November 1985, Röntgen noticed that a piece of a cardboard coated with barium platinocyanide showed a faint, flickering, greenish light when electrical discharge took place in a Hittorf-Crookes tube near the cardboard. This observation, which was actually the X-ray detection, led to many inventions and developments of nowadays radiation detectors. The detection capabilities have widened to effectively detect charged particles such as betas, protons and alphas, or even neutral particles such as neutrons which are relatively harder to detect. In particular, the neutron detection has increased its importance in many aspects. One important example is the application of neutron detection in national security. Since plutonium (Pu) and uranium (U) are main ingredients for nuclear weapon enrichment, it becomes international threats if the materials are in terrorist possessions. Since these two elements are amongst other heavy elements that often undergo nuclear fission processes and emit neutrons, one possible and effective way to detect these heavy elements is to have a detector capable of detecting neutrons with high neutron detection efficiency while limiting false alarms from gamma detection. In the past, a helium-3 (^{3}He)-based detector was commonly used due to its high neutron absorption cross section, which leads to high neutron detection efficiency. ^{3}He is a byproduct from the beta decay of tritium (^{3}H) and is separated from ^{3}H as part of the tritium purification process for refurbishment and dismantlement of the nuclear stockpile. Lately, due to international agreement to reduce numbers of existing nuclear weapons and nuclear stockpiles, the production of ^{3}He from ^{3}H decay has significantly declined and could harm the supply of ^{3}He for ^{3}He-based neutron detector. On the other hand, after the terrorist attack on September 11, 2001 in USA, the demand for ^{3}He has increased

[1] To whom any correspondence should be addressed.

significantly as ^{3}He-based neutron detectors were rapidly deployed around the world to detect illegal nuclear materials. These two phenomena lead to the reduction in production and the increase in demand of ^{3}He. It is estimated that the total ^{3}He demand is approximately 65,000 liters/year, while the total supply is approximately 15,000 liters/year. These numbers show the disparity between supply and demand of ^{3}He and a prediction of a more severe shortage of ^{3}He. In order to maintain national security at the highest level, ^{3}He-free neutron detectors must be invented to replace the ^{3}He-based neutron detectors.

In addition to the need of neutron detectors for national security purposes, many scientific and industrial researches also require high-performance neutron detectors. Some examples of research areas that require the uses of neutron detectors include

- Material sciences: The neutron detectors can be used to study and measure the substructures of materials and biological organs in the angstrom level by measuring elastic and inelastic scattering of neutrons off interested materials.
- Reactor instrumentation: The neutron detectors are used in monitoring power of nuclear power reactor and research reactor since the reactor power is linearly proportional to neutron flux.
- Astrophysics: The neutron detectors can be used to detect secondary neutrons, which are parts of the particle showers produced in Earth's atmosphere by cosmic rays.
- Medical imaging: Detection of neutron radiation can be used for medical imaging of patients' organs and tissues.

2. The Gas Electron Multiplier (GEM) detector

The Gas Electron Multiplier (GEM) detector is a gaseous charge amplification structure, invented by F. Sauli and his research group at CERN, Switzerland, in 1997 [1]. The main component of the GEM detector is a thin multilayer foil with many high precision holes, which is known as a GEM foil. The GEM foil consists of a thin insulating foil (polyimide). Polyimide foil has an excellent dielectric strength of 150-300 kV/mm for a thickness of 25-125 μm. The usual thickness for polyimide used in GEM foils is 50 μm and has metal clads (usually copper) on both sides and perforated with a regular matrix of holes. The holes are very closely spaced and have a diameter ~70 μm, and a pitch or a distance from one center of the hole to the center of the closest neighboring holes of ~140 μm.

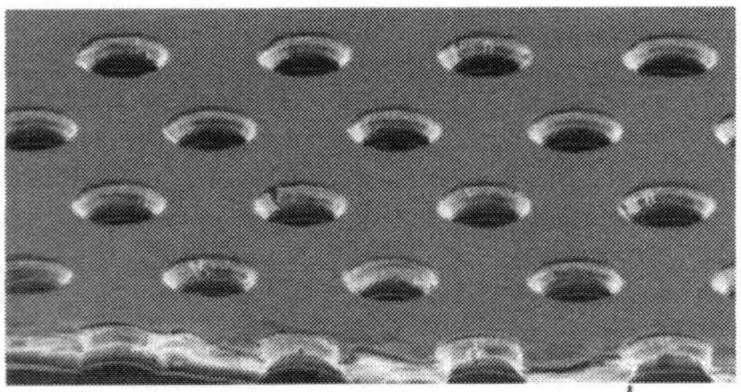

Figure 1. A magnified image of a GEM foil shows a regular matrix of holes. Holes have ~70 μm in diameter and a pitch of ~140 μm [2].

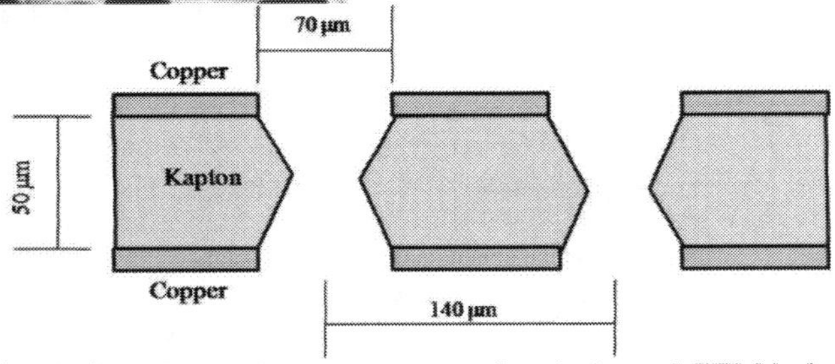

Figure 2. A figure shows a cross sectional view of GEM holes

Another important component of the GEM detector is the gas flushing through the detector. In principle, the GEM detector can be operated in a pure noble gas such as argon (Ar). However, in order to increase the detector stability and to reduce discharge probability, a mixture of gasses is used. The standard gas mixture used in most applications is a mixture of Ar/CO_2 in a ratio of 70/30, where Ar acts as the main ionizing gas and CO_2 acts as a quencher.

The main operating principles of the GEM detector are as the following steps.

- By applying a potential difference about 300-400 V between the top and the bottom metal clads, strong electric fields develops inside the holes.
- A particle passes through the drift region of the detector and ionizes gas molecules inside the detector. These ionized electrons will then travel through the strong electric fields inside the GEM holes, gain more energy, and ionize other gas molecules creating electron avalanches.
- In the last amplification stage, the amplified electron avalanches will be captured by readout pads or readout strips depending on the applications. The moving of electron avalanches between the last GEM foil and the readout electrode will create detectable signals, which will be processed through an appropriate data acquisition (DAQ) system.

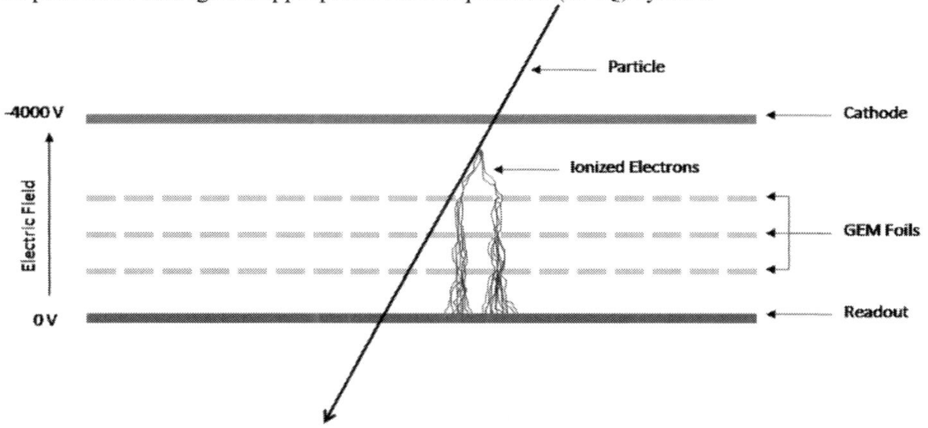

Figure 3. The picture shows the three amplification stages inside the GEM detector. The last electron avalanches will be captured by the readout pads or the readout strips. The moving of electron avalanches between the last GEM foil and the readout electrode will create detectable signals and will be processed through DAQ system.

3. GEM detector as a neutron detector

Although the GEM-based detector is mostly used to detect charged particles and low-energy photons, the detector can be modified by adding appropriate neutron converters to the detector such that it is now able to detect neutrons. The ability to detect neutrons relies on a conversion mechanism that yields charges in the drift volume of the detector. In the case of thermal neutrons, conversion takes place via a nuclear reaction with an appropriate isotope, while higher energy neutrons may cause nuclear recoils. To increase the probability of the conversion, an appropriate neutron converter is added to the detector. Examples of commonly used neutron converters include Helium-3 (^3He), boron-10 (^{10}B), lithium-6 (^6Li) or hydrogen-rich materials such as polyethylene ($(C_2H_4)_nH_2$) and polypropylene ($(C_3H_6)_n$). The nuclear reaction products are an alpha (α) particle of ~2 MeV and either a triton (for ^6Li) or lithium ion (for ^{10}B), isotropically emitted in opposite directions. Example of the nuclear reaction with ^{10}B can be shown as the following:

$$^{10}B + n \longrightarrow {}^7Li + \alpha + 2.79\ \text{MeV}\ (6\%)$$
$$^{10}B + n \longrightarrow {}^7Li^* + \alpha + 2.31\ \text{MeV}\ (94\%)$$

where $^7Li^*$ is an excited particle. The $^7Li^*$ spontaneously emits a 0.48 MeV gamma ray and returns to the ground state. These emitted alpha particles will ionize gas molecules inside the GEM detector, which will later be amplified in the amplification stages and get captured by readout pads. The efficiency of neutron detection depends on the neutron reaction rate which also depends on various factors. The neutron reaction rate is shown in (3.1).

$$R = \int \emptyset(E) N\sigma(E) dE \tag{3.1}$$

where $\emptyset = \int \emptyset(E)dE$ is the neutron flux, N is the number of ^{10}B atoms in the neutron converter, and $\sigma(E)$is the neutron cross section. The value of neutron cross section is the most important quantity that defines the efficiency of the detector and is shown in figure 4.

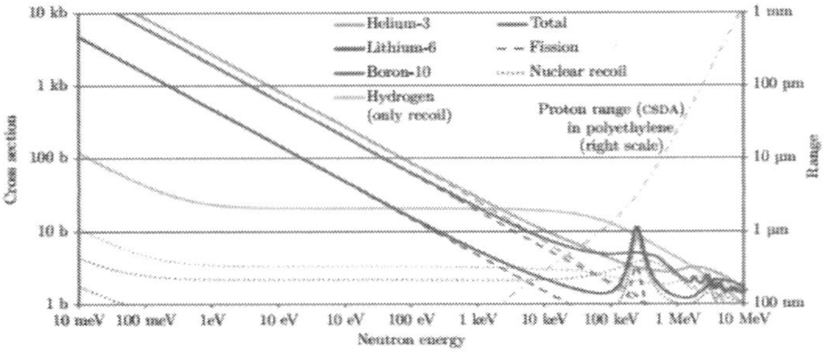

Figure 4. Cross sections of selected elements used as neutron converters. Dashed and dotted curves decompose the total interaction cross section from nuclear reaction and nuclear recoil reactions [3].

As shown in figure 4, ^3He has the highest cross section and has been used widely for neutron detectors in the past. Due to a crisis in the shortage of ^3He and the sharp increase in the price of ^3He, alternatives to ^3He-based neutron detectors are urgently needed to maintain the highest level of national security and researches. These include the uses of thin film coated with ^{10}B or ^6Li, or filling the detector with hydrogen-rich gas.

4. Recent development and result

Since the invention of the GEM detector, the growing interests in its capabilities have motivated scientists and researchers to develop and improve the detector for various applications. GEM-based neutron detector is one of applications that have been vastly developed in recent days. Some of the notable developments and results will be discussed in this section.

4.1 Manufacturing of boron-coated GEM foil

Since the performances of the GEM-based neutron detector especially the efficiency of the detection depend heavily on the neutron converting process occurred at the converters, the development of methods to apply converters (^{10}B) to drift cathode and GEM foils is very important. As shown by Se-Hwan Park *et al.* [4,5], ^{10}B was coated on the aluminized Mylar, which was used as a drift cathode, using electron-beam evaporator. The coated boron became stable with time. However, ^{10}B did not adhere to the GEM foil very well and could be peeled off as well as cracked easily from the GEM foil. The picture of borated GEM foil that peeled off is shown in figure 5.

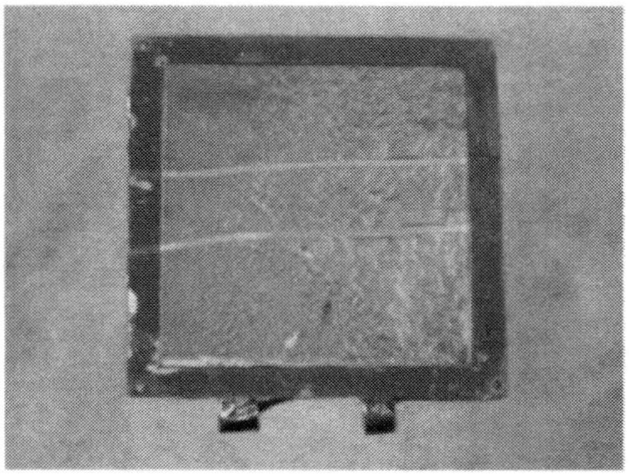

Figure 5. Boron coated on the GEM foil using electron-beam evaporator done by Se-Hwan Park *et al.* could be easily peeled off after the process. Image was taken from [5].

Se-Hwan Park *et al.* had solved the problem by, instead of coating one full piece of ^{10}B film on the GEM foil, the ^{10}B film was segmented into small squares. To achieve that, the main additional procedure included the introduction of a stainless mesh placed in front of the substrate so that the ^{10}B films were divided into ~1.5 x 1.5 cm^2 squares and had the gap between the adjacent squares of 0.3 mm. Pictures of drift cathode and GEM foil that were coated with the new procedure are shown in figure 6.

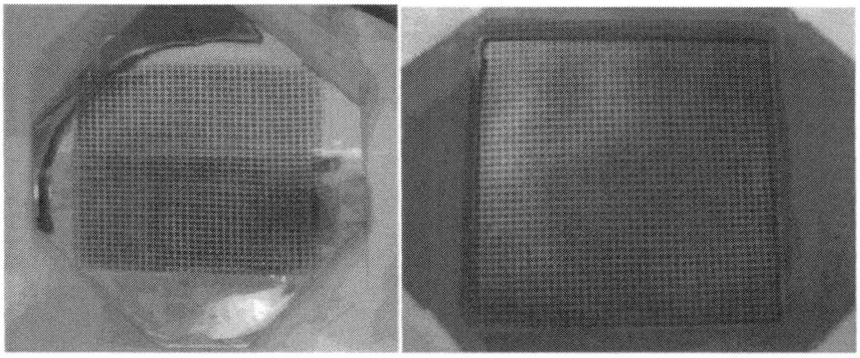

Figure 6. Pictures of ^{10}B-coated drift cathode (GEM foil) is shown on the left (right). Images were taken from Se-Hwan Park *et al.* [4,5].

The idea of segmenting ^{10}B film into small squares reduced the stress on the film surface significantly and even if the crack occurred, the crack remained in the broken square and did not spread to other parts. The technology has improved the manufacturing of the ^{10}B-coated GEM foil and significantly improved qualities of GEM-based neutron detector. Furthermore, the technique has increased the working duration of the foil and the detector.

4.2 Improvement on the configuration and layout of the converter

Since the ability to detect neutrons depends on the ability to convert neutrons to charged particles such as alphas or protons, the way the converter is introduced to the detector is highly important to the performances of the detector. In a traditional configuration, the converters such as ^{10}B and polyethylene are coated directly onto the drift cathode and the GEM foils. However, this method of applying converters limits the conversion process to only at the surface of the detector. Furthermore, even if the thickness of the converter increases which would increase the conversion probability, charged particles produced cannot penetrate the thick converter. Hence, the thickness of the converter is limited to only

2-3 μm in the case of ^{10}B. A. Pietropaolo *et al.* [6] has designed a new GEM-based neutron detector to remove these constraints by introducing layers of borated glass sheets as converters that are perpendicular to the drift cathode as shown in figure 7.

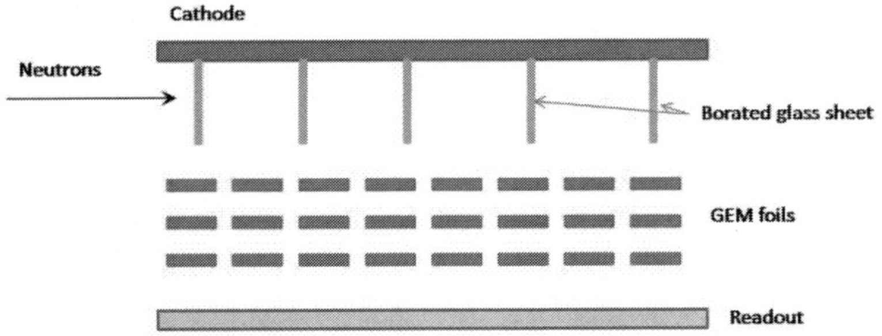

Figure 7. Schematic of GEM-based neutron detector designed by A. Peitropaolo *et al.*: The borated glass sheets were applied perpendicular to the drift cathode [6].

The main purpose of having layers of borated glass sheet aligned as in figure 7 is to loosen up the requirement that neutrons must be converted to charged particles at the very first μm-thick surface of the detector. The improvement consequently increases the probability of the conversion and, hence, the efficiency of the detector. The efficiency measured by A. Pietropaolo *et al.* showed that the efficiency of the neutron detection came out to be 4.8(5) % which was much better than previously reports of traditional configurations, which had the efficiency of less than 1%. A similar design was also used in the research conducted by M. Cortesi *et al.* [7] for fast neutron detection, which used polyethylene (HPDE) layers as converters. The reports showed that the estimated efficiency that could be achieved by this configuration was in the range of 5-8 % for 2.5 MeV neutrons.

5. Possible improvement and application

Although growing number of tests and researches have been conducted on the GEM-based neutron detector, there are still rooms for further improvement such as:

- Conducting extensive researches on various neutron converters including gaseous converters such as He-4 and other solid converters (^{6}Li), or a mix of both solid and gaseous converters used in the detector.
- Designing new layouts of converters such that there are more possibility of conversion. This might include adding more layers of borated glass sheets to the detector.

Since many reports have shown the excellences of the GEM-based neutron detector, more researches should be aimed to adapt the detector for applications. One example of applications was done by S. Uno *et al.* The GEM-based neutron detector was used to perform a 2-dimensional imaging of metal content, which could clearly show content of gold in a Japanese oval gold coin from Edo period. Furthermore, the detector could also be used to perform a 2-dimensional imaging of crystallite size of iron bars after bending [8][9]. These examples clearly show the wide areas of applications of the GEM-based neutron detectors and there are still plenty of rooms to explore.

6. Conclusion

The GEM detector has been development significantly since the invention in 1997. From the ability to detect charged particles and photons, the GEM detector has been adapted to be used in various applications. Neutron detector based on the GEM detector is one of the detection technologies that greatly benefit educational, industrial, medical, and national security sectors. Many research projects have been conducted to improve the performances of the detector. These include the attempts to improve the method to apply converters to the drift cathode and the GEM foils such that the converters could adhere to the surface better and increase the lifetime of the detector. Also, researchers have designed new arrangements of the converters to improve the chances of the conversion by adding layers of borated glass sheets or HDPE perpendicular to the drift cathode. This design has increased efficiency of the detector to be in the range of 5%, which improves from previously reports of ~1%. Furthermore, attempts to adapt the GEM-based neutron detector to be used in various applications such as in material analysis, medical imaging, and national security should be emphasized and invested.

Acknowledgments
This work was supported by the Thailand Research Fund (TRG5780292) and Kasetsart University Research and Development Institute. Additional support was provided by Faculty of Science, Kasetsart University. I also would like to express my appreciation to Dr. Nilanga K. Liyanage (University of Virginia) and Dr. Thiraphat Vilaithong (Thailand Center of Excellence in Physics) for their useful advices.

References

[1] Sauli F 1996 A new concept for electron amplification in gas detectors *Nucl. Instr. and Meth. A.* **386** 531–34
[2] Sauli F 2004 Progress with the gas electron multiplier *Nucl. Instr. and Meth. A.* **522** 93-98
[3] Pinto S D 2013 GEM applications outside high energy physics *Mod. Phys. Lett. A.* **28** 1340025
[4] Park S H, Kim Y K and Kim J K 2005 Neutron detection with a GEM *IEEE Transactions on Nuclear Science.* **52**(5) 1689–92
[5] Park S H, Kim H S, Ha J H, Kim Y K and Kim J K 2006 Fabrication of a GEM-based gas detector for thermal neutron measurement *J. Korean Phys. Soc.* **49**(5) 1939–43
[6] Pieotropaolo A *et al*2013 A new ^{3}He-free thermal neutrons detector concept based on the GEM technology *Nucl. Instr. and Meth. A.* **729** 117–26
[7] Cortesi M, Zboray R, Adams R, Dangendorf V and Prasser H M 2012 Concept of a novel fast neutron imaging detector based on THGEM for fan-beam tomography applications *JINST* **7** C02056
[8] Uno S *et al* 2012 Two-dimensional neutron detector with GEM and its applications *Physics Procedia* **26** 142–52
[9] Uno S *et al* 2012 Development of a two-dimensional gaseous detector for energy-selective neutron radiography *Physics Procedia* **37** 600–05

Improvement of portable computed tomography system for on-field applications

K. Sukrod[1], P. Khoonkamjorn and C. Tippayakul
Thailand Institute of Nuclear Technology, Bangkok, Thailand

E-mail: kokiat@tint.or.th

Abstract. In 2010, Thailand Institute of Nuclear Technology (TINT) received a portable Computed Tomography (CT) system from the IAEA as part of the Regional Cooperative Agreement (RCA) program. This portable CT system has been used as the prototype for development of portable CT system intended for industrial applications since then. This paper discusses the improvements in the attempt to utilize the CT system for on-field applications. The system is foreseen to visualize the amount of agarwood in the live tree trunk. The experiments adopting Am-241 as the radiation source were conducted. The Am-241 source was selected since it emits low energy gamma which should better distinguish small density differences of wood types. Test specimens made of timbers with different densities were prepared and used in the experiments. The cross sectional views of the test specimens were obtained from the CT system using different scanning parameters. It is found from the experiments that the results are promising as the picture can clearly differentiate wood types according to their densities. Also, the optimum scanning parameters were determined from the experiments. The results from this work encourage the research team to advance into the next phase which is to experiment with the real tree on the field.

1. Introduction

The gamma Computed Tomography (CT) is an area of interest at Thailand Institute of Nuclear Technology (TINT) for analyzing industrial process units and other similar applications. This technique is considered complementary to the other NDT techniques such as radiotracer or gamma scanning and may provide useful information to diagnose the plant problems. Researches in this area have been conducted at TINT and experimental setups have been carried out throughout the years. The ultimate purpose of these researches is to develop a portable CT system which can be deployed easily at the site in the plant process environment. Applications such as analyzing corrosion under insulation or monitoring multiphase flows are anticipated examples of using such portable CT system. However, the portable CT system has a number of challenges to develop due to its complexity of the construction and operation. For example, the portable CT system shall be flexible enough to align with the process unit in different orientations and it shall be light-weighted such that the setup can be done without assistance from lifting equipment which may not be available or accessible in the installation area. This paper discusses the effort to improve and prepare a portable CT system available at TINT to analyze one particular challenging problem – to visualize the amount of resin in agarwood. This portable CT system was received from IAEA as part of the Regional Cooperative Agreement (RCA) program and it is being used for research work. Uses of different radioactive sources for the portable

[1] To whom any correspondence should be addressed.

CT system were tested in this research. The capability of the portable CT system to differentiate timber densities was also studied. Finally, test of a live tree was conducted to verify the operability of the portable CT in the field which should have environment similar to the actual utilization.

2. Description of Portable Computed Tomography System

TINT received a portable CT system (named "eGorbit") from IAEA in 2010 as part of the Regional Cooperative Agreement (RCA) program. This eGorbit system has been used as the prototype for development of the portable CT system intended for industrial applications since then. Basically, this portable CT system consists of the mechanical gantry, the controller unit, the measurement unit, the hardware interface software and the image processing software. The portable CT system is operated by a computer running the hardware interface software which controls the rotations and the movements by use of two servo motors. Figure 1 shows the picture of the portable CT system (eGorbit).

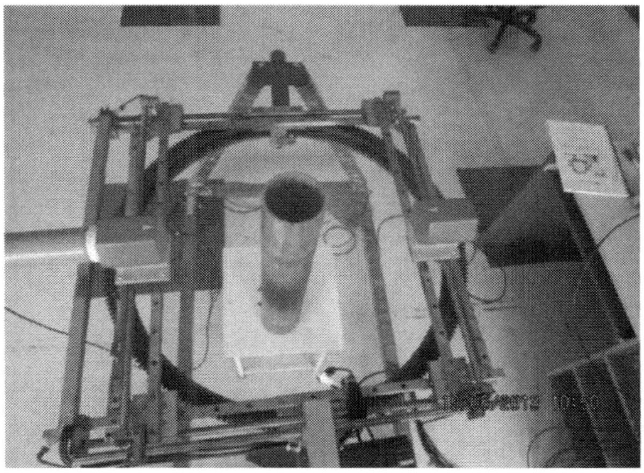

Figure 1. Picture of the portable CT system.

This portable CT system is actually the first generation CT scanner, that is, it performs parallel scanning in which the source, emitting pencil beam of radiation, and the detector are coupled together such that the detector is always facing the source and they both travel in the same direction during the scan [1]. The components and connections among the components in the portable CT system are described in Figure 2.

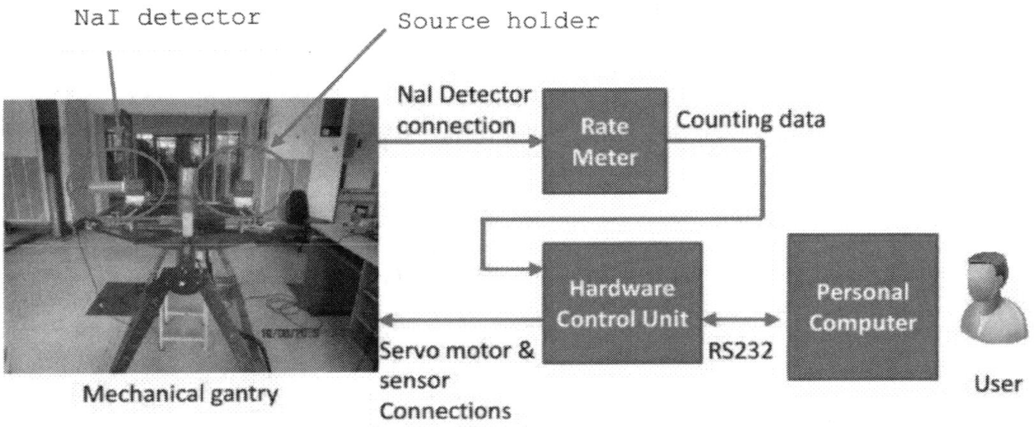

Figure 2. Components and Connections in the eGorbit system.

The portable CT system consists of 2"x2" NaI detector and gamma source holder as shown in figure 2. The radiation source and the NaI detector are collimated by use of lead collimators having the aperture size of approximately 4.5 mm and the thickness of 2 inches. The detector is connected to Ludlum rate meter (measurement unit) which transmits counting data to hardware control unit of the eGorbit system. The hardware control unit controls two servo motors and communicates with the personal computer via RS-232 port. The first servo motor moves the detector and the source along the X direction while the second one rotates the detector and source in radial direction. The typical radiation source used with this eGorbit system is Cs-137.

3. Improvements of the CT system

In this work, various activities have been carried out to optimize and improve the portable CT system. In fact, the following main tasks were performed in this research.

3.1. Stability improvements

After the eGorbit system was received and assembled, it had been tested for operation with some samples in laboratory. It was, however, found that the operation of the eGorbit system had experienced some malfunctions from time to time. The eGorbit system was then examined for its instability. The problem of internal wiring within the hardware control unit was identified as one of the causes for the abnormal operation. This causes unknown electrical noises into the counting data as well. Another problem was found that the cable connectors were not in good contact. The research team rectified this problem by rewiring the internal cables and the cable connectors of the eGorbit hardware control unit.

3.2. Adopting Am-241 source

The typical radiation source for use in the eGorbit system is Cs-137. However, Cs-137 which has single gamma energy of 661 keV may not be suitable for all applications. One of the current applications of interest is to use the eGorbit system for scanning live agarwood. It was hypothesized that Cs-137 is not best for using in this application since the gamma energy is relatively high which does not have good attenuation property in agarwood. Therefore, a new radiation source was adopted in this work. The Am-241 was selected because it has low gamma energy (main gamma energy of 59.7 keV) [2] and it is considered to have good attenuation property in wood. Comparison of using the eGorbit system with Am-241, Cs-137 and Co-60 was conducted. The activities of the Am-241, Cs-137 and Co-60 sources used in this experiment are approximately 200 mCi, 25 mCi and 14 mCi respectively.

3.3. Experiments in preparation for on-field application

It is expected that the eGorbit system will be used for scanning objects on-field. Laboratory experiments as described in the followings were designed and conducted to test the feasibility prior to such on-field work:

3.3.1. Test specimen. A test specimen was prepared for testing with the eGorbit system. The test specimen was made of timbers with different densities (approximately 20% difference). The purpose of this test was to observe whether the eGorbit system can differentiate the pattern of wood types in the specimen. The scanning test was conducted with different scanning views to observe the effects of the scanning views to the picture quality as well.

3.3.2. Live tree. Another test performed in preparation to use of the eGorbit system for on-field application was to scan a live tree. The tree was actually located near the laboratory building in TINT. The tree with diameter of approximately 13 cm. was selected for the test. The purpose of this test was to observe the overall performance of the eGorbit system under on-field environments where test conditions were not controlled (i.e., close to the real use). Figure 3 shows the setup of the portable CT system for scanning a live tree.

Figure 3. Setup of the portable CT system for live tree.

4. Experiment results

4.1. Adopting Am-241 source

The tests of using the eGorbit system with Am-241, Cs-137 and Co-60 for scanning a tree trunk were conducted. It shall be noted that the collimator of 4" thickness in total has the transmission factor of approximately 0.001% for Cs-137 while that of Co-60 is approximately 0.4%. These transmission factors show that only small fraction (less than 1%) of uncollimated radiation beam can enter the detector. Therefore, the collimator shall be sufficiently thick enough for all sources used in this experiment. The pictures from the tests are shown in comparison in figure 4.

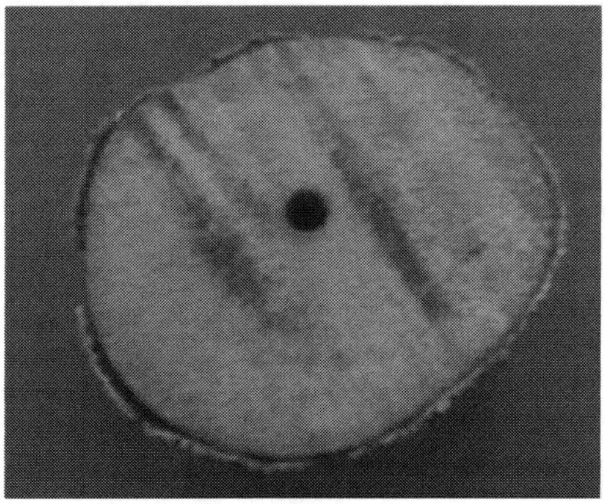

Figure 4a) Test specimen.

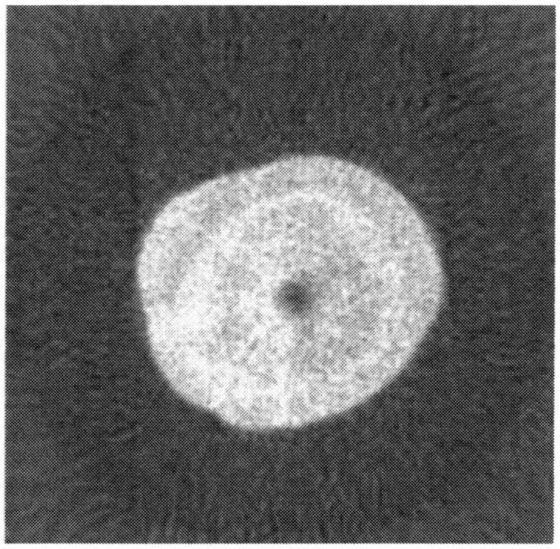

Figure 4b) Am-241 source.

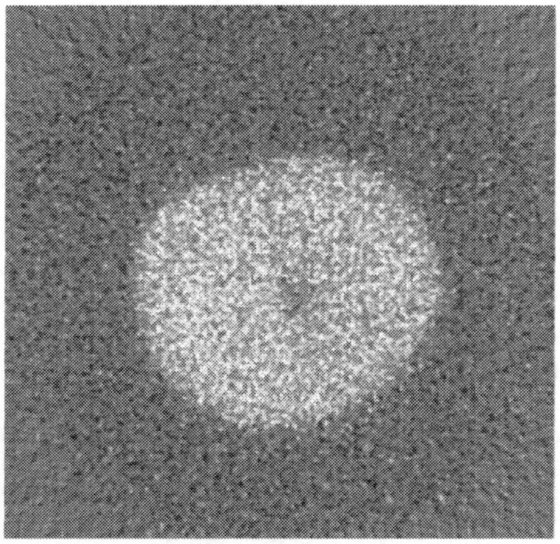

Figure 4c) Cs-137 source.

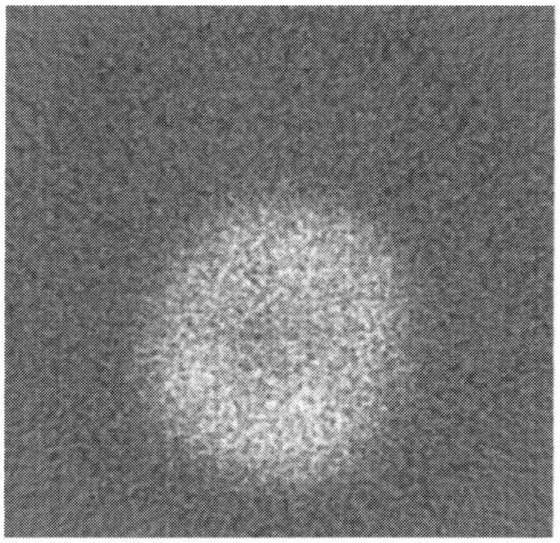

Figure 4d) Co-60 source.

As it can be clearly seen from the comparison in Figure 4, the picture quality scanned by use of different radiation sources varies. For tree trunk which is made of wood, radiation source with high gamma energy would not interact very well with the material. Therefore, the picture quality is not as good as that of using lower gamma energy. Thus, the Am-241 source produces best picture quality of all and the Co-60 source produces the worst picture quality of all.

4.2. Test specimen
The results from scanning the test specimen using the eGorbit system are shown in Figure 5.

Journal of Physics: Conference Series **611** (2015) 012017 doi:10.1088/1742-6596/611/1/012017

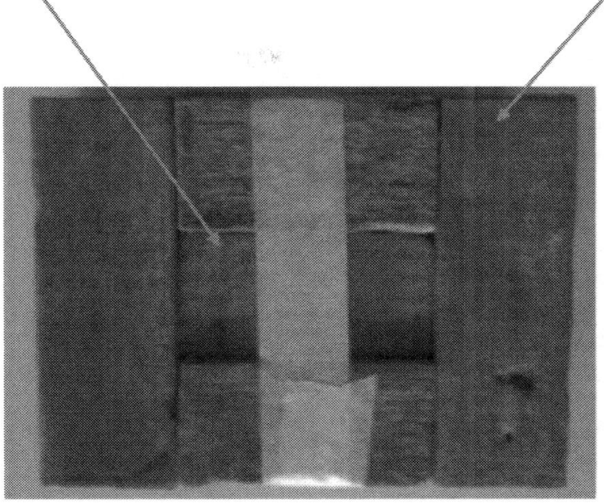

Figure 5a) Test specimen.

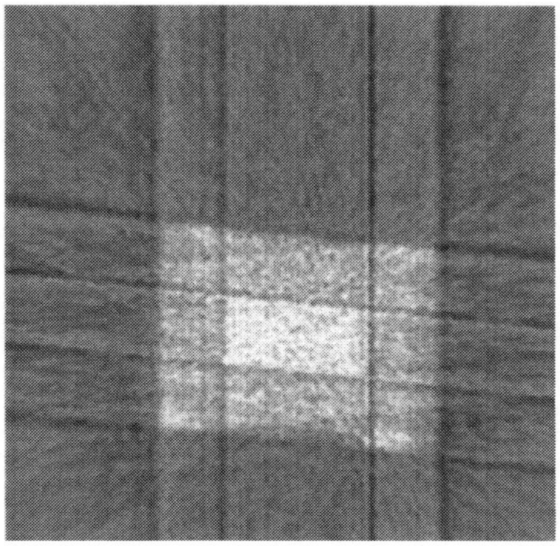

Figure 5b) 16 views.

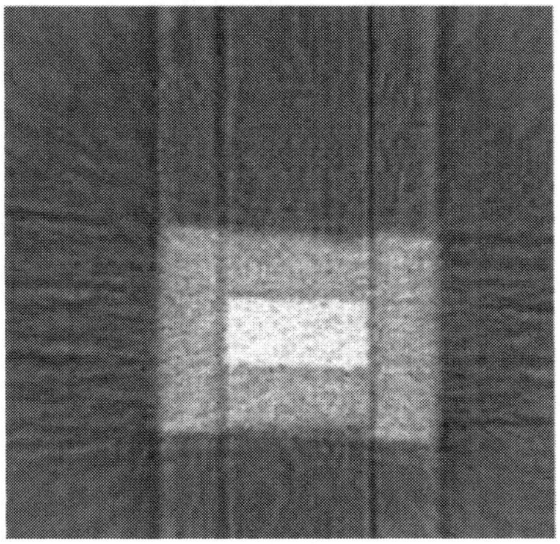

Figure 5c) 32 views.

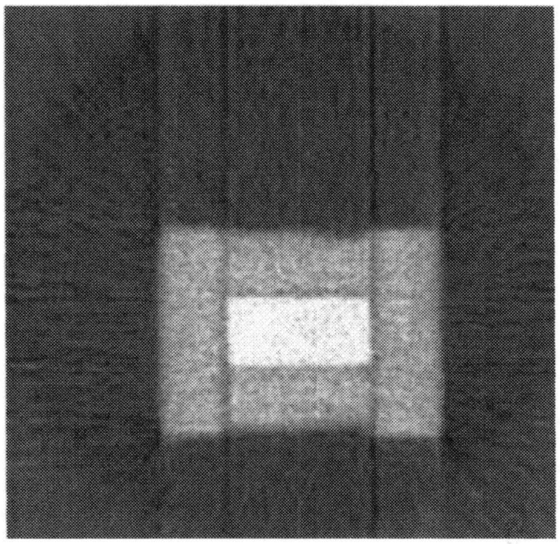

Figure 5d) 64 views.

It can be seen from the pictures shown in figure 5 that the eGorbit system using Am-241 source is able to clearly differentiate the types of wood which have different densities. Moreover, the effect of picture quality when increasing the number of scanning views is compared. It can be observed that using 64 scanning views provides better picture quality compared to using 32 scanning views. However, the scanning time is proportional to the number of scanning views. In this test specimen, the scanning time of approximately 7 hours is typical when using 32 scanning views. This scanning time doubles when using 64 scanning views. Thus, the nominal scanning view of 32 is selected to compromise between practical scanning time and the picture quality.

4.3. Live tree

The scanning of the live tree takes approximately 6 hours. The test of the eGorbit system in the environment similar to the actual one was completed successfully without malfunctions of the hardware and hence it fulfils the purpose of the test. However, since the tree was not cut down to compare with the scanned picture. There is no picture comparison for this test.

5. Conclusion

This research has been conducted to realize the use of the portable CT system available at TINT for on-field applications. It was found at the beginning of the research that the hardware of the portable CT system had some problems and it was rectified by rewiring the cables within the hardware control unit. Moreover, laboratory experiments were conducted to evaluate the feasibility to use the portable CT system for visualizing the amount of resin in agarwood. In order to do that, Am-241 source was adopted to replace the typical Cs-137 source in the portable CT system. The experiment showed that Am-241 source provides better picture quality for wood than Cs-137 or Co-60. Additionally, the use of the portable CT system for differentiating wood densities was performed. It was found that the portable CT system with Am-241 source is able to clearly differentiate the wood density difference of 20%. Finally, the portable CT system was tested with a live tree in order to observe the overall performance under on-field environments and the test was completed successfully. The results from this work encourage the research team to advance into the next phase which is to experiment with the real agarwood.

References

[1] Goldman L W (2007) Principles of CT and CT Technology *J. Nucl. Med. Technol.* **35** 115-28

[2] Strain J E and Leddicottee G W 1962 The preparation, properties, and uses of Americium 241, alpha-, gamma- and neutron sources *ORNL-3335* (Oak Ridge: Oak Ridge National Laboratory)

Preliminary study of the use of radiotracers for leak detection in industrial applications

S Wetchagarun[1], A Petchrak and C Tippayakul
Reactor Center, Thailand Institute of Nuclear Technology (Public Organization), Bangkok 10900, Thailand

E-mail: saensuk@tint.or.th

Abstract. One of the most widespread uses of radiotracers in the industrial applications is the leak detection of the systems. This technique can be applied, for example, to detect leak in heat exchangers or along buried industrial pipelines. The ability to perform online investigation is one of the most important advantages of the radiotracer technique over other non-radioactive leak detection methods. In this paper, a preliminary study of the leak detection using radiotracer in the laboratory scale was presented. Br-82 was selected for this work due to its chemical property, its suitable half-life and its on-site availability. The NH_4Br in the form of aqueous solution was injected into the experimental system as the radiotracer. Three NaI detectors were placed along the pipelines to measure system flow rate and to detect the leakage from the piping system. The results obtained from the radiotracer technique were compared to those measured by other methods. It is found that the flow rate obtained from the radiotracer technique agreed well with the one obtained from the flow meter. The leak rate result, however, showed discrepancy between results obtained from two different measuring methods indicating further study on leak detection was required before applying this technique in the industrial system.

1. Introduction

The radiotracer technique has been widely used for troubleshooting and process optimization in industry. According to IAEA, many developing countries have radiotracer application groups [1]. Thailand Institute of Nuclear Technology (TINT) is also interested in deploying the radiotracer applications especially in the petrochemical industry. The applications of interest include determining leak rate detection in heat exchangers or similar plant units or determining Residence Time Distribution (RTD) in mixing reactors. The economic benefits of using radiotracer in these applications are significant. However, research work shall be conducted to develop the techniques as well as engineering the operation equipment to be suitable for harsh environments in the chemical processes. This paper discusses the preliminary work at TINT in attempt to develop radiotracer technique for leak rate detection. An available experimental setup at TINT was used for conducting the leak detection measurement. This paper provides the setup details of the experiment. The obtained measured data from the experiment were processed using the IAEA software and the result was compared with the actual measurement.

[1] To whom any correspondence should be addressed.

2. Experimental description

The existing radiotracer experimental setup at Nuclear Technology Service Centre, Thailand Institute of Nuclear Technology (TINT) had been used in this work for the laboratory experiment. The results obtained were compared with calculating results. The details of the experimental setup and the calculating methods are as follows.

2.1. Laboratory experiment

The experimental setup consists of three main parts: 1) the water tanks and piping system, 2) the radiotracer injection system, and 3) the liquid radiotracer. In this work, only the water tank #2 was used to simulate the leakage in the heat exchanger. A small opening valve was used to mimic the leakage of water from the primary loop to the secondary shell side of a shell and tube heat exchanger. Figure 1 shows the diagram of the experimental setup.

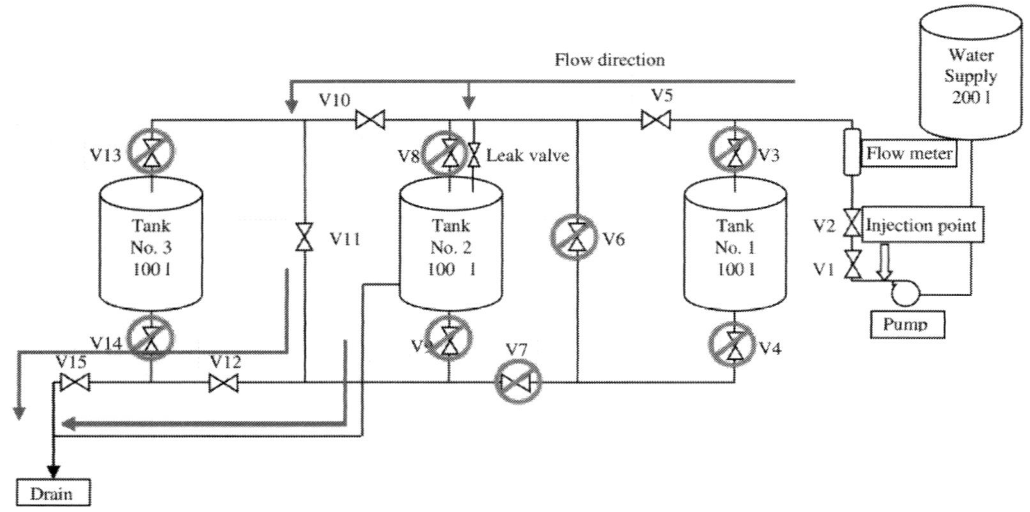

Figure 1. Layout of the radiotracer laboratory setup.

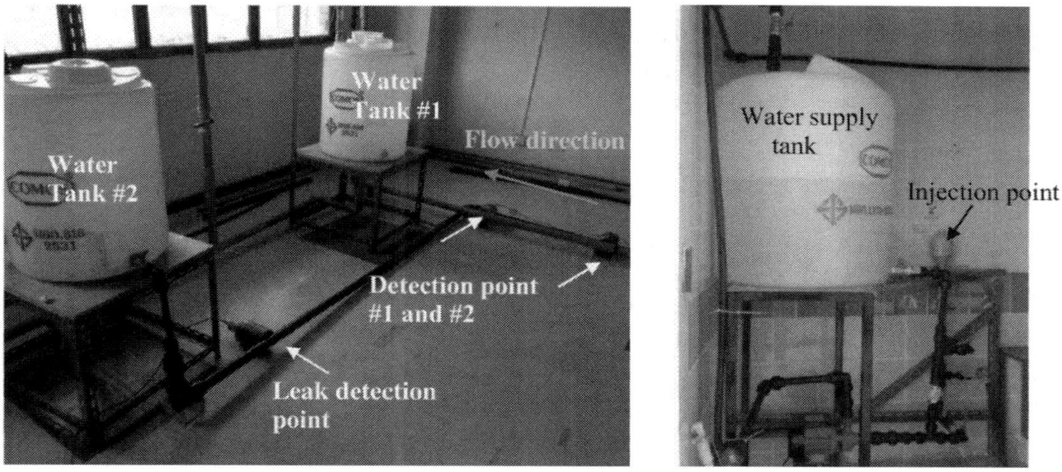

(a) (b)

Figure 2. Injection and detection points.

The radiotracer, Br-82 (approximately 10 mCi), in the form of NH_4Br aqueous solution was injected into the test system at the injection point as shown in Figure 2b. Three NaI detectors were placed along the pipeline to measure the radiation signals (see Figure 2a). The first two detectors were located along the straight pipe between the pump outlet and the glass flow meter to measure the flow rate of the system. The third detector was setup at the outlet pipe of the water tank #2 in order to measure the flow leakage.

2.2. Calculation method

The radiation signals attained from the experiment were further processed for the flow rate and the amount of leakage from the main flow. The flow rate was calculated by the software called RTD program version 1.0 distributed by International Atomic Energy Agency (IAEA). The ratio of the leakage to the main flow was computed from the areas under curves of the radiation signals measured by the leak detector and the detector located at the main pipe flow [2], e.g., detector #1.

3. Results and Discussion

3.1. Flow rate measurement

The glass flow meter shows approximately 5 LPM during the experiment. The pipe diameter at the measuring locations for detector #1 and #2 is about 5 cm. Therefore, the velocity of water in the pipe is approximately 4.24 cm/s.

By using the RTD program, the radiation signals from detector #1 and #2 were used as the input to the software. The program calculated the mean residence time between the two detectors, tau or τ, as 15.29 s. The two detectors were located 100 cm apart, so the velocity of the water in the pipe is about 6.54 cm/s. Besides the optimal mean resident time, tau, the RTD program also provided the optimal value of the Peclet number, Pe, which is the ratio of the advective transport rate to the diffusive transport rate of the flow. Figure 3 shows the calculation results of τ and Pe by the RTD program.

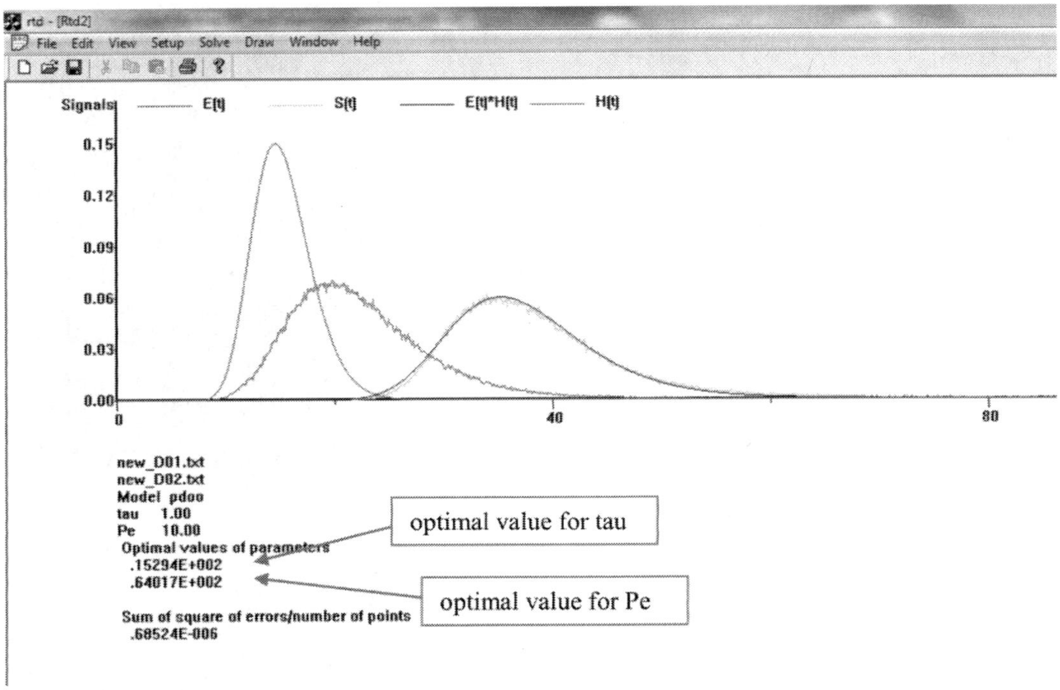

Figure 3. Flow rate calculation using RTD program.

It is found that the difference of the flow rate obtained from the glass flow meter and that from the RTD program is about 54%. This discrepancy might be a result of the coarse scale of the glass flow meter and the fluctuation of the flow during the experimental time. Since the flow rate in the experiment is quite low, only a slight deviation of the value read from the flow meter can cause a high percentage difference compared to the flow rate itself. For example, if the actual flow rate is 7 LPM, but 5 LPM was read from the flow meter (due to coarse meter scale), in this case, it is approximately 29% error. If the flow rate is 7 LPM, the pipe flow velocity is around 5.94 cm, which is only 10% discrepancy from what computed by the RTD program. Therefore, to lessen this discrepancy, more accurate/finer scale flow meter and/or higher operating flow is needed for the experiments. However, if high discrepancy still exists even though more accurate flow meter is applied, mathematical model in the RTD program used in this case may not be suitable for the flow field of interest.

3.2. Leakage measurement

Before injecting the radiotracer into the system, the leak rate was measured and compared to the main flow rate. It is found that the leakage is about 0.5% of the main flow.

By using numerical integration (Trapezoidal rule), it is found that the areas under curve of the radiation signal of the main flow (detector #1, D01) and that of the leakage (detector #3, D03) are 20671.7 and 1696.3, respectively. This leads to the leakage of about 8% of the main flow. Figure 4 illustrates the main flow and leakage signals.

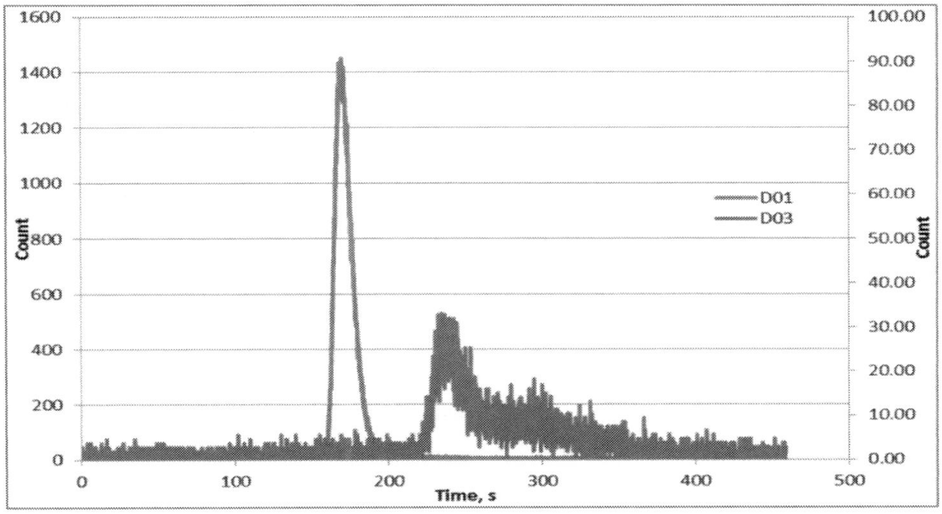

Figure 4: Radiation signal from the main flow detector #1(D01) and the leak detector #3(D03).

It can be seen that the leakage computed from the ratio of area under curves of the radiation signal measured from the leakage to that from the main flow are around 16 times more than the direct measurement of the leak itself. This large difference may be the result of the size difference of the measuring pipes. The sizes of the leakage and the main pipes are different which may affect the measuring efficiency of the detectors. Moreover, due to the geometry of the setup, there can be a significant accumulation of the radiotracer at the reducing pipe for the leakage loop. This may lead to higher radiotracer detected over time which may cause misinterpretation for higher leakage than reality.

To further investigate the cause of this large discrepancy, it was hypothesized that the accumulation of the radiotracer in the experiment setup was the major contributor. To test this hypothesis, first, the noise is filtered out from the signals of interest with the median filtering technique. Then, the effect of radiotracer accumulation at the leakage measurement point (signal from D03) is analyzed. The long tail of the leak signal reveals the possibility of radiotracer accumulation at the leak setup. If there were no radiotracer build-up at the pipe reducer, the peak of the leak signal should be almost symmetry similar to that of the signal obtained from the main flow (D01). If we consider only the area under peak for the main flow and assuming symmetry shape for the leak signal, peak areas are 19771 and 554 for the main flow and for the leakage, respectively. The percentage leakage in this case is approximately 2.8 which is about 5.6 times larger than the result from the direct measurement. The filtered signals are shown in Figure 5.

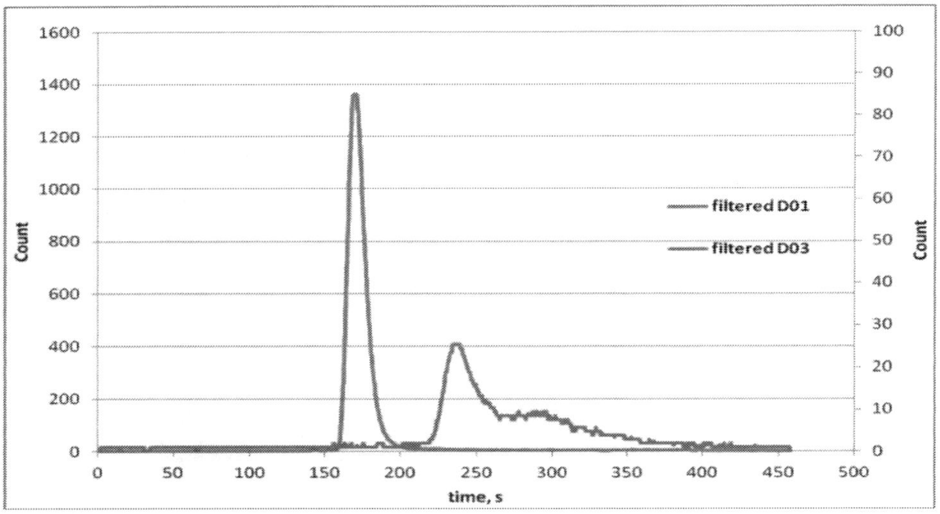

Figure 5: Filtered signals from the main flow detector #1(D01) and the leak detector #3(D03).

When comparing to the discrepancy of 16 times in the previous result, this analysis shows that, without the accumulation in the experiment setup, the measured leakage is more accurate. The analysis encourages the possibility of the hypothesis that radiotracer accumulation at the leakage measurement point is one cause for the discrepancy of the results from the radiotracer method and the direct measurement. Future experiments will be performed to identify other sources of discrepancies.

In order to study the factors that cause the result discrepancy in more details, the effect of the pipe size to the detector measuring efficiency should be studied. In the future, the setup for leak study should also be modified to lessen or, if possible, eliminate the radiotracer accumulation effect.

4. Conclusions

The radiotracer experiment using Br-82, in the form of NH_4Br aqueous solution, was performed at TINT. Two flow properties were studied in this work, i.e., flow rate and leakage from the main flow. The radiation signals were detected by three NaI detectors placed in three different locations in order to measure those flow properties. Direct measuring methods were also conducted to measure flow rate and leakage from the main flow. This information was used to compare with the results obtained by the radiotracer methodology. In this work, the velocity and percentage leakage obtained from the radiotracer method were found to be different from those measuring from the direct measurements. The plausible causes of the discrepancy are the coarse scale of the measuring equipment, the effect of pipe size on the measuring efficiency of detectors and the significant accumulation of radiotracer at the reducing pipe of the leakage control valve. If the effects of those causes can be lessen or eliminate, it is highly possible that the radiotracer method can provide a fairly consistent results with the one obtained from the direct measuring methods. In that case, the radiotracer method will be a very useful way for on-line measuring of flow properties where direct measuring method cannot be performed.

5. References

[1] International Atomic Energy Agency 2004 *Radiotracer Applications in Industry - A Guidebook, Technical Reports Series 423* (Vienna: International Atomic Energy Agency)

[2] International Atomic Energy Agency 2009 *Leak detection in heat exchangers and underground pipelines using radiotracers: material for education and on-the-job training for practitioners of radiotracer technology, Training course series No.38* (Vienna: International Atomic Energy Agency)

Innovative neutron shielding materials composed of natural rubber-styrene butadiene rubber blends, boron oxide and iron(III) oxide

C Jumpee and D Wongsawaeng[1]

Department of Nuclear Engineering, Faculty of Engineering, Chulalongkorn University, Bangkok, Thailand, 10330

E-mail: doonyapong.w@chula.ac.th

Abstract. Optimized flexible and lightweight neutron shielding materials were designed using the Monte Carlo N-Particle (MCNP) code. Thicknesses of 10 mm and 100 mm were tested for neutron shielding performances. Simulation results indicated that the 10 mm shielding material of natural rubber (NR) and styrene butadiene rubber (SBR) blend (1:1) with 60 part per hundred rubber (phr) boron oxide (B_2O_3) and 100 mm shielding material with four alternating layers of NR with 100 phr iron (III) oxide (Fe_2O_3) and of NR and SBR blend (1:1) with 10 phr B_2O_3 were most suitable for thermal neutron shielding and all-energy neutron shielding, respectively. Experimental results verified the shielding efficiency of these optimal designs and ease of fabrication.

1. Introduction

Neutron radiation is widely utilized in medical, agricultural and industrial applications. As neutrons have no electrical charge, they can pass through an electron cloud to interact directly with the atomic nucleus and transfer most of their energy to the nucleus with a mass close to that of a neutron. As the human body has a high water content, there is a high density of hydrogen atoms present. The mass of the hydrogen nucleus is close to that of the neutron. Therefore, exposure to free neutrons can be hazardous, since the interaction of neutrons with molecules in the body can cause disruption to molecules and atoms as well as cause reactions which give rise to other forms of radiation such as protons with high Linear Energy Transfer (LET) [1], [2]. Neutron shielding requires materials with an atomic mass close to that of the neutron. For example, hydrogen-rich materials are often used to shield against neutrons, since ordinary hydrogen both scatters and slows down neutrons, after which neutrons may be absorbed by an isotope exhibiting high affinity for adsorbing slow neutrons such as boron.

There have been several studies of polymeric composition containing materials with high neutron cross section to determine their thermal neutron shielding properties. Materials made of polyethylene or epoxy resin mixed with boron compound are widely used as neutron shielding materials, but they usually exhibit sizable weight and volume, which results in low-flexibility performances [3], [4].

[1] To whom any correspondence should be addressed.

Flexible polymers such as natural rubber (NR) have been used as a matrix for neutron shielding materials, but they only shield against thermal neutrons [5], [6].

There is considerable natural rubber cultivation in many areas of Thailand. This rubber is a hydrocarbon polymer containing a large number of hydrogen atoms and can be easily fabricated into flexible materials. In addition, borate minerals such as boracite, or boron compounds, are found in the northeast of Thailand. Thus, not only is the production of neutron shielding materials comprising natural rubber/ SBR blends and a borate material useful for neutron-related work, but it also utilizes domestic resources.

The most effective neutron shielding material can be obtained by appropriately mixing high hydrogen-content materials, heavy elements and thermal neutron absorbers. High hydrogen-content materials can undergo elastic scattering with fast and intermediate-energy neutrons. Heavy elements can undergo inelastic scattering with fast neutrons and can attenuate the secondary gamma ray emission. Finally, neutron absorbers function to greatly reduce the number of thermal neutrons.

In this experimental work, rubbers, which are low-cost hydrocarbon polymers that can easily be fabricated into many different products and shapes, were selected as the substrate of the composite together with a very low-cost and often disregarded neutron absorbing material, B_2O_3. Fe_2O_3 (commonly known as rust), a very low-cost and usually disregarded material, was also utilized as the heavy element in the composite.

The method of calculation for neutron shielding can be found in references [7], [8]. This research incorporated inelastic scattering and shielding of secondary gamma ray emissions into the simulation. The compositions of neutron shielding materials were optimized by Monte Carlo simulations using the MCNP5 code to examine the neutron flux encompassing the entire range of interaction energy and accounting for secondary gamma ray emissions.

Neutron attenuation experiments were carried out to verify neutron and secondary gamma ray shielding performances of the designed neutron shielding materials using a simulated neutron source (energy $1\text{x}10^{-8}$ – 100 MeV) and an Am-Be neutron source.

2. Materials and Methods

2.1. Material selection and MCNP simulation

The main raw materials were selected from simple elements or compound materials, taking into consideration cost, radiation shielding performances and physical and thermal properties. The main compositions of the selected neutron shielding material to shield against the entire range of neutron energy and secondary gamma ray emission are listed in Table 1. They consisted of natural rubber (NR), styrene butadiene rubber (SBR-1502 from BST elastomer Co., Ltd. Thailand) and neutron absorber/scattering agent [boron oxide (B_2O_3), iron(III) oxide (Fe_2O_3) and lead dioxide (PbO_2)]. 40 phr of each agent was used for each sample because preliminary fabrication indicated that at 40 phr, rubber still exhibited desirable properties. The lead or iron content is desirable for inelastic neutron scattering to effectively bring the neutron energy down to lower-energy regions. In addition, lead and iron are used for attenuation of gamma radiation. Neutrons in the lower-energy regions can be moderated by hydrogen and other light elements by elastic scattering. In the thermal energy region, boron (as B_2O_3) is used for neutron absorption. Because the interaction between a thermal neutron and boron-10 produces a secondary gamma ray from the following interaction:

Thermal neutron + B-10 → He-4 + Li-7 (478 keV gamma ray),*

without the gamma ray attenuating material (Pb or Fe), this neutron shielding material would itself become a significant source of gamma radiation.

For each sample in table 1, the input card of the MCNP5 transport code was appropriately prepared for testing and comparison of shielding properties of different materials. A simplified model was employed: a single radiation source, an isotropic emission of simulated neutron source with energy $1\text{x}10^{-8}$ – 100 MeV, was located at the center of a sphere filled with the shielding material as shown in figure. 1. The spherical shielding material was considered for shielding the neutron dose equivalent rate [(rem/hr)/source strength] and the secondary gamma dose equivalent rate [(rem/hr)/source strength]. The spherical model of radius 10 mm or 100 mm was chosen for neutron shielding in the thermal region or in the entire-energy range, respectively, to ensure that all neutron particles interacted with the shielding material at least one time before leaving the system. For each shielding material, the dose equivalent rate per source strength was estimated on the outer surface of the sphere.

Table 1. Main compositions of considered neutron shielding materials
(40 phr each of B_2O_3, PbO_2 and Fe_2O_3)

Sample No.	No. of layers	Material composition	Density (g/cm^3)	Element content (part per hundred rubber; phr)						
				H	C	B-10	B-11	O	Fe	Pb
0	0	No material (air)	0.00120	Air *						
1	1	NR+SBR	0.987	10.378	89.622	-	-	-	-	-
2	1	NR+SBR+B_2O_3	1.210	10.378	89.622	2.460	9.963	27.577	-	-
3	1	NR+SBR+PbO_2	1.343	10.378	89.622	-	-	5.351	-	34.649
4	1	NR+SBR+Fe_2O_3	1.301	10.378	89.622	-	-	12.023	-	13.988
5	1	NR+SBR+B_2O_3+PbO_2	1.273	10.378	89.622	1.230	4.982	16.464	-	17.325
6	1	NR+SBR+B_2O_3+Fe_2O_3	1.254	10.378	89.622	-	-	19.800	-	27.977

Multi-layer samples

		Layer 1	Layer 2	Layer 3	Layer 4	Layer 5
7	4	Sample 2	Sample 3	Sample 2	Sample 3	-
8	4	Sample 2	Sample 4	Sample 2	Sample 4	-
9	4	Sample 3	Sample 2	Sample 3	Sample 2	-
10	4	Sample 4	Sample2	Sample4	Sample 2	-
11	5	Sample 2	Sample 3	Sample 2	Sample 3	Sample 2
12	5	Sample 2	Sample 4	Sample 4	Sample 4	Sample 2
13	5	Sample 3	Sample 2	Sample 3	Sample 2	Sample 3

*[8] Air's composition (weight fraction) ; Carbon : Nitrogen : Oxygen : Argon
0.000124 : 0.755268 : 0.231781 : 0.012827

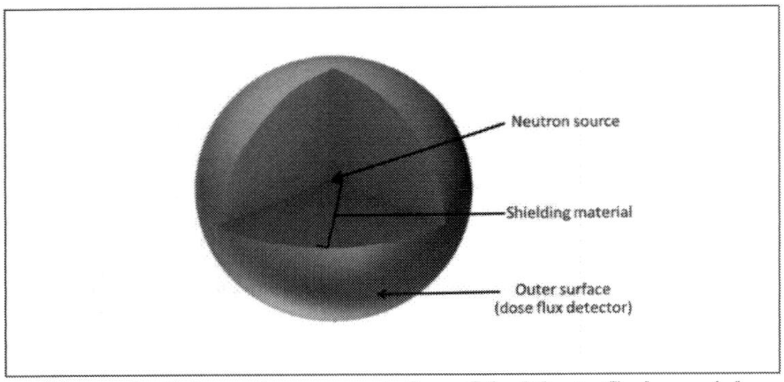

Figure 1. Schematic representation of the Monte Carlo model

Table 2. Ingredients for fabrication of most appropriate rubber compounds

Ingredient (phr)	Samples						
	1	2	4	10			
				Layer 1	Layer 2	Layer 3	Layer 4
NR	50	50	100	100	50	100	50
SBR-1502	50	50	0	0	50	0	50
ZnO	5	5	5	5	5	5	5
Steric acid	2	2	2	2	2	2	2
Wingstay-L	1	1	1	1	1	1	1
B_2O_3	0	60	0	0	10	0	10
Fe_2O_3	0	0	5	100	0	100	0
MBTS [a]	1.2	1.2	1.2	1.2	1.2	1.2	1.2
TMTD [b]	1	1	1	1	1	1	1
DPG [c]	0.5	0.5	0.5	0.5	0.5	0.5	0.5
Sulfur	2.5	2.5	2.5	2.5	2.5	2.5	2.5
Compression molded time (min.)	8	30	4	4	15	4	15

[a] Dibenzothaizyledisulphide, [b] Tetramethylthigramdisulphide , [c] Diphenyl Guanidine

2.2. Fabrication of the shielding materials

Composite materials shown in table 2, which were the most appropriate rubber compounds after MCNP simulation of all compounds listed in table 1, were fabricated by trial and error, as it was unpredictable which composition would result in a stable compound. All rubber mixtures were prepared in a two-roll mill of 8″ x 20″ in diameter with the working distance of 300 mm (speed of slow roll 18 rpm) and the gear ratio of 1:4. Powders of B_2O_3, PbO_2 and Fe_2O_3 were appropriately introduced into the blend according to the recipe of compositions given in table 2. Mixing time was about 25 minutes. After compounding, the stock was left for 24 hrs to mature. The sheets were cut into slabs and then compression molded by an electrically-heated hydraulic press at 160±2 °C and 4 MPa for a suitable duration of each sample.

The homogeneity of each shielding material sample was verified by neutron radiography at the Thai Research Reactor TRR-1/M1, which is a TRIGA-type reactor. A neutron imaging plate (ND 2040) attached to the shielding material by an aluminium tape was exposed to the neutron beam for 4 minutes. The neutron flux and the Cd ratio at the position of the shielding material were approximately 9×10^5 n cm^{-2} s^{-1} and 100, respectively. The neutron imaging plate was processed by an image reader (BAS 2500).

2.3. Experimental work on neutron attenuation

The neutron attenuation experiment was carried out at the Radiation Measurement Laboratory at the Nuclear Engineering Department, Faculty of Engineering, Chulalongkorn University to assess neutron shielding performances of the selected shielding materials shown in table 2. Shielding materials were slab shaped with thicknesses of 10 mm and 100 mm and dimensions of 150 mm x 150 mm. These slabs were stacked together to a height of 20 cm. The Am-Be neutron source whose strength was 3.08 $\times 10^3$ n cm^{-2} s^{-1} was placed at the bottom of the polyethylene (PE) source holder with a distance of 40 mm from the neutron shielding material, as illustrated in figure 2. The neutron measurement was performed using a NE-905 glass scintillation neutron detector connected to a counting system.

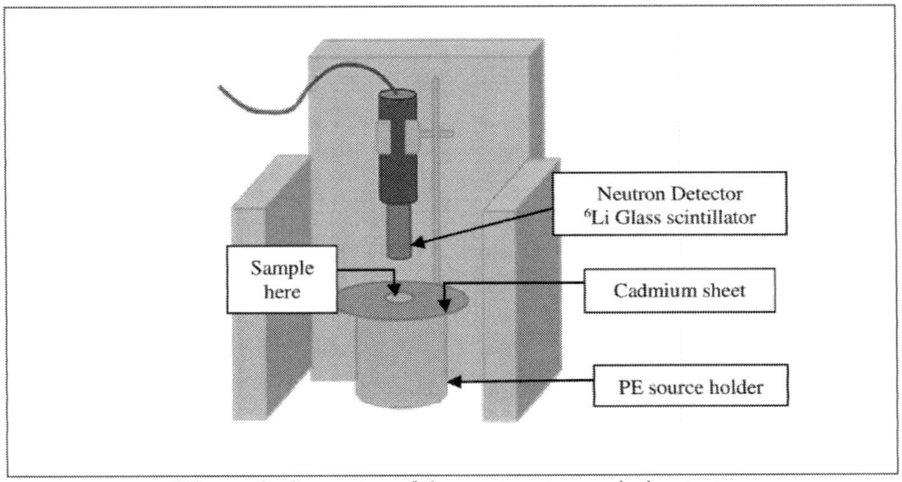

Figure 2. Geometry of the neutron transmission test.

3. Results and Discussion

3.1. MCNP Simulation

Results from the MCNP transport code revealed that the 10-mm thick sample #2 (NR+SBR+B_2O_3) and 100-mm thick sample #10 (4 alternating layers of NR+Fe_2O_3/NR+SBR+B_2O_3) exhibited excellent neutron and secondary gamma ray shielding performances, as seen in figures 3 and 4. For sample #2, because the thickness of 10 mm was smaller than the mean free path of fast neutrons in rubber, only thermal neutrons could be absorbed by this material, which also had the highest boron content. For the 100-mm thick sample #10, fast neutrons could undergo inelastic scattering with iron and elastic scattering with hydrogen present in the rubber to become thermal neutrons and finally be absorbed by boron. Thus, sample #2 and sample #10 are suitable for thermal neutron shielding and all-energy neutron shielding, respectively.

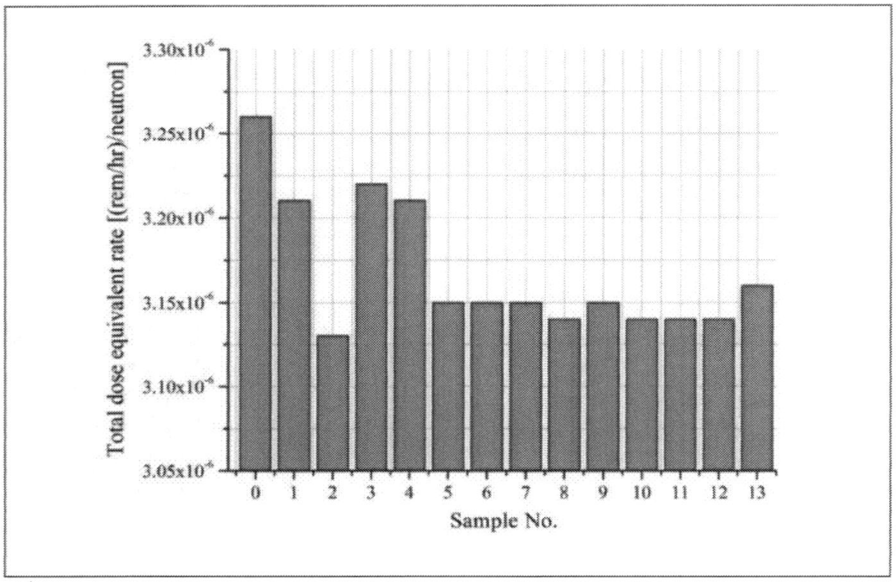

Figure 3. Total dose rate equivalent [neutron ($1x10^{-8}$ – 100 MeV) and secondary photon] on the outer surface of 10-mm radius sphere shielding sample.

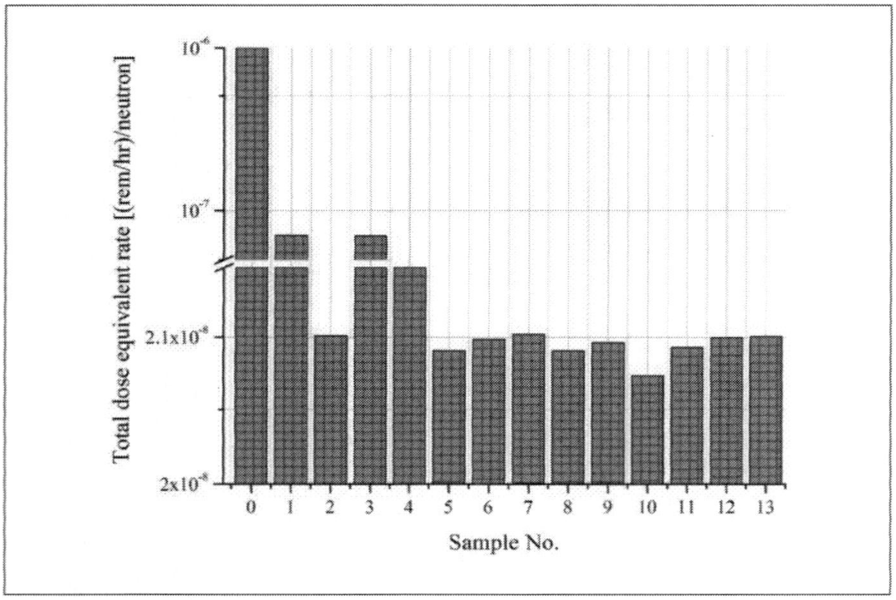

Figure 4. Total dose rate equivalent [neutron ($1x10^{-8}$ 100 MeV) and secondary photon] on the outer surface of 100-mm radius sphere shielding sample.

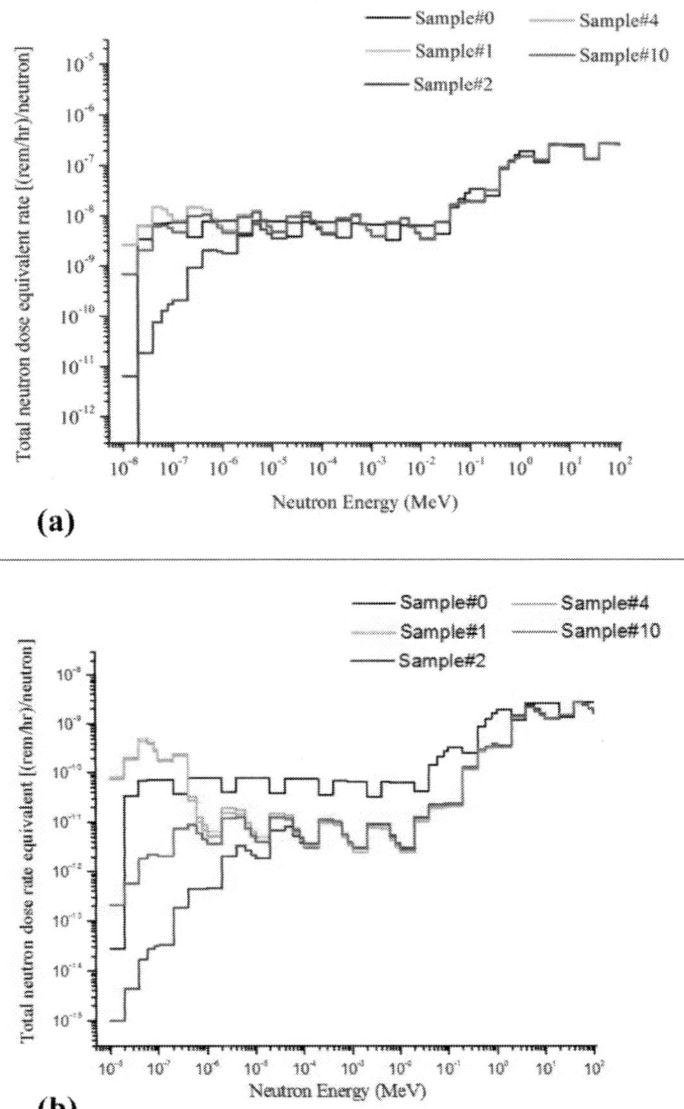

Figure 5. Total neutron dose rate spectra for various materials for 10 mm thickness (a) and for 100 mm thickness (b).

Figures 5(a) and 5(b) illustrate the neutron shielding efficiencies at different energy regions. For sample #2, the best shielding efficiency occurred in the thermal energy range ($1x10^{-8}$ – $1x10^{-2}$ MeV) because boron exhibits a high thermal neutron absorption cross-section. In the fast neutron regime ($1x10^{-1}$ – $1x10^{2}$ MeV), for 100-mm thick shielding designs, although not clearly visible in Figure 5(b),

Journal of Physics: Conference Series **611** (2015) 012019 doi:10.1088/1742-6596/611/1/012019

sample #10 outperformed sample #2 because the total neutron dose rate equivalent was lower in the entire fast neutron energy range. Moreover, for sample #10, it also performed better than the NR/SBR rubber compound (sample #1) and NR/SBR/Fe_2O_3 (sample #4) in the thermal energy range due to the presence of the boron compound.

To further determine the optimized amount of B_2O_3 and Fe_2O_3 additions, a simulation of a 10 mm thick sample #2 with varying amounts of B_2O_3 added from 0 to 70 phr was performed. Results are shown in figure 6(a). As more B_2O_3 was added, the total dose rate equivalent became reduced and appeared to saturate after about 60 phr. Thus, the most appropriate B_2O_3 addition for sample #2 was chosen to be 60 phr. Moreover, a simulation of a 100 mm thick sample #10 with varying amounts of B_2O_3 added from 0 to 70 phr and Fe_2O_3 added from 0 to 100 phr was performed. The reason why the amount of Fe_2O_3 added was limited to 100 phr was because during a trial fabrication test, the rubber became very rigid at 100 phr and more, making it much less flexible. Results in figure 6(b) revealed that the most appropriate amounts were 100 phr of Fe_2O_3 and 10 phr of B_2O_3.

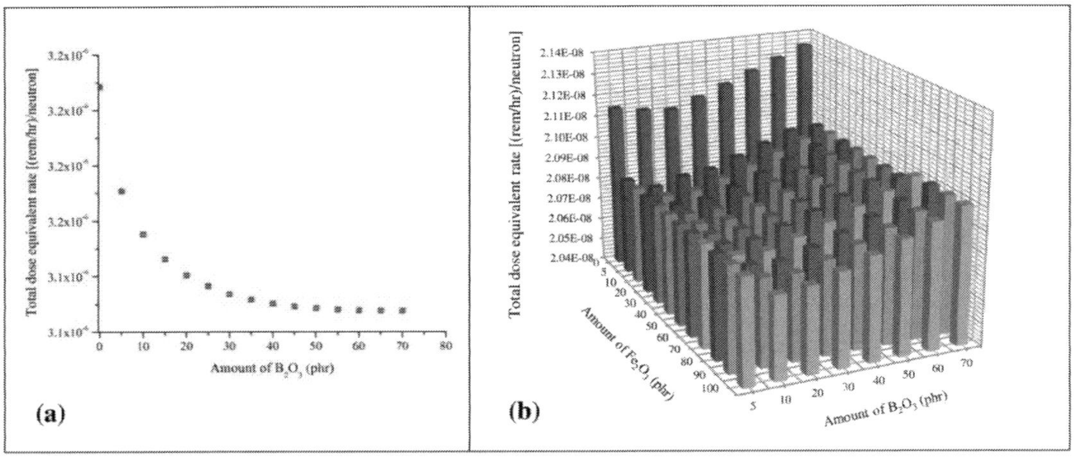

Figure 6. Total dose rate on outer surface of sphere neutron shielding at different mass ratios of neutron absorber contents (Part per hundred rubber; phr) for (a) 10 mm of sample#2 and (b) 100 mm of sample#10.

3.2. Fabrication of rubber compound

The optimized designs according to the MCNP simulation were fabricated by trial and error. The successfully-fabricated shielding materials with slab geometry are shown in figures 7(a) and 8(a). Homogeneity was one of several factors determining the suitability of the fabricated shielding materials. In reality, for the rubber compounds of samples #2 and #10, if they were left exposed to moisture in room condition, B_2O_3 would quickly react with water vapour in the atmosphere, turning into boric acid crystals, causing a problem with homogeneity. Thus, the samples must be completely encased in a thin layer of conventional rubber compound. Therefore, figure 7(a) represents the rubber compound encased in the green conventional rubber. In figure 8(a), because the rubber with iron oxide didn't require any encasement, the red slab represented the actual rubber compound. The corresponding neutron radiographs displayed in figures 7(b) and 8(b) exhibiting a very even color tone indicated that the neutron absorbing/scattering materials (B_2O_3 or Fe_2O_3) can be homogenously dispersed in the rubber compounds of NR and SBR.

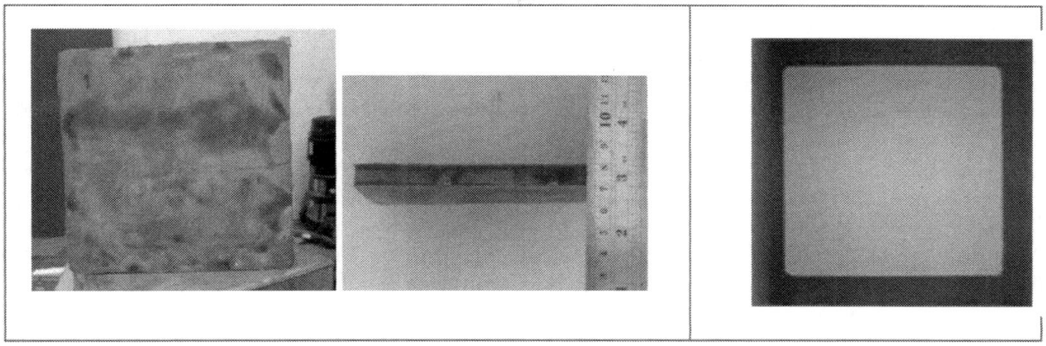

Figure 7. (a) Front view and (b) side view of 10 mm (NR+SBR+B_2O_3 60 PHR: sample#2) shielding material samples trial-produced to fabricate slab shape. (c) Neutron radiograph of this shielding material (front view).

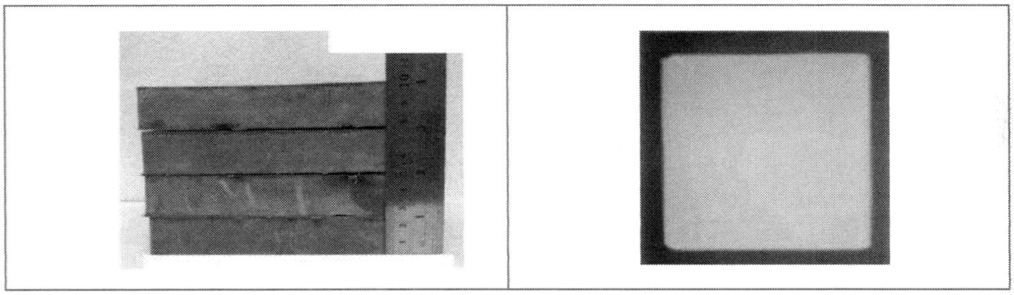

Figure 8. (a) Side view of 100 mm (4 layers: sample#10) shielding material samples trial-produced to fabricate slab shape. (b) Neutron radiograph of this shielding material.

3.3. Neutron transmission test

From the neutron transmission test, the 10 mm rubber compound with 60 phr of B_2O_3 can reduce the neutron dose from the Am-Be source by 53.91%. The 100 mm rubber compound composed of 4 alternating layers with 100 phr Fe_2O_3 and 20 phr B_2O_3 can reduce the neutron dose by 72.56%.

4. Conclusion

Optimized flexible and lightweight neutron shielding materials were designed using the MCNP code. The neutron shielding materials with thicknesses of 10 mm and 100 mm were examined for neutron shielding performances. Simulation results indicate that the 10 mm shielding material of NR and SBR blend (1:1) with 60 phr B_2O_3, and the 100 mm shielding material having 4 alternating layers of NR with 100 phr Fe_2O_3 and of NR and SBR blend (1:1) with 10 phr B_2O_3 were most suitable for thermal neutron shielding and all-energy neutron shielding, respectively. Experimental results verified the shielding efficiency of these optimal designs and the ease of fabrication. The designed shielding materials are highly suitable for applications in nuclear science and technology.

Acknowledgments

This work was financially supported by the Graduate School, Chulalongkorn University and The Commission on Higher Education under Strategic Scholarships for Frontier Research Network for the Thai Ph.D. Doctoral Degree Program.

References

[1] Kenneth S J and Faw R 2000 *Radiation Shielding* (Illinois: American Nuclear Society)
[2] Knoll G 1979 *Radiation Detection and Measurement* (Toronto: John Wiley & Sons)
[3] Koichi O 2005 Neutron shielding material based on colemanite and epoxy resin *Radiat Prot Dosimetry* **115** 258–61
[4] Gwaily S E, Badawy M M, Hassan H H and Madani M 2002 Natural rubber composites as thermal neutron radiation shields II — H_3BO_3/NR composites *Polym Test* **21** 513-17
[5] Gwaily S E, Badawy M M, Hassan H H and Madani M 2002 Natural rubber composites as thermal neutron radiation Shields I. B_4C/NR composites *Polym Test* **21** 129-33
[6] Huasi H *et al* 2008 Study on composite material for shielding mixed neutron and gamma rays *IEEE Trans Nucl Sci* **55** 4
[7] Elbio C, Florian G, Burkhard S, Harald T 2011 Reusable shielding material for neutron- and gamma-radiation *Nucl Instrum Methods Phys Res A* **651** 77-80
[8] McConn Jr R J, Gesh C J, Pagh R T, Rucker R A and Wlliams III R G 2011 *Radiation portal monitor project: Compendium of material composition data for radiation transport modelling* (Washington: Pacific Northwest National Laboratory)

Journal of Physics: Conference Series **611** (2015) 012020 doi:10.1088/1742-6596/611/1/012020

An industrial radiography exposure device based on measurement of transmitted gamma-ray intensity

C Polee[1], N Chankow, S Srisatit and D Thong-Aram

Department of Nuclear Engineering, Faculty of Engineering, Chulalongkorn University, Bangkok, Thailand

E-mail: bansak@hotmail.com

Abstract. In film radiography, underexposure and overexposure may happen particularly when lacking information of specimen material and hollowness. This paper describes a method and a device for determining exposure in industrial gamma-ray radiography based on quick measurement of transmitted gamma-ray intensity with a small detector. Application software was developed for Android mobile phone to remotely control the device and to display counting data via Bluetooth communication. Prior to film exposure, the device is placed behind a specimen to measure transmitted intensity which is inversely proportional to the exposure. Unlike in using the conventional exposure curve, correction factors for source decay, source-to-film distance, specimen thickness and kind of material are not needed. The developed technique and device make radiographic process economic, convenient and more reliable.

1. Introduction

Industrial radiography is widely used for the nondestructive testing (NDT) of materials. Film is still the most popular image recorder due to economic, technical and practical reasons. In general practice, exposure is obtained from the exposure curve at the thickness of specimen equivalent to steel. Correction factors for kind of specimen material, source-to-film distance and source decay are required [1-4]. In case of lack of information of specimen material and hollowness, film density may be out of range resulting in waste of time and money. In this research, a technique and a device were developed for determination of the optimum exposure by measurement of the transmitted gamma-ray intensity. The appropriate exposure can then be determined without information of the specimen thickness and specimen material. Moreover, it is independent on source activity and source distance, which makes radiographic process economical, convenient and more reliable. The gamma-ray detector is essentially small so that the transmitted intensity can be measured at the desired parts of the specimen. This research, therefore, selected PIN photodiode which is small and sensitive to x-rays and gamma-rays particularly at high dose [5-8]. The device is remotely controlled using an Android mobile phone making the procedure even more convenient and safer [8]. Bluetooth communication allows up to 10 meters to control the device and to send counting data to the phone. Upon appropriate calibration, the device can also be used to monitor dose rate for safety purpose in industrial radiography and any other radiation practices.

[1] To whom any correspondence should be addressed.

2. Experimental

2.1. Radiation counting device using PIN photodiode

The Si Positive-Intrinsic-Negative or PIN photodiode is a semiconductor detector type, which creates electron-hole pairs from interaction of an incident gamma-ray with P-N junction. By supplying the reverse-biasing voltage, charges can be then collected [5]. The basic structure of the PIN photodiode is shown in figure 1. The PIN photodiode is widely used in the radiation measurement [9, 10] due to the above mentioned properties [6] as well as its high-speed response, high sensitivity, low noise and compactness.

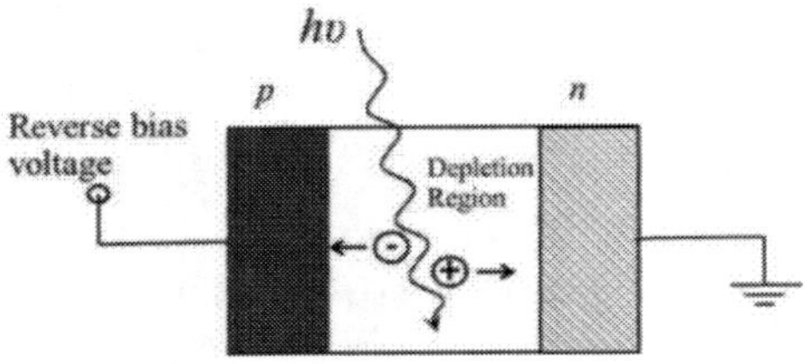

Figure 1. Basic structure of PIN photodiode.

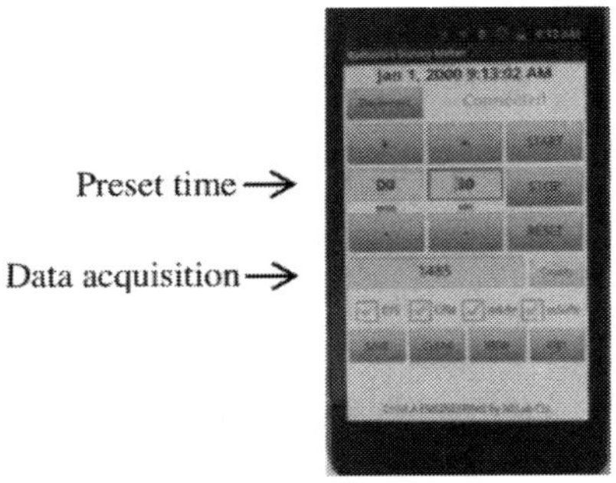

Figure 2. Developed interface.

The in-house developed radiation counting device is composed of four main parts; a PIN photodiode radiation detector, a pulse shaping circuit, a microcontroller board and a Bluetooth module. The developed counting system can be used for a maximum counting rate of 1×10^5 cps and an exposure rate is up to 300 mR/hr [8]. The android interface software was also developed to control the counting process and data communication as well as to display the counting results.

The software was designed to be simple and user-friendly and its interface is shown in figure 2. The preset time can be set by touching screen on the menu + and - for increasing and decreasing time, respectively. For the menu START, STOP and RESET, user can remotely controlled to start, to stop and to reset the system via Bluetooth technology. Then the acquisition data will be displayed on the screen. Note that the interface is not only designed for this purpose but also for other applications.

To investigate the relationship between the transmitted gamma-ray intensity and the thickness of specimen, a steel step wedge was used as a standard sample. The thicknesses of a steel step wedge are 5.48, 6.98, 8.14, 10.72, 11.55, 12.79, 14.30, 16.60, 18.10, 20.00, 21.06, 22.56, 24.00 and 25.33 mm. The PIN photodiode detector Hamamatsu model S3590-08 with an active area of 1 cm × 1 cm was set behind a steel step wedge and far from the 10 Ci Ir-192 radiation source of 80 cm-distance. Small size of the detector enables detection at high dose, and increase the precision of the inspection in a small area. The obtained signals from the PIN photodiode were counted by the pulse shaping circuit and the results were then delivered to a smartphone. The experimental setup and the schematic diagram of the developed counting device are shown in figure 3 (a) and (b), respectively. The relationship between the steel thicknesses and the transmitted counts was carried out.

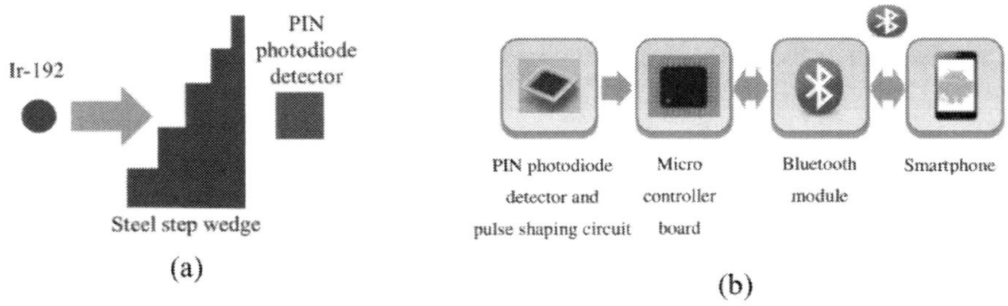

Figure 3. The experimental setup (a) and the schematic diagram of the developed counting device (b).

2.2. Film optical density (OD) for various steel thicknesses

For the industrial x-ray and gamma-ray radiography with film technique, the film optical density or OD is a conventional factor which indicates image darkness. The OD is defined as the logarithm of inverse of light transmittance. In practice, the OD of the radiographic image should fall in the range of 1 – 3 to keep the image contrast at maximum value [3].

To determine the film OD as a function of the specimen thickness and the exposure time, a steel step wedge was radiographed using Kodak AA 400 x-ray film at exposure times of 5, 10, 15 and 20 minutes, respectively. The experimental setup is shown in figure 4. After film processing, the film OD at each steel thickness was measured by a densitometer of X-rite model 301. The relationship between the exposure time and the film OD were then plotted for each steel thickness. Subsequently, the film OD calibration chart was created for determination of the appropriated exposure time.

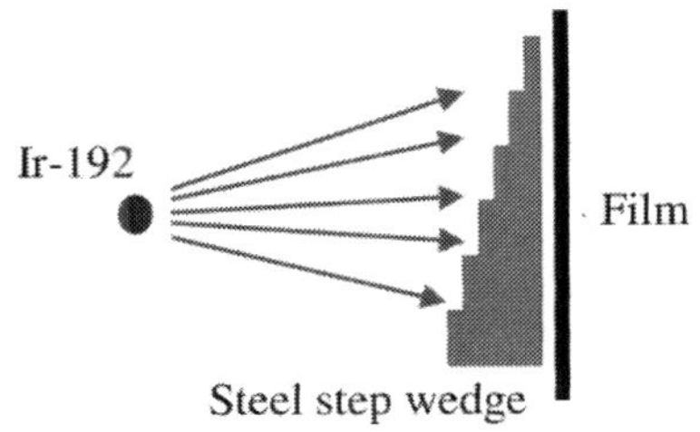

Figure 4. Schematic diagram of the experimental setup for determining the film OD.

2.3. Testing the developed device for determination of exposure time

To test the developed device and method, various types of specimen including a welded steel plate, a cast iron and a lock were radiographed. For the preliminary testing, we aimed to obtain the film OD to be 2.0. The developed device was set behind a specimen to measure the transmitted gamma-ray intensity. The investigated areas of the samples were indicated in figure 8. The obtained counting data were used to determine the appropriated exposure time of each specimen from the calibration chart. The three specimens were finally radiographed using the obtained exposure times. After the film chemical processing, the OD was measured at the position where the transmitted gamma-ray intensity was counted to compare with the expected OD of 2.0.

3. Results and discussions

3.1. Transmitted gamma-ray intensity versus steel thickness

The transmitted gamma-ray intensity was successfully measured by the developed device. The obtained transmitted gamma-ray intensity versus steel thickness could be plotted as shown in figure 5.

From the results, the transmitted gamma-ray intensity decreased with increasing steel thickness. The minimum count of all obtained counting data was found to be approximately 37,000 counts per 30 second. The standard deviation was therefore predicted from the square root of 37,000 which was approximately 192. The precision (3σ) was about 577 which was better than 2% at 99.7% confidence. The results showed that the uncertainty of gamma-ray counting is lower than 2%. However, it can be decreased with increasing of the counting time.

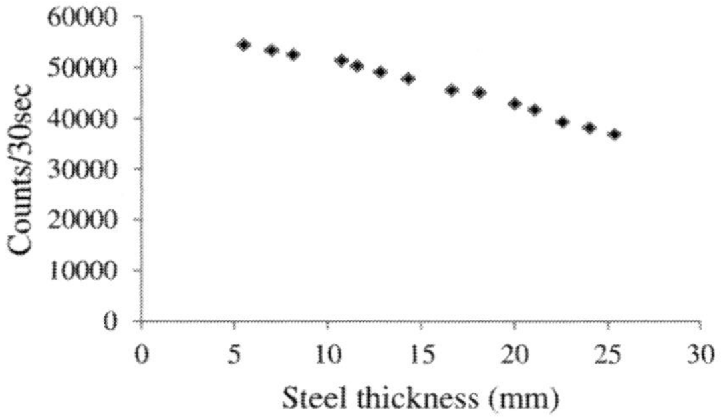

Figure 5. Relationship between the gamma-ray intensity measured by the developed counting device and the steel thickness.

3.2. Film optical density versus steel thickness

By varying the exposure times, the film OD increased with increasing the exposure time whereas it decreased with increasing the steel thickness. The obtained data were fitted with the linear function as shown in figure 6. To avoid overlapping of the fitted curves as shown in figure 6, only 10 out of 14 data sets from our experiments were selected to present.

The relationships of the transmitted gamma-ray count rates and the calculated exposure times at various thicknesses were created by fitting the logarithmic function. Finally, the calibration chart from the developed method was achieved as shown in figure 7. The calibration chart obviously shows that the impact of counting uncertainty from the in-house developed device is small in determining the exposure time. The obtained calibration chart was then employed to determine the appropriate exposure time using the transmitted gamma-ray intensity of and the test specimens, regardless of the thickness and kind of material.

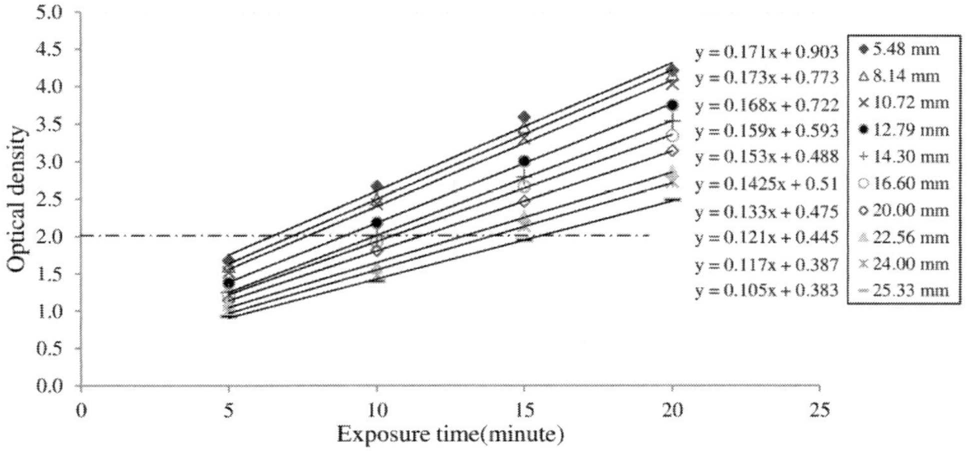

Figure 6. Relationships between the exposure times and the ODs at various thicknesses.

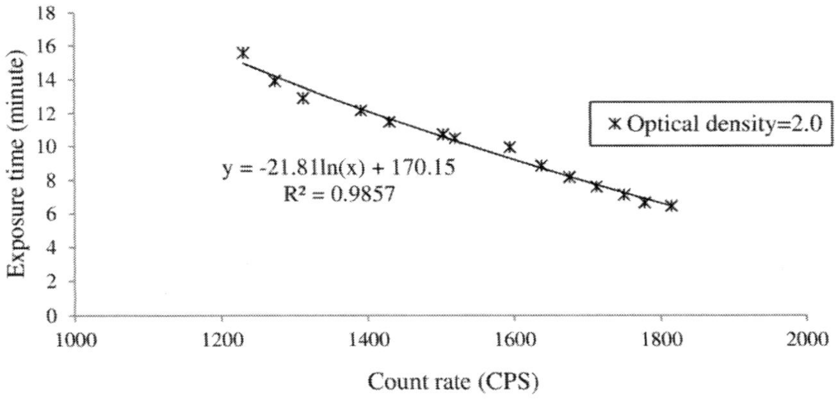

Figure 7. The calibration chart for the film OD at 2.0.

3.3. Determination of exposure using the developed technique and device

By applying the developed method, the appropriate exposure time for the OD is 2.0 could be successfully determined using the measurement of the transmitted gamma-ray intensity. The radiographic images were shown in figure 8. The film ODs were measured at the positions indicated by the squares in figure 8 where the PIN photodiode was placed to measure transmitted gamma-ray intensity. The film ODs of a welded steel plate, a cast iron and a lock radiograph were found to be 2.05, 1.98 and 1.92, respectively, as shown in table 1. The discrepancies of the film OD were found to be very small compared to the expected film OD of 2.0. The obtained ODs were still in the acceptable range. Furthermore, the results indicated that the developed counting system and the calibration method could give appropriate exposure time without specimen information and any correction factor. However, film chemical quality of film processing should be strictly controlled.

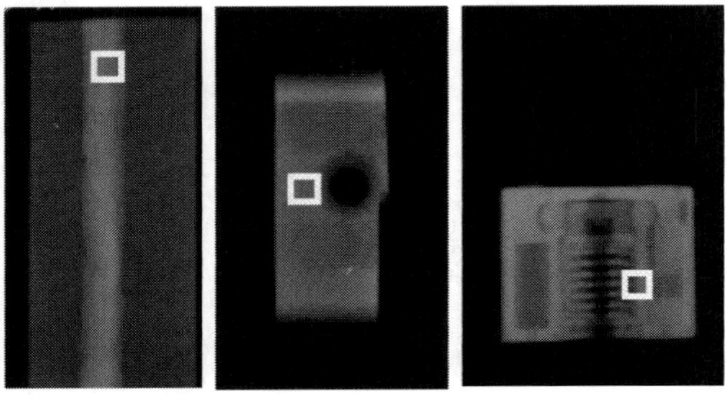

Steel welded seam Cast iron Lock

Figure 8. Radiographic image of the test specimens.

Table 1. The film OD of the radiographic testing.

Sample	Thickness (cm)	Gamma-ray intensity (cps)	Exposure time (min)	Optical density (OD)	Discrepancy
Welded steel plate	0.8	1713	7.68	2.05	+0.05
Cast iron	1.5	1530	10.14	1.98	- 0.02
Lock	2.0	1355	12.78	1.92	- 0.08

4. Conclusions

Nowadays, industrial radiography is actually not only used to radiograph industrial specimens. It is also widely used to inspect ancient objects and other large specimens like concrete structures and large animals. The latter is due to too low energy medical x-rays. Radiographers often have difficulty in determining appropriate exposures due to lack of appropriate calibration and specimen information The in-house developed gamma-ray counting device combining with a smartphone and Bluetooth technology was successfully tested in this work. It can facilitates the experiment is more convenient and safer. The calibration chart for the gamma-ray radiography using film with Ir-192 source was successfully created by the development method. The new technique based on measurement of transmitted gamma-ray intensity accompany with the calibration method could provide the appropriate exposure time with an expected optical density. Therefore, the method can help the radiographer to reduce time and cost loss from the trial and error of unknown specimen.

Acknowledgments
We would like to thank the Thai Non-destructive Testing Public Company Limited for providing us the Ir-192 Gamma source and the NDT division of Nuclear Technology Service Centre of TINT for allowing us to use the film chemical processing laboratory.

Journal of Physics: Conference Series **611** (2015) 012020 doi:10.1088/1742-6596/611/1/012020

References

[1] Rigagu Denki (n.d) *Radiography Testing Training Course Lecture Note*

[2] International Atomic Energy Agency 1992 *Industrial radiography manual of the syllabi contended in IAEA TECDOC-628 training guidelines for non-destructive testing techniques* (Vienna: International Atomic Energy Agency)

[3] Raad J A and Kuiper A 2007 *Industrial Radiography Image Forming Techniques* (General Electric Company, Inspection Technologies)

[4] International Atomic Energy Agency 2009 *Lecture script for the IAEA/RCA regional training course on digital industrial radiology and computed tomography applications in industry (Kajang, Malaysia, 2 – 6 November 2009)*

[5] Hamamatsu Photonics K.K 1993 *Silicon photodiodes and charge sensitive amplifiers for scintillation counting and high energy physics* (Solid State Division catalog June 1993)

[6] Hamamatsu Photonics K.K 2014 *Silicon photodiodes* (Solid State Division catalog February 2014)

[7] Ramírez-Jiménez F J 2003 Radiation detectors of PIN type for X-Rays *API* **674** 313

[8] Watanapanich P and Thong-Aram D 2014 Development of a smartphone based radiation survey meter *The engineering institute of Thailand under H.M. The King's Patronage* **25** 105

[9] Han S K 2009 Fabrication and performance characteristics of a CsI(Tl)/PIN diode radiation sensor for industrial applications *Appl. Radiat. Isotopes.* **67** 7

[10] Lucia A 2014 Comparison between a silicon PIN diode and a CsI(Tl) coupled to a silicon PIN diode for dosimetric purpose in radiology *Nucl. Instrum. Meth. A* **762** 11

Specific activity of natural and anthropogenic radionuclides and radiological hazard assessment in surface soil samples collected along the Andaman sea coast in southern region of Thailand

P Kessaratikoon[1,3], R Boonkrongcheep[1], S Benjakul[1] and S Udomsomporn[2]
[1]Nuclear and Material Physics Research Unit (NuMPRU), Department of Physics, Faculty of Science, Thaksin University, Muang, Songkhla, Thailand.
[2]Office of Atoms for Peace (OAP), 16 Vibhavadi Rangsit Rd, Chatuchak, Bangkok, Thailand.

E-mail: prasong_mi@hotmail.com

Abstract. The specific activities of natural (^{40}K, ^{226}Ra and ^{232}Th) and anthropogenic radionuclides (^{137}Cs) in 314 surface soil samples collected from three provinces (Phuket, Phang-Nga and Krabi) along the Andaman sea coast in southern region of Thailand were studied and evaluated. Experimental results were obtained by using a high-purity germanium (HPGe) detector and gamma spectrometry analysis system. It was found that the mean values of specific activities of ^{40}K, ^{226}Ra, ^{232}Th and ^{137}Cs were 2859.63 ± 209.83, 157.10 ± 8.06, 137.16 ± 7.26 and 4.88 ± 2.34 Bq/kg, respectively. Furthermore, four radiological hazard indices which are absorbed dose rate in air (D), radium equivalent activity (Ra_{eq}), external hazard index (H_{ex}) and annual effective dose rate (AED_{out}) in the area under consideration were also calculated and came out to be 277.11 ± 16.96 nGy/h, 575.16 ± 34.67 Bq/kg, 1.55 ± 0.09 and 0.34 ± 0.02 mSv/y, respectively. The experimental results were also compared and found to be comparable with national and global radioactivity measurements and evaluations. Moreover, the radioactive contour maps of the investigated area were also created and presented in this paper.

1. Introduction

Naturally occurring radioactive material (NORM), which potentially includes all radioactive elements found in the environment. The acronym TENORM, or technologically enhanced NORM, is often used to refer to those materials which the amount of radioactivity has actually increased or concentrated as a result of industrial processes. The issues related to NORM and TENORM attracted a lot of public interest and mass media in the late 1970s. Long-lived radioactive elements such as uranium, thorium and potassium and any of their decay products, such as radium and radon are examples of NORM [1-3]. These elements are always present in the Earth's crust and atmosphere. Many scientists and researchers around the world are studying and measuring the level of NORM concentration in soil and beach sand samples in some countries such as in Brazil, India, China, Turkey, France and Thailand [4-11]. Along the Andaman sea coast in the southern region of Thailand, the most common activities

[3] To whom any correspondence should be addressed.

related to TENORM are industrial related to tin mining and its subsequent processing. For this reason, some areas especially in Phuket, Phang-Nga and Krabi provinces could be affected by an increase in naturally occurring radioactive materials [12].

Caesium-137 (^{137}Cs) is a radioactive alkali metal created and globally distributed by the atmospheric testing of atomic weapons during the period 1945-1963. This fission radioactive material was deposited on the earth's surface as fallouts associated with precipitation. The other prime source of ^{137}Cs has been accidental releases by accidents of nuclear power plant (e.g., Three Mile Island in 1979, Chernobyl in 1986, and Fukushima Dai-Ichi in 2011). Baseline monitoring of ^{137}Cs in the landscape is used to assess the present health risk from gamma (γ)-radiation emitted following ^{137}Cs decay and to monitor for new releases of ^{137}Cs. With a half life of 30.17 years, ^{137}Cs has a long residence in the environment and is a health hazard if found in sufficient concentrations.

Furthermore, on Sunday, 26 December 2004, the 2004 Indian Ocean earthquake was an undersea megathrust earthquake, with an epicentre off the west coast of Sumatra, Indonesia. The resulting tsunami hit Thailand's southwestern coast along the Andaman Sea, causing death and destruction. The hardest hit areas in terms of loss of life and property destruction were in Phuket, Phang-Nga and Krabi provinces because they were the most developed and the most densely populated areas along the coast. Much of Phuket province's west coast was heavily damaged by the tsunami, and most homes, hotels, restaurants and other structures on low ground required significant repairs or rebuilding. Some areas, including Khao Lak in Phang-Nga province, were almost completely wiped out by the waves. Beyond the heavy toll on human lives, the Indian Ocean earthquake has caused an enormous environmental impact that will affect the region for many years to come [13-14]. It has been reported that severe damage has influence on ecosystems such as mangroves, coral reefs, forests, coastal wetlands, vegetation, sand dunes and rock formations, animal and plant biodiversity and groundwater. In addition, the spread of solid and liquid waste and industrial chemicals, water pollution and the destruction of sewage collectors and treatment plants threaten the environment even further, in uncountable ways. The environmental impact will take a long time and significant resources to evaluate [15].

For this reason, the area in the southwestern coast of Thailand or the Andaman sea coast (Phuket, Phang-Nga and Krabi provinces) should be the best place to study and evaluate the specific activities of NORM/TENORM (^{40}K, ^{226}Ra and ^{232}Th) and anthropogenic radionuclides (^{137}Cs). This paper reports the range and mean values of natural (^{40}K, ^{226}Ra and ^{232}Th) and anthropogenic radionuclides (^{137}Cs) in surface soil samples collected from three provinces (Phuket, Phang-Nga and Krabi) along the Andaman sea coast in the southern region of Thailand. By using the resulting data, radiological hazard could be studied and evaluated for the investigated area and compared to national and global radioactivity measurement and evaluations. Moreover, the radioactive contour maps in the monitored area could be also created and discussed.

2. Experimental

2.1 Sample collection and preparation

Surface soil samples (314 samples) were collected (from a ground surface down to 10-15 cm in depth) from 19 districts of Phuket (3), Phang-Nga (8) and Krabi (8) provinces in the southern region of Thailand. The geographical locations of the study areas were determined using the geographical positioning system (GPS) and located between latitude 7° 28' and 9° 21' N and longitude 98° 12' and 99° 24' E. After collection, each sample was dried up at room temperature and sieved through a 2 mm mesh-sized sieve to remove stones, pebbles and other macro-impurities. All samples were oven dried at a temperature of 100 °C for 3 hours before the analysis to remove the moisture. The homogenized sample was placed in a 290 cm^3 PVC containers. The container was sealed hermitically and externally using a cellophane tape and kept aside for about a month to ensure equilibrium between ^{226}Ra and its daughters before being taken for gamma spectrometric analysis.

2.2 Measurement and analysis

The specific activities of natural (^{40}K, ^{226}Ra and ^{232}Th) and anthropogenic(^{137}Cs) radionuclides in all surface soil samples were determined by employing a high-purity germanium detector (HPGe, CANBERRA Model GC 2018) and gamma spectrometry analysis system at Nuclear and Material Physics Laboratory, Department of Physics, Faculty of Science, Thaksin University. The detector was enclosed in a massive 10 cm thick lead shielding. Gamma ray radioactive standard sources ^{60}Co, ^{137}Cs and ^{133}Ba were used to calibrate the measurement system up to about 2 MeV. Geometric efficiency for beach sand matrices in the container was determined by KCl, IAEA/RGU-1, IAEA/RGTh-1 and IAEA/SL-2 reference materials (International Atomic Energy Agency IAEA, Vienna, Austria). The spectra were analyzed using the program GENIE 2000. The specific activity of ^{40}K was determined from its 1460.8 keV γ-line. The specific activities of ^{226}Ra and ^{232}Th were determined by their decay products ^{214}Pb (351.9 keV) and ^{208}Tl (583.2 keV), respectively. The specific activity of ^{137}Cs was also determined from its 661.7 keV γ-line. The counting time interval was 10,800 s. The background spectrum was recorded immediately after or before the sample counting.

2.3 Equations and formulae

By using the specific activity values of ^{40}K, ^{226}Ra, and ^{232}Th from this study, the absorbed dose rate (D) in outdoor air at 1 m above the ground was calculated using the conversion factors published in [7] and is given below

$$D = 0.461C_{Ra} + 0.623C_{Th} + 0.0414C_{K}, \quad (1)$$

where C_{Ra}, C_{Th} and C_K are the specific activities of ^{226}Ra, ^{232}Th and ^{40}K in Bq/kg, respectively. Furthermore, the radium equivalent activity (Ra_{eq}) was calculated through the following relationship [5, 16]:

$$Ra_{eq} = C_{Ra} + 1.43C_{Th} + 0.077C_{K} \quad (2)$$

Moreover, the external hazard index (H_{ex}) was also evaluated by using the equation which was defined as [5]:

$$H_{ex} = C_{Ra}/370 + C_{Th}/259 + C_{K}/4810 \leq 1 \quad (3)$$

Using the absorbed dose rates in air (D) obtained from the specific activities of natural nuclides in beach sand, adopting the conversion factor of 0.7 Sv/Gy to convert from the absorbed dose in air to the effective dose received by adults and considering that people in Thailand, on average, spend approximately 20% of their time outdoors, the annual effective dose rate (AED_{out}) were calculated using the following equation [6 -7]:

$$AED_{out}\,(mSv/y) = D\,(nGy/h) \times 8760\ h \times 0.2 \times 0.7\,(Sv/Gy) \times 10^{-6} \quad (4)$$

All four calculated and average values were evaluated and compared with the recommended values proposed by UNSCEAR [1-3].

3. Results and discussion

3.1 Specific activities of radionuclides in southern Thailand

Specific activities of natural (^{40}K, ^{226}Ra and ^{232}Th) and anthropogenic (^{137}Cs) radionuclides in 314 surface soil samples collected from the investigated areas could be analysed and evaluated from the measured gamma ray energy spectrum. Hence, the ranges and average values of those specific activities were also calculated and shown in table 1.

Table 1. Ranges and average values of specific activities of ^{40}K, ^{226}Ra, ^{232}Th and ^{137}Cs in 314 surface soil samples collected from Phuket, Phang-Nga and Krabi provinces (Thailand).

Locations	Range and Average Values of Specific Activity (Bq/kg)			
	^{40}K	^{226}Ra	^{232}Th	^{137}Cs
Phuket Province (103)	202.30 – 18803.28[a] 4896.45 ± 288.74[b]	24.12 – 830.82 240.11 ± 10.79	20.23 – 1014.21 210.68 ± 9.89	2.09 – 11.26 5.91 ± 2.94
Phang-Nga Province (97)	251.50 – 15740.34[a] 2879.87 ± 225.50[b]	15.21 – 791.42 165.44 ± 8.44	18.14 – 854.34 160.11 ± 7.92	1.48 – 16.91 5.76 ± 2.49
Krabi Province (114)	120.54 – 6586.33[a] 802.58 ± 115.25[b]	3.05 – 930.59 65.75 ± 4.95	4.83 – 194.07 40.69 ± 3.96	0.95 – 11.05 2.96 ± 1.59
Ranges Average Values	120.54 – 18803.28[a] 2859.63 ± 209.83[b]	3.05 – 930.59 157.10 ± 8.06	4.83 – 1014.21 137.16 ±7.26	0.95 – 16.91 4.88 ± 2.34

[a] Ranges
[b] Average values

From the result in table 1, the ranges and average values of specific activities of all radionuclides in surface soil samples collected from Phuket province were the highest values compared to Phang-Nga and Krabi provinces.

3.2 Comparison of the specific activities to research data in the South of Thailand and worldwide means

The average values of specific activities of natural and anthropogenic radionuclides in 314 surface soil samples collected from the inquired area were compared to data from other research reports, the national and global radioactivity measurements as shown in table 2. The national radioactivity data for Southern Thailand were previously reported by Office of Atoms for Peace, Thailand [23]. The global data were reported by UNSCEAR [3].

Table 2. Comparison of the average value of specific activities of ^{40}K, ^{226}Ra, ^{232}Th and ^{137}Cs in 314 surface soil samples collected from Phuket, Phang-Nga and Krabi provinces (Thailand) with research data in the South of Thailand and worldwide mean.

Data Sources	Specific Activities (Bq/kg)			
	^{40}K	^{226}Ra	^{232}Th	^{137}Cs
Songkhla [17]	3562.14 ± 223.56	107.28 ± 6.74	51.53 ± 5.02	3.05 ± 1.58
Nakhon Si Thammarat [18]	4313.18 ± 148.16	108.55 ± 31.19	73.95 ± 4.15	4.84 ± 2.59
Phatthalung [19]	3573.35 ± 203.89	135.89 ± 6.71	76.34 ± 5.32	4.00 ± 1.92
Chumphon [20]	2135.69 ± 168.87	57.32 ± 5.19	56.98 ± 4.68	2.30 ± 1.38
Surat Thani [21]	2119.10 ± 171.72	75.72 ± 5.75	47.39 ± 4.76	3.85 ± 1.67
Yala [22]	3607.70 ± 235.48	128.94 ± 7.42	85.93 ± 6.13	4.53 ± 2.24
Phuket	4896.45 ± 288.74	240.11 ± 10.79	210.68 ± 9.89	5.91 ± 2.94
Phang-Nga	2879.87 ± 225.50	165.44 ± 8.44	160.11 ± 7.92	5.76 ± 2.49
Krabi	802.58 ± 115.25	65.75 ± 4.95	40.69 ± 3.96	2.96 ± 1.59
South of Thailand [23]	511.04 ± 7.04	171.55 ± 3.13	211.19 ± 1.98	1.13 ± 0.49
Worldwide Mean [3]	400	35	30	-

From the result in table 2, the average specific activity of ^{40}K in surface soil samples collected from Phuket province was the highest but the average from Krabi province was the lowest compared to Songkhla, Nakhon Si Thammarat, Phatthalung, Chumphon, Surat Thani and Yala provinces. The average specific activities of ^{226}Ra, ^{232}Th and ^{137}Cs in surface soil samples collected from Phuket and

Phang-Nga provinces were higher than in Songkhla, Nakhon Si Thammarat, Phatthalung, Chumphon, Surat Thani and Yala provinces. Furthermore, the average specific activities of ^{226}Ra, ^{232}Th and ^{137}Cs in surface soil samples collected from Krabi province were comparable with all data of Songkhla, Nakhon Si Thammarat, Phatthalung, Chumphon, Surat Thani and Yala provinces. We could also see that the average specific activities of ^{40}K in Phuket, Phang-Nga and Krabi provinces were always higher than the research data from the south of Thailand and worldwide mean. Moreover, the average specific activities of ^{226}Ra and ^{232}Th were lower than the research data from the south of Thailand (except only the average specific activity of ^{226}Ra in Phuket province which was higher) but higher than the worldwide means. The average specific activity of ^{137}Cs in surface soil samples collected from all three investigated provinces were higher than the research data from the south of Thailand, previously reported.

3.3 Radiological hazard evaluation and comparison

By using the specific activity values of ^{40}K, ^{226}Ra and ^{232}Th as shown in table 1 and table 2, the gamma absorbed dose rate in air (D), the radium equivalent activity (Ra_{eq}), the external hazard index (H_{ex}) and the annual effective dose rate (AED_{out}) could be evaluated by using equations (1) to (4), compared to UNSCEAR data and presented in table 3.

Table 3. Gamma absorbed dose rate in air, radium equivalent activity, external hazard index, and annual effective dose rate of 314 surface soil samples collected from Phuket, Phang-Nga and Krabi provinces (Thailand).

Locations	**D (nGy/h)**	**Ra_{eq} (Bq/kg)**	**H_{ex}**	**AED_{out} (mSv/y)**
Phuket	52.12 – 1793.31	113.13 – 3728.99	0.31 – 10.07	0.06 – 2.20
(103)	446.92 ± 23.19	922.99 ± 47.37	2.49 ± 0.13	0.55 ± 0.03
Phang-Nga	35.84 – 1319.88	75.07 – 2709.96	0.20 – 7.32	0.04 – 1.62
(97)	295.24 ± 18.16	616.14 ± 37.13	1.66 ± 0.10	0.36 ± 0.02
Krabi	16.81 – 575.97	33.83 – 1249.45	0.09 – 3.38	0.02 – 0.71
(114)	89.16 ± 9.53	186.35 ± 19.50	0.50 ± 0.05	0.11 ± 0.01
Ranges	16.81 – 1793.31	33.83 – 3728.99	0.09 – 10.07	0.02 – 2.20
Average values	277.11 ± 16.96	575.16 ± 34.67	1.55 ± 0.09	0.34 ± 0.02
UNSCEAR [1-3]	55	370	1	0.48

From table 3, we saw that the average values of all four radiological hazard indices in Phuket province were higher than the UNSCEAR recommended values. In Phang-Nga province, the gamma absorbed dose rates in air (D), the radium equivalent activity (Ra_{eq}) and the external hazard index (H_{ex}) were higher than those recommended values but the annual effective dose rate (AED_{out}) in this area was lower. There was only one radiological index which was the absorbed dose rates in air in Krabi province that was higher than the referenced value. Furthermore, the average values of the gamma absorbed dose rates in air, the radium equivalent activity and the external hazard index in all three provinces were 5.04, 1.55 and 1.55 times higher than the UNSCEAR recommended values. Moreover, the calculated annual effective dose rate with average value 0.34 mSv/y was lower than the worldwide average value (0.48 mSv/y).

3.4 Contour maps of radiological hazard indices

Furthermore, the contour maps of all four radiological hazard indices which were the gamma absorbed dose rates in air (D), the radium equivalent activity (Ra_{eq}), the external hazard index (H_{ex}) and the annual effective dose rate (AED_{out}) from 314 surface soil samples collected from Phuket, Phang-Nga and Krabi provinces (Thailand) were constructed by using a computer program and shown in figure 1 - 4.

Journal of Physics: Conference Series **611** (2015) 012021 doi:10.1088/1742-6596/611/1/012021

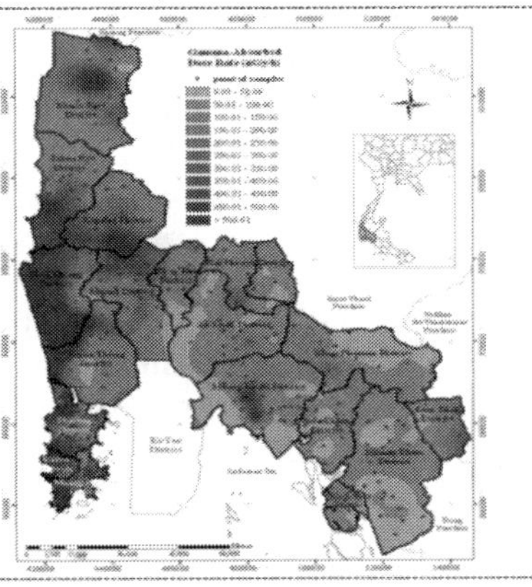

Figure 1. The contour map of the absorbed dose rates in air (D) from 314 surface soil samples collected from Phuket, Phang-Nga and Krabi provinces (Thailand).

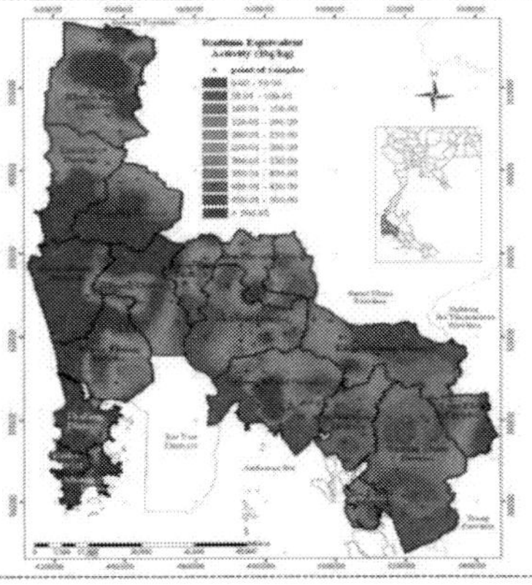

Figure 2. The contour map of the radium equivalent activity (Ra_{eq}) from 314 surface soil samples collected from Phuket, Phang-Nga and Krabi provinces (Thailand).

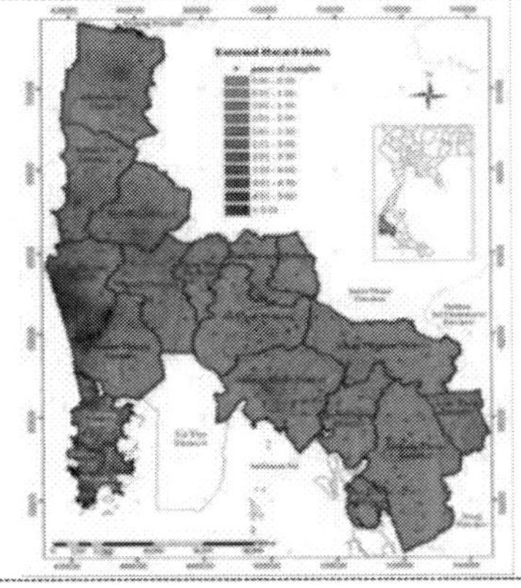

Figure 3. The contour map of the external hazard index (H_{ex}) from 314 surface soil samples collected from Phuket, Phang-Nga and Krabi provinces (Thailand).

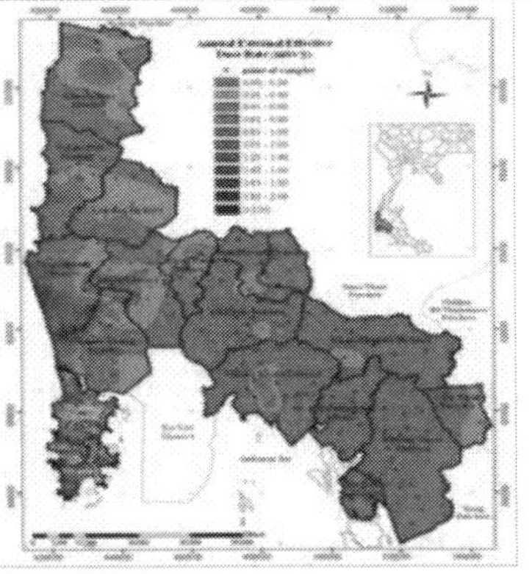

Figure 4. The contour map of the annual effective dose rate (AED_{out}) from 314 surface soil samples collected from Phuket, Phang-Nga and Krabi provinces (Thailand).

4. Conclusion

Specific activities of natural (^{40}K, ^{226}Ra, ^{232}Th) and anthropogenic (^{137}Cs) radionuclides in 314 surface soil samples collected from three provinces (Phuket, Phang-Nga and Krabi) along the Andaman sea coast in southern region of Thailand ranged from 120.54 – 18803.28, 3.05 – 930.59, 4.83 – 1014.21 and 0.95 – 16.91 Bq/kg with the average values of 2859.63 ± 209.83, 157.10 ± 8.06, 137.16 ±7.26 and 4.88 ± 2.34 Bq/kg, respectively. Furthermore, four radiological hazard indices which were the gamma absorbed dose rate in air (D), the radium equivalent activity (Ra_{eq}), the external hazard index (H_{ex}), and the annual effective dose rate (AED_{out}) in the monitoring areas were evaluated to be 277.11 ± 16.96 nGy/h, 575.16 ± 34.67 Bq/kg, 1.55 ± 0.09 and 0.34 ± 0.02 mSv/y, respectively. It was found that the average value of the gamma absorbed dose rate in air obtained in this study (277.11 nGy/h) was 5.04 times higher than the global average (55 nGy/h). The average value of the radium equivalent activity (575.16 Bq/kg) was 1.55 times higher than 370 Bq/kg, which was acceptable for safe use [1-3]. The average values of the external hazard index obtained in this study was found to be 1.55 which was higher than unity. We could see that three out of four radiological hazard indices were greater than the values recommended by UNSCEAR [1-3], especially in Phuket and Phang-Nga provinces. This indicated a possibility that the radiation hazard might be a concern for the population living in the investigated area. According to the calculations, the average annual effective dose rate in the area was 0.34 mSv/y, which was still lower than the worldwide average value (0.48 mSv/y). For this reason, the radiation hazard was considered insignificant for the population living in the studied area. Nonetheless, from the contour maps, we could see that populations living in the northern and western parts, especially in Phang-Nga and Phuket provinces, could be exposed to a higher background radioactivity than those in the central and eastern regions of the investigated area.

Acknowledgments
The investigation was fully supported by Research and Development Institute, Thaksin University (RDITSU). The authors would like to thank Department of Physics, Faculty of Science, Thaksin University, for allowing us to use all instruments and radioactive standard sources at Nuclear and Material Physics Laboratory. The authors wish to thank Assoc. Prof. Dr. Tripop Bhongsuwan, Department of Physics, Faculty of Science, Prince of Songkhla University (PSU), Hat Yai Campus, for his generous permission to let us use the standard reference materials (KCl, IAEA/RGU-1, IAEA/RGTh-1 and IAEA/SL-2) for calculations and analyses in this study. The authors are particularly indebted to some of undergraduate and graduate students for their works and patience in sample collections, measurements and analyses of parts of the data.

References
[1] United Nations Scientific Committee on the Effects of Atomic Radiation 1988 Source, Effects and Risks of Ionizing Radiation *UNSCEAR 1988 Report to the General Assembly, with Scientific Annexes* (New York :United Nations)
[2] United Nations Scientific Committee on the Effects of Atomic Radiation 1993 Sources and Effects of Ionizing Radiation *UNSCEAR 1993 Report to the General Assembly, with Scientific Annexes* (New York :United Nations)
[3] United Nations Scientific Committee on the Effects of Atomic Radiation 2000 Sources and Effects of Ionizing Radiation *UNSCEAR 2000 Report to the General Assembly, with Scientific Annexes* (New York :United Nations)
[4] Alencar A S and Freitas A C 2005 *Radiat. Meas.* **40** 76-83
[5] Veiga R *et al.* 2006 *Radiat. Meas.* **41** 189-96
[6] Mohanty A K, Sengupta D, Das S K, Vijayan V and Saha S K 2004 *Radiat. Meas.* **38** 153-65
[7] Singh S, Rani A and Mahajan R K 2005 *Radiat. Meas.* **39** 431-39
[8] Yang Y X, Wu X M, Jiang Z Y, Wang W X, Lu J G, Lin J, Wang L M and Hsia Y F 2005 *Appl. Radiat. Isot.* **63** 255-59
[9] Kurnaz A, Kucukomeroglu B, Keserb R, Okumusoglu N T, Korkmazb F, Karahanc G and Cevika_ U 2007 *Appl. Radiat. Isot.* **65** 1281-89
[10] Vassas C, Pourcelot L, Vella C, Carpena J, Pupin J P, Bouisset P and Guillot L 2006 *J. Environ. Radioac.* **91** 146-59
[11] Kessaratikoon P, Boonkrongcheep R, Benjakul S and Youngchauy U 2013 *J. Radioanal. Nucl. Chem.* **297** 247-55
[12] International Atomic Energy Agency (IAEA) 2003 Extent of Environmental Contamination by Naturally Occurring Radioactive Material (NORM) and Technological Options for Mitigation *Technical Report Series No. 419* (Vienna: IAEA)
[13] Malain D, Regan P H, Bradley D A, Matthews M, Al-Sulaiti H A and Santawamaitre T 2012 *Appl. Radiat. Isot.* **70** 1467-74
[14] Kessaratikoon P, Choosiri N, Kongratana P, Boonkrongcheep R and Youngchauy U 2013 Frequency Distribution and Radiological Hazard Assessment of Specific Activities of

Natural and Anthropogenic Radionuclides in Beach Sand Samples from Patong and Naiyang Beaches in Phuket Province, Thailand *Proc. The 39th Congress on Science and Technology of Thailand "Innovative Sciences for a Better life" (STT39)* October 21-23, 2013 Bangkok Thailand pp 55-62

[15] Wilkinson C, Souter D and Goldberg J 2006 *Status of Coral Reefs in Tsunami Affected Countries: 2005* (Townsville: Australian Institute of Marine Science (AIMS))

[16] Beretka J and Mathew P J 1985 *Health. Phys.* **48** 87-95

[17] Ratanasumniang C 2009 *Measurement of Specific Activity of Natural and Anthropogenic Radionuclides in Soil Samples in Songkhla Province (Thailand)* (dissertation) (Songkhla:Thaksin University)

[18] Thongna T 2009 *Natural Radioactivity Assessment in Nakhon Si Thammarat Province* (dissertation) (Songkhla:Thaksin University)

[19] Daoh M 2010 *Measurement of Specific Activity of Natural (^{40}K, ^{226}Ra and ^{232}Th) and Anthropogenic Radionuclides (^{137}Cs) in Soil from Phathalung Province (Thailand) Using Gamma Ray Spectrometry* (dissertation) (Songkhla:Thaksin University)

[20] Chauymanee S 2012 *Assessment of Natural and Anthropogenic Radioactivity in Soil and Beach Sand Samples in Chumphon Province, Thailand* (dissertation) (Songkhla:Thaksin University)

[21] Kessaratikoon P, Boonkrongcheep R, Benjakul S and Youngchauy U 2013 Assessment and Radioactive Contour Maps of Specific Activities of Natural (^{40}K, ^{226}Ra and ^{232}Th) and Anthropogenic (^{137}Cs) Radionuclides in Surface Soil Samples from Surat Thani Province, Thailand *The 8th Annual Conf. of the Thai Physics Society "Thai Physics Society on the Road to ASEAN Community"(SPC 2013)* March 11-13, 2013 Chiang Mai Thailand

[22] Jahlee M 2012 *Quantitative Measurement of Natural and Anthropogenic Radionuclide (^{137}Cs) in Soil Samples from Yala Province, Thailand* (dissertation) (Songkhla:Thaksin University)

[23] Office of Atoms for Peace (OAP) 1994 – 2002 *Office of Atoms for Peace Annual Reports* (Bangkok: Ministry of Science and Technology)

Tolerance and accumulation of cesium and strontium in saprothophic fungi

P Seeprasert, M Yoneda, Y Shimada and Y Matsui
Department of Environmental Engineering, Graduate School of Engineering, Kyoto University, Kyoto, 615-8540, Japan

E-mail: noon@risk.env.kyoto-u.ac.jp

Abstract. Soil contamination by nuclear accidents has led to a resurgence of interest in microbe-radionuclide interactions. Soil fungi accumulate radioactive elements from contaminated soil, and it has been hypothesized that this may alter the availability of radionuclides to plants and alter their movement in particular areas. This study intended to demonstrate how soil saprotrophic fungi accumulated Cs and Sr isotopes in both stable and radioactive solution forms. An experiment to determine the tolerance of fungal growth under extremely inhibitory concentrations of Cs and Sr stable isotopes was conducted. The results showed that fungal cells were more sensitive to Cs than to Sr. Accumulations of Cs and Sr were examined through sorption mechanisms using resting cells in the solutions under various conditions. The sorption capacity was indirectly determined by analysing the elements, which decreased in the solution. The equilibrium data were fitted with sorption isotherms to show the best fit with the Langmuir isotherm for both elements, assuming that the sorption sites form a surface monolayer. In addition, pH was examined to investigate its effect on the sorption capacity of Cs and Sr.

1. Introduction

Microbiota (which include fungi, bacteria, actinomycetes, protozoa, microfauna, and algae) are one of the most important components that play a role in radionuclide bioavailability and cycling. Some studies focusing on the interaction of microorganisms with cesium (Cs) and strontium (Sr) removal were reported several years ago. Unicellular organisms (e.g., *Euglena* and *Chlorella*) were the first species of microorganisms to be investigated for their ability to accumulate Cs [1]. Research on freshwater algae, Cyanobacteria, and bacteria (e.g., *Synechocystis* spp., *Rhodococcus* spp. and *Streptomyces* spp.) isolated from the natural environment were also carried out to investigate their accumulation capacity. Moreover, yeasts (e.g., *Saccharomyces cerevisiae*, *Rhodotorula rubra* and *Candida albicans*) [2-5] showed greater accumulation in the logarithmic phase in comparison to the stationary phase of kinetic growth [6]. Fungi are often the major component of soil microbiota that play a critical role in soil ecological function. Fungal biomass can function as a bio-accumulator of soluble forms of inorganic pollutants, including radionuclides. Higher fungi (e.g., lichens and mushrooms) have been investigated as indicators of Cs contamination in the environment and have presented accumulation in their fruitbodies or hyphae. The mechanism has been explained as Cs being immediately and non-specifically adsorbed onto the negatively charged cell surfaces and then taken up into the cytoplasm [5]. Soil fungi have the potential to immobilize radioactive Cs, which may alter availability to plants and radionuclide movement in surface soil layers and reduce migration through the soil [7]. Mycorrhizal fungi are an important group of soil fungi that may have similar effects, thus influencing the uptake of inorganic pollutants including radionuclides [8]. Saprotrophic fungi are a group of fungi that are free-living in the soil and have an ability to decompose dead organic residues. *Trichoderma* is a genus of saprotrophic fungi that produces various kinds of secondary metabolites. Radionuclides may attach to these fungal cells extracellularly or be actively taken up intracellularly.

Radioactive isotopes, especially Cs and Sr, are present in the soil in their cationic forms, and there are no general differences between their radioactive and stable isotopes with respect to their chemical elements or their behavior in the environment. Cs is a rare type of alkali metal. Its stable isotope is ^{133}Cs, while the fission products include four main Cs isotopes. It exists in the environment as a monovalent cation (Cs^+). Based on the long half-life of radioactive Cs (2 years for ^{134}Cs and 30 years ^{137}Cs), they have become the dominant radionuclides contaminating the environment. In nature, Sr occurs only in the Sr^{2+} oxidation state. Radioactive isotopes can be produced by fission reactions. In particular, ^{90}Sr is a problematic environmental contaminant because of its long half-life (29 years), high transferability, and high fission yield. Currently, soil contamination caused by radioactive elements resulting from nuclear accidents is leading to a resurgence of interest in microbe-radionuclide interactions. However, information on the role of the fungi as a storage organism has been scarce. This study intends to clearly understand how a soil saprotrophic fungi *Trichoderma* spp. accumulates Cs and Sr as both stable and radioactive isotopes.

2. Materials and methods

2.1. Soil saprotrophic fungi

Trichoderma spp. is the one of the soil saprotrophic fungi representatives that was isolated from a rhizosphere soil that had aboundant microbial activity. The sampling sites were located in a conifer forest at Takizawa Research Forest of Iwate University, Japan (39°46'32N, 144°9'25E). *Trichoderma* spp. was isolated by making a dilution series of soil solution from which 1 mL of each dilution was spread onto a potato dextrose agar (PDA) plate. The plates were incubated for 7 days at 25 – 30 °C. Colonies were selected and re-cultured on fresh PDA plates, and DNA extraction and PCR amplification were conducted. Entire internal transcribed spacer (ITS) regions that were potentially useful in rapid and accurate diagnosis of fungal isolation were sequenced [9-10]. ITS1 (5′-TCC GTA GGT GAA CCT TGC GG-3′) and ITS4 (5′-TCC TCC GCT TAT TGA TAT GC-3′) were used as forward and reverse primers. Sequences of DNA fragments were identified using Basic Local Alignment Search Tool (BLAST; National Center for Biotechnology Information, USA).

2.2. Cell preparation

2.2.1. Active cell preparation

Trichoderma spp. was cultured in potato dextrose broth (PDB) with shaking at 110 rpm at 25°C until it reached the late exponential growth phase.

2.2.2. Resting cell preparation

Trichoderma spp. cells at the late exponential growth phase derived from the active cells were washed three times with sterile purified water to stop cell growth.

2.3. Growth kinetics and tolerance

To investigate the effect of Cs and Sr on the growth kinetics of *Trichoderma* spp., Cs^+ and Sr^{2+} were each prepared in 50 mL of fresh PDB with concentrations of 5, 10, 25, 50, and 100 mM. 1 mL of the active cell culture was then inoculated into the PDB solution. After that, they were incubated with shaking at 110 rpm at 25°C until the experiment was completed (168 h). Approximately 5 mL of the suspension was sampled at each time interval with a sterilized syringe. All the experiments were conducted in duplicate. The cell concentration was determined indirectly by measuring the optical density at 600 nm (OD_{600}), using a spectrophotometer (PhotoLab 6100 VIS, Wissenschaftlich-Technische Werkstätten, Germany).

2.4. Accumulation

To investigate the effect of Cs and Sr in both stable and radioactive isotopic forms on sorption, the resting cells were resuspended in 25 mL of sterile purified water. The experimental conditions are shown in table 1. The cells were incubated with shaking at 110 rpm at 25°C for 3 h. They were then filtrated through dried filter paper (Whatman no.1). The remaining Cs and Sr in the aliquot and dry weight of cell pellets were measured.

Table 1. Experimental Conditions.

No.	Elements	Concentration	pH
1	^{133}Cs	1, 10, 20, 25, and 50 ppb	4[a]
2	^{88}Sr	1, 10, 20, 25, and 50 ppb	4[a]
3	^{133}Cs/ ^{88}Sr	2.5,5,10, 25, and 50 ppb	4[a]
4	^{134}Cs/ ^{85}Sr	^{85}Sr; 20, 40, 80, 200, and 400 Bq[b] ^{134}Cs; 100, 200, 400, 1000, and 2000 Bq[b]	4[a]
5	^{133}Cs/ ^{88}Sr	10 ppb	3, 5, 7 and 9
6	^{134}Cs/ ^{85}Sr	^{85}Sr; 80 Bq[a]; and ^{134}Cs; 400 Bq	3, 5, 7 and 9

[a] pH was not adjusted. The initial pH was approximately 4.
[b] Activity was measured from 25 mL of the samples.

2.5. Radioactive materials

^{134}Cs has a radioactive half-life of 2.06 years. It is a β-emitter, and it also emits γ-ray with an energy level of 604.7 keV (97.5%). ^{85}Sr has a radioactive half-life of 64.85 days; it decays by electron capture, emitting γ-ray with an energy level of 514.0 keV (98.3%). The ^{134}Cs and ^{85}Sr stock solutions contained an activity concentration of approximately 2.5 kBq mL^{-1}, the pH was adjusted to pH 7. The radiotracer solution was added to each experiment at 50, 100, 200, 500, and 1000 μL to achieve an activity of approximately 20, 40, 80, 200, and 400 Bq for ^{85}Sr and 100, 200, 400, 1000, and 200 for ^{134}Cs.

2.6. Measurement

^{133}Cs and ^{88}Sr are stable isotopes. They were measured by inductively coupled plasma mass spectrometry (ICP-MS; XSeries 2, Thermo Scientific, USA). Sample preparation followed the standard methods ISO17294-2 and EPA 6020a [11]. ^{134}Cs and ^{85}Sr are radioactive isotopes. The γ-ray spectrum was measured for 30 to 60 min using a gamma spectrometer with a HPGe detector (GMX series, EG&G, Ortec, USA) and a channel analyser.

3. Results and Discussion

3.1. Soil saprotrophic fungi

Colonies on the PDA plates that represented those with major growth and different morphology were selected using the hyphal tip isolation technique. This technique isolates single colonies that represent a strain of soil fungus. These strains were then re-cultured on fresh PDA plates. The ITS regions from the amplified DNA samples were compared against the nucleotide collection database in GenBank by using BLAST. The result showed greater than 95% similarity to *Trichoderma* spp. (data not shown). Therefore, it was concluded that the representative fungus used in this study belonged to *Trichoderma* spp., which is a genus of saprotrophic fungi that produces various kinds of secondary metabolites [12].

3.2. Growth kinetics and tolerance

The effect of Cs and Sr on the kinetic growth of *Trichoderma* spp. was investigated under extreme concentrations (5, 10, 25, 50 and 100 mM) and compared with a control condition under which no elements were added (0 mM). Growth curves were plotted for the logarithm of the number of cells (N), which were indirectly measured using optical density in the unit of $\ln(N/N_0)$ against time in the unit of hours. The results are presented in figure 1.

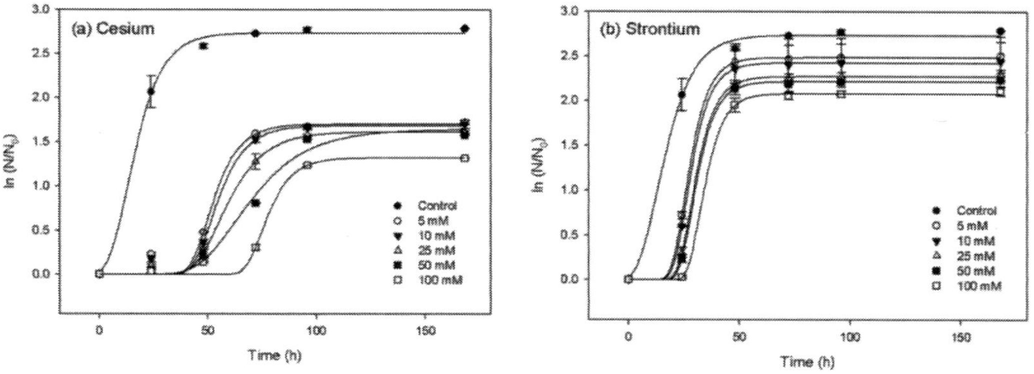

Figure 1. *Trichoderma* spp. cell growth kinetics under various concentrations of (a) cesium and (b) strontium fitted with Gemperts model.

Trichoderma spp. yielded a higher number of cells at lower Cs and Sr concentrations and the number of cells decreased as the concentration increased. *Trichoderma* spp. growth mostly exhibited an exponential phase up to 48 h. However, when the Cs concentration was increased to 50 mM, the exponential phase extended to 72 h before reaching the stationary phase. Cell growth at the stationary phase decreased maximally to 52% for Cs and 24% for Sr compared with the control at 100 mM for both elements. These results showed that *Trichoderma* spp. was more sensitive to Cs than Sr. Similar results were also observed by various kind of microorganisms in previous studies. For instance, the filamentous actinomycetes also could not grow in the presence of Cs concentrations higher than 25 mM, which is more sensitive to Cs than planktonic bacteria [13]. Cyanobacterium demonstrated considerable toxicity, as shown by changes in various parameters such as dry weight, protein and carbohydrate content as well as acid phosphatase activity, which also decreased with increased Sr concentrations [14]. In addition, Sr inhibited the growth of *Micrococcus* spp. and *Cupriavidus metallidurans* when the concentration reached 300 mM [15].

The phases of the growth curve can be described by three parameters. Maximam specific growth rate (μ_m) is defined as the specific growth rate in a certain period of time, resulting in values for lag time (λ) and the asymptote (A) which is defined as the specific growth rate at time ∞; $A = \ln(N_\infty/N_0)$. Those parameters were obtained for each concentration. Gemperts model is a sigmoidal mathematical model that has been used to estimate these parameters. The Gemperts function (equation 1) was rewritten by substituting the mathematical parameters with biological parameters: μ_m, λ, and A (equation 2) [16].

$$y = a\,.exp[-exp(b - cx)] \quad (1)$$

$$y = A\,.exp\left\{-exp\left[\frac{\mu_m . e}{A}(\lambda - t) + 1\right]\right\} \quad (2)$$

The data-fitting equations obtained using Gemperts model were statistically evaluated for the residuals and the goodness of fit (table 2). Serial correlations for the residuals were examined using the Durbin-Watson (DW) statistics, which was used to test residuals for their independence to each other. The DW values were obtained by fitting the models to the growth data. If the residuals do not correlate, the

DW values will be in the range 1.5 to 2.5, which is obtained when the serial correlation is small and the residuals are randomly distributed around the zero line when plotted against time. There were fewer curves for DW values, indicating that the errors were positively autocorrelated which was shown by DW values that were lower than the lower bound or negatively autocorrelated that were indicated by DW values higher than the upper bound. There were more DW values, indicating that the errors were not autocorrelated. The R^2 values for most curves were greater than 0.98, indicating that this statistics can be used as a basis for model evaluation.

Table 2. Statistics and parameter values obtained when the models were fitted to the growth data.

	Cesium (Cs)					Strontium (Sr)				
	DW	R^2	A	λ	μ_m	DW	R^2	A	λ	μ_m
Control	1.77	0.997	2.74	0.05	8.90	1.77	0.997	2.74	0.05	8.90
5 mM	2.01	0.983	1.70	0.62	5.11	1.50	1.000	2.48	0.55	5.44
10 mM	2.02	0.990	1.68	0.55	5.44	1.51	1.000	2.42	0.55	5.36
25 mM	1.99	0.996	1.61	0.33	6.85	1.50	1.000	2.27	0.61	5.06
50 mM	3.16*	0.986	1.64	0.14	10.72	1.53	1.000	2.21	0.65	4.82
100 mM	1.50	0.989	1.32	1.14	3.72	1.50	1.000	2.08	0.81	4.40

*DW value was higher than the upper bound

3.3. Accumulation

Data were gathered using several sorption isotherms. The best fitting model was indicated by high values of the correlation coefficient of determination (figure 2).

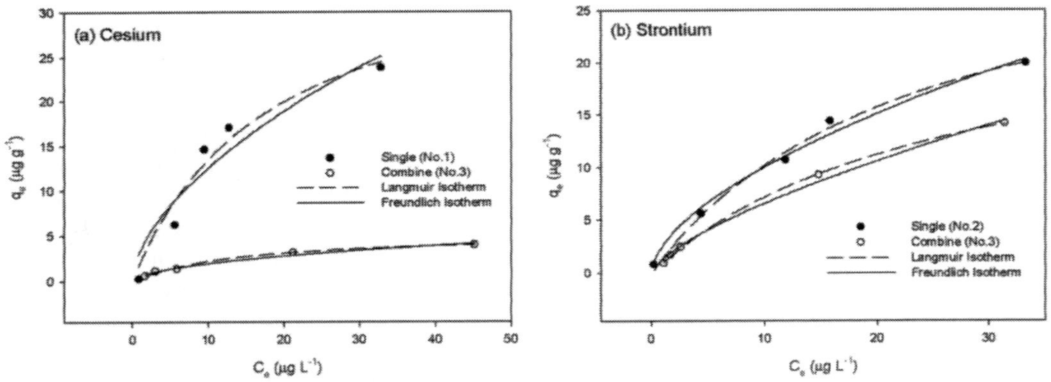

Figure 2. Data fitting with the Langmuir and Freundlich isotherm for (a) cesium and (b) strontium.

The results indicated that Cs and Sr were adsorbed under all conditions in agreement with the Langmuir isotherm ($R^2 = 0.96 - 0.99$), which was theoretically derived supposing no interaction between the elements and the cell surface and that adsorption occurred on fixed homogenous adsorption sites forming a monolayer surface coverage (equation 3).

$$q_e = \frac{q_m K_L C_e}{1 + C_e} \quad (3)$$

where q_e is the amount of elements adsorbed (µg g^{-1}), C_e is the amount of element concentration in the solution (µg L^{-1}), q_m is the maximum amount adsorbed (µg g^{-1}) and K_L is an equilibrium constant representing the affinity between adsorbate and adsorbent.

However, for Sr, there was also strong agreement with the Freundlich isotherm for both single and combined conditions ($R^2 = 0.99$). The Freundlich isotherm is an empirical equation based on the adsorption on the heterogeneous surface (equation 4).

$$q_e = K_F C_e^{1/n} \tag{4}$$

where q_e is the amount of elements adsorbed (μg g^{-1}), C_e is the amount of element concentration in the solution (μg L^{-1}), K_f is a constant that is related to sorption capacity and n is an empirical parameter that varies with the degree of heterogeneity.

The results show the same trend for both elements. For the stable isotopes (^{133}Cs and ^{88}Sr), the sorption capacity increased when the initial concentration of elements was increased and showed a specific trend pertaining to approaching a plateau. Lan *et al.* (2014) reported a similar result that assumed that the biosorption of Cs onto *Rhodosporidium fluviale* (initial Cs concentration varied from 0.01 to 2.0 mg/L) was a monolayer adsorption [17]. Furthermore, ^{88}Sr was also found to highly agree with the Freundlich isotherm, which assumes that heterogeneity occurred and that the adsorption capacity of ^{88}Sr was higher than ^{133}Cs. The effect of the presence of competitive ions occurred under combined conditions whereby each element was less adsorbed than under the single condition. This suggests that Sr^{2+} and Cs^{+} ions were adsorbed onto similar sites. A similar result was observed by Ofomaja *et al.* (2013), who studied the effect of Cs sorption in the presence of Na^{+} and Ca^{2+} [18], which have an ionic radious of 0.99 Å, and was found to be similar to that for Sr^{2+} 1.12 Å. To determine the mechanistic parameters associated with stable isotope (^{133}Cs and ^{88}Sr) sorption, results were obtained from sorption experiments for both single and combined conditions. The isotherm parameters are shown in table 3.

Table 3. Isotherm parameters for cesium and strontium sorption onto *Trichoderma spp.*

Element	Langmuir constants			Freundlich constants		
	q_m	K_L	R^2	K_f	1/n	R^2
^{133}Cs (single)	37.72	17.87	0.9604	3.32	0.58	0.9174
^{133}Cs (combined)	5.10	12.96	0.9914	0.63	0.49	0.9758
^{88}Sr (single)	34.01	23.38	0.9940	2.58	0.59	0.9901
^{88}Sr (combined)	25.85	26.15	0.9995	1.34	0.69	0.9925

Figure 3 demonstrates the effect of the initial activity of Cs and Sr, which was derived from the experiments in combined conditions between ^{134}Cs and ^{85}Sr to investigate the sorption capacity (Q), which was calculated based on equation 5.

$$\mathrm{Q}\ (Bq/mg) = \frac{(A_i - A_f)}{M} \tag{5}$$

where A_i and A_f are the initial and final activities in the solution (A) and M is the mass of *Trichoderma* spp. (mg).

Both Cs and Sr showed the same trend (figure 3): the sorption capacity of Sr was higher than Cs at the same initial activity, which corresponded with the stable isotopes mentioned above. However the ability to accumulate Cs and Sr was dependent on the different species of microorganisms, according to reports from previous studies. For example *Pseudomonas fluorescens* and *Rhodococcus spp.* isolated from the soil were grown in the presence of Cs, whereas no detectable Cs accumulation was observed for *Pseudomonas fluorescens*[4].

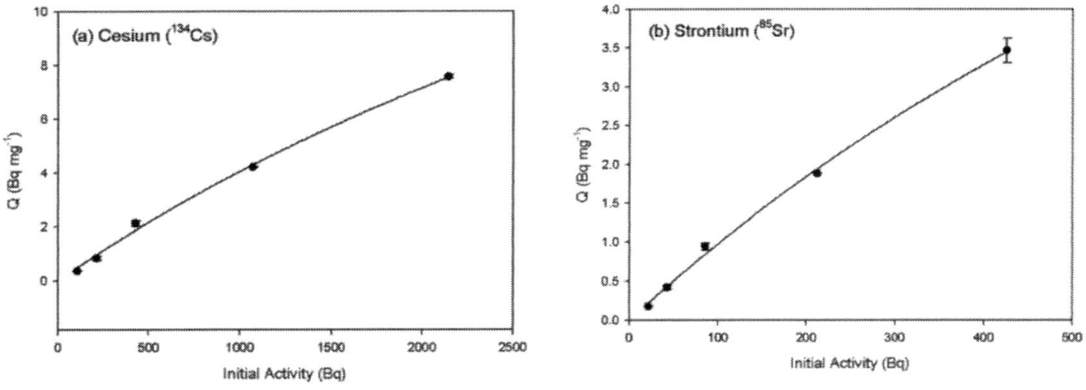

Figure 3. Effect of the initial activity of (a) cesium (^{134}Cs) and (b) strontium (^{85}Sr) on the sorption capacity.

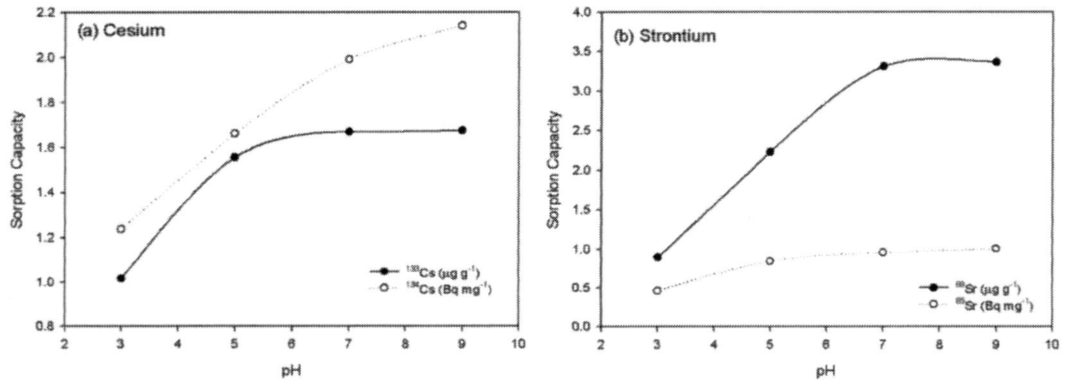

Figure 4. Effect of pH on the sorption of (a) cesium and (b) strontium.

External physico-chemical factors, such as pH, can be a major factor that influences sorption of Cs and Sr. Figure 4 shows the effect of pH on the sorption of Cs and Sr onto *Trichoderma* spp. where pH was plotted against the sorption capacity in the unit of $\mu g\ g^{-1}$ for stable isotopes and $Bq\ mg^{-1}$ for radioactive isotopes. The results demonstrated that pH had a greater effect on the sorption of both elements with the same trend for stable and radioactive isotopes. When the pH was increased from 3 to either 7 or 9, the sorption capacity onto *Trichoderma* spp. greatly increased for both Cs and Sr. Meanwhile, at lower pH (pH 3) the sorption capacity decreased to its lowest capacity, probably because H^+ also interacts with sites mediating the monovalent and divalent cation transport system. Then, at the lower pH, binding of H^+ inhibits the sorption of other cations such as Cs^+ and Sr^{2+} [19-20]. Thus, during the experiment, the sorption ratio of Cs and Sr onto *Trichoderma* spp. increased when pH was increased from 3 to 9. In addition, other species, for example the marine bacteria, *Vibrio alginolyticus* and yeast (*Saccharomyces cerevisiae*), have also shown the same mechanism [6,19, 21].

4. Conclusion

Trichoderma spp. was isolated from the soil. Experiments were conducted with two main approaches: growth kinetics and accumulation. An experiment to investigate the growth kinetics was conducted under extreme concentrations of stable Cs and Sr isotopes. The results showed that *Trichoderma* spp. was more sensitive to Cs than to Sr. For Cs, *Trichoderma* spp. grew at lower concentrations (5 to 25 mM) and an inhibition of growth appeared at 50 mM, which became more significant when the concentration increased to 100 mM. The growth rate decreased evenly when the concentration of Sr increased. The accumulation of Cs and Sr were conducted under various conditions using both stable and radioactive isotopes. The results showed the same trend for both elements and both isotopes; the sorption capacity increased when the initial concentration and initial activity increased. Then, the effect of pH, showed a decrease in sorption capacity at lower pH because H^+ also interacts with sites mediating the monovalent and divalent cation transport system. The radioactive accumulation in fungi was found to be the most difficult to predict. This work did not intend to provide a comprehensive result of biological processes, but rather to study the possibility of soil saprotrophic fungi to accumulate the radioactive elements. The experimental results indicated that *Trichoderma* spp. effectively contributed substantially to the long-term retention of Cs and Sr in an organic layer which is essential to reduce the migration in soil layers by directly bind fungal cells extracellularly or actively taken up intracellularly which should be completed by future studies.

Acknowledgement
This work was partially supported by the Human Security Engineering Education Program, Graduate School of Engineering, Kyoto University, Japan. The authors would like to thank the reviewers for their relevant and helpful comments. Thanks are extended to the staff at Kyoto University Research Reactor Institute (KURRI), who provided us with valuable advice throughout the study.

References
[1] Louis G W 1960 *Limnol. Oceaogr.* **5** 301–11
[2] Harvey R S and Patrick R 1967 *Biotechnol. Bioeng.* **9** 449–56
[3] Avery V S, Codd G A and Gadd G M 1991 *J. General. Microbiol.* **137** 405–13
[4] Tomioka N, Uchiyama H and Yagi O 1994 *Appl. Environ. Microbiol.* **60** 2227–31
[5] Kuwahara C, Fukumoto A, Nishina M, Sugiyama H, Anzai Y and Kato F 2011 *J. Environ. Radioact.* **102** 138–44
[6] Perkins J and Gadd G M 1993 *Mycol. Res.* **97** 717–24
[7] Dighton J, Clint G M and Poskitt J 1991 *Mycol. Res.* **95** 1052–56
[8] Riessen T K and Brunner I 1996 *Environ. pollut.* **93** 1–8
[9] Turenne C, Sanche S, Hoban D, Karlowsky J and Kanabi A 1999 *J.Clin.Microbial* **37** 1846–51
[10] White T J, Bruns T D, Lee S B and Taylor J W 1990 Amplification and direct sequencing of fungal ribosomal RNA genes for phylogenetics *PCR Protocols: A Guide to Methods and Applications* (New York: Academic Press) pp 315-22
[11] van de Wiel H J 2003 *Determination of Elements by ICP-AES and ICP-MS* (Bilthoven: National Institute for Public Health and the Environment)
[12] Mukherjee P K, Nautiyal C S and Mukhopadhyay A N 2008 *Soil Biology* **15** 243–62
[13] Kuwahara C, Fukumoto A, Ohsone A, Furuya N, Shibata H, Sugiyama H and Kato F 2005 *Sci. Total Environ.* **345** 165–73
[14] Khalil Z 1997 *Turkis J. Mar.Sci.* **3** 65–81
[15] Salem I B, Sghaier H, Trifi H, Heni S, Khwaldia K, Saidi M and Landoulsi A 2012 *African J. Microbiol. Res.* **6** 851–8
[16] Zwietering M H, Jonggenburger I, Rombouts F M and Riet K 1990 *Appl. Environ. Microbiol.* **56** 1875–81
[17] Lan T, Feng Y, Liao J, Li X, Ding C, Zhang D, Yang J, Zeng J, Yang Y, Tang J and Liu N 2014 *J. Environ. Radioact.* **134** 6–13
[18] Ofomaja A E, Pholosi A and Naidoo E B 2013 *J. Taiwan Inst. Chem.* E **44** 943–51
[19] Avery V S 1995 *J. Ind. Microbiol.***14** 76–84
[20] Jones R P and Gadd G M 1990 *Enzyme Microb. Technol.* **12** 402–18
[21] Nakamura T, Tokuda H and Unemoto T 1982 *Biochim. Biophys. Acta* **692** 389–96

Assessment of radionuclides and heavy metals in marine sediments along the Upper Gulf of Thailand

S Khuntong[1,6], C Phaophang[2] and W Sudprasert[3]
[1] Faculty of Science at Si Racha, Kasetsart University Si Racha campus, 199 Sukhumvit Rd., Si Racha, Chonburi 20230 Thailand
[2] Radioisotope Center, Thailand Institute of Nuclear Technology, Vibhavadi Rangsit Rd., Chatuchak, Bangkok, 10900 Thailand
[3] Department of Applied Radiation and Isotopes, Faculty of Science, Kasetsart University, Ngam Wong Wan Rd., Chatuchak, Bangkok 10900 Thailand

E-mail: srcstk@ku.ac.th

Abstract. Due to the Fukushima Daiichi nuclear disaster in 2011 and the development of nuclear power plant in neighboring countries such as Vietnam in the near future, radionuclide assessment in marine sediment during 2010 - 2011 may be useful as background levels for radiation protection in Thailand. Marine sediments (10 samples) were collected approximately 1 km away from the coastline along Chonburi to Pattaya, Chonburi Province. The sediments were ground and sieved through 2-mm test sieve after air drying. Radionuclides were measured with a gamma spectrometer equipped with a well-calibrated HPGe detector. The samples were prepared in the same geometry as the reference material. The optimal counting time was 60,000 – 80,000 s for statistical evaluation and uncertainties. No contamination of ^{137}Cs as an artificial radionuclide was found. Naturally-occurring radionuclides including ^{238}U, ^{232}Th and ^{40}K were found. The mean specific activities of ^{238}U, ^{232}Th and ^{40}K were 44 ± 10, 59 ± 17 and 463 ± 94 Bq/kg in the rainy season (2010); 41 ± 6, 50 ± 9 and 484 ± 83 Bq/kg in the winter (2010), and 39 ± 6, 41 ± 7 and 472 ± 81 Bq/kg in the summer (2011), respectively. The mean specific activities were higher than the values in the UNSCEAR report of 35, 30 and 400 Bq/kg for ^{238}U, ^{232}Th and ^{40}K, respectively. From the measured specific activities, the absorbed dose rate, radium equivalent activity, external hazard index and annual external effective dose rate were calculated in order to assess the health risk. No radiation hazards related to the radioactivity in the sediment were expected. The accumulation of radionuclides varied with the particle size and the organic matter content in the sediment. The accumulation of heavy metals showed similar results to that of the radionuclides in the sediment.

1. Introduction

Radionuclides such as U, Th, Ra, and Am [1] constitute a category of heavy metals in marine environment as they become and cause serious risks to human health and ecosystem [2]. They are generated from two different sources: 1) natural sources which yield a constant activity level and 2) artificial sources which are caused by nuclear weapon tests, nuclear accidents, nuclear power plants, nuclear waste and so on causing a variation in the radioactivity level. Increasing utilization of radioactive

[6] To whom any correspondence should be addressed.

materials for industrial and medical purposes may cause contamination of radionuclides in the marine environment. Daughter decay products (Rn, Ra, Pu and Pb) from natural radioactive series Th-232, U-235 and U-238 and the others (K-40 and Ru-87) are mobilized and accumulated in the marine sediments. Components of the marine environment such as the appearance of the sediment, the organic content, the grain size and geochemical properties (properties of seawater), the benthos and the marine organisms could be affected by the accumulation of radionuclides in marine sediments.

Because of their high molecular weight, the chemical properties of several radionuclides may be similar to heavy metals. They are normally bound to particulate matters and finally settled down and incorporated into the marine sediments when they are released into the aquatic system [3]. Marine sediments are then considered as the secondary source originated from the sinking of the pollutants. Resuspension, bioturbation and diagenesis of sediments naturally can attenuate the pollutants in the sediments [4], [5]. Similar to heavy metals, radionuclides can accumulate in sediments over a long time [6]. The physical, chemical and biological properties of interstitial water and sediments bring about various forms of metals. The mechanisms of metal accumulation in the sediments lead to five geometric forms: (1) exchangeable, (2) bound to the carbonate phase, (3) bound to iron-manganese-oxide, (4) bound to the organic matters and (5) residual metal phase [7].

The study area in this report was located along the Chonburi coastline of Thailand from Ban Suan to Bang Lamung spreading over a 50 km distance. This area constituted mostly the industrialized area, the shipping area and the tourism area. Radioactivity levels from the natural radionuclides and the fission products before Fukushima Daiichi disaster especially I-131, Cs-134 and Cs-137 were evaluated. The data may be used as a guideline for background levels of radionuclides in the marine sediments.

2. Material and method

2.1. Sampling locations and sample preparation

Marine sediments were collected along Chonburi coastline from Ban Suan to Bang Lamung district (part of the upper Gulf of Thailand, a semi-enclosed gulf) over about 50 km distance. Each sampling point was located approximately 1 km away from the shore in three seasons (rainy: September, 2010, winter: December, 2010, and summer: March, 2011). Coordinates were determined by the global positioning system (figure 1). The physical parameters consisting of temperature, electrical conductivity, salinity and dissolved oxygen were measured at each half depth below the sea level by using YSI-60 multimeter. The sediment samples were collected by Ekman dredge grab and some samples were taken by scuba diving. The depth of the sediment cores were between 0 – 10 cm.

After air drying, the dried sediment was ground and sieved through a 2 mm test sieve. The ground sediment was then filled into a plastic cylindrical container with the same geometry as IAEA Soil 6 reference material (7.5 x 7 cm^2).

2.2. Classification of sediment texture and organic content

The texture of the sediments and their organic content were characterized since they might affect the deposition of radionuclides and heavy metals in the sediments. Dried sediment (5 g) was heated at 350-440 °C for 24 h to remove organic matter. The percentage of organic content was determined by the weight difference [8].

2.3. Benthos sampling and identification

Fractions of the sediment samples were washed and collected for benthos by sequential sieving through 2 mm, 850 µm and 425 µm sieves. The benthos were removed by forceps and immediately placed under a stereomicroscope for taxonomic identification. If the benthos could not be immediately identified, they were stored in 10% formalin until identification.

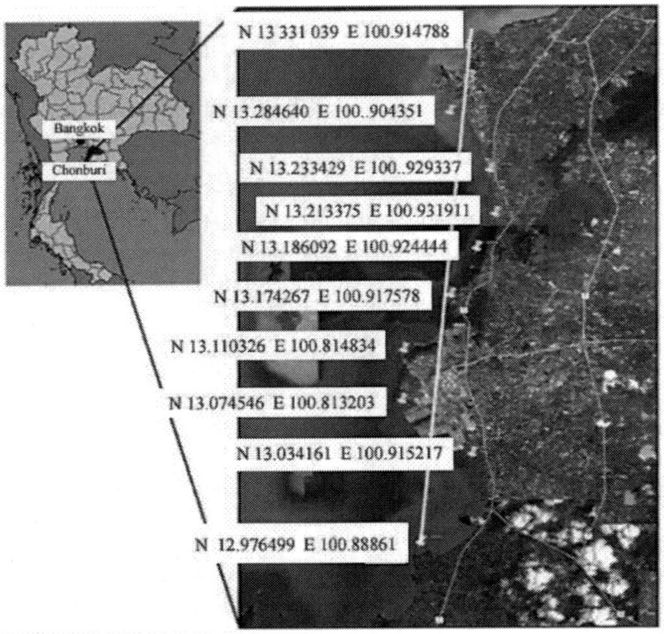

Figure 1. The sampling locations: (1) site 1: Fish dock, Chonburi; (2) site 2: Wornapa beach; (3) site 3: Wastewater discharge, Bangpra; (4) site 4: Bangpra community; (5) site 5: Si Racha Fisheries Research Station; (6) site 6: Si Racha pier; (7) site 7: Si Racha Shipping area; (8) site 8: Leam Chabang pier; (9) site 9: Bang Lamung; (10) site 10: The Sanctuary of Truth. The geographical coordinates and the corresponding yellow pins marked sites 1 – 10 in order from top to bottom.

2.4 Measurement of radioactivity and determination of health risks

Radionuclides in the sediments were measured with a gamma spectrometer equipped with HPGe detector (Ortec GWL-120230 MJ-GWL). The energy and efficiency of the detector were calibrated with IAEA Soil 6 reference material [9].

For statistical confidence, the counting time of marine sediment was optimized to 60.000-80,000 s. The radioactivity was calculated by:

$$A = \frac{R_n}{\varepsilon P_\gamma W} \quad (1)$$

where A is the activity of the radionuclides (Bq/kg); R_n is the counting rate (cps); ε is the detector efficiency; P_γ is the absolute probable decay of gamma energy ≤ 1, and W is the sediment weight (g).

To assess the radiation hazards associated with the radioactivity in the sediments, the absorbed dose rate in air (D), the radium equivalent activity (Ra_{eq}), the external hazard index (Hex), and the annual external effective dose rate (AED_{out}) were calculated from activities of U-238, Th-232 and K-40 using equations 2 - 5 [10].

$$D\ (nGy/h) = 0.0417C_K + 0.462C_U + 0.604C_{Th} \quad (2)$$

$$Ra_{eq}\ (Bq/kg) = 0.077C_K + C_U + 1.43C_{Th} \quad (3)$$

$$Hex = (C_K/4810) + (C_U/370) + (C_{Th}/259) \quad (4)$$

$$AED_{out}\ (mSv/y) = D \times 8760 \times 0.2 \times 0.7 (Sv/Gy) \times 10^{-6} \quad (5)$$

where C_K, C_U and C_{Th} are specific activities (Bq/kg) of K-40, U-238 and Th-232, respectively.

2.5 Analysis of heavy metals by ICP-AES

Marine sediment was digested according to USEPA [11]. 1 g of dried sediments was digested in 10 mL of 8 M nitric acid over a hot plate at 92 – 95 °C for 10 min. After cooling, 5 mL of concentrated nitric acid was added with continuous heating for 30 min. Deionized water (2 mL) and 30% hydrogen peroxide (3 mL) were added; the digested sediment was then heated until bubbles disappeared. Hydrochloric acid

(6 M, 10 mL) and 2 mL of deionized water were then added; the mixture was heated for 10 min, then left to cool and filtered through Whatman paper no. 42. The solution was adjusted to 100 mL with deionized water.

Concentrations of heavy metals As, Cd, Cr, Cu, Hg, Ni, Pb and Zn were determined by a pre-calibrated ICP-AES (JY 2000, Jobin Yvon) with detection limits of 2.82, 0.32, 4.49, 0.002, 0.05, 0.12, 0.007 and 0.05 ppm, respectively.

Benthos samples were digested and analyzed in the same manner as the sediments, following classification of the benthic communities. All experiments were performed in three replications.

3. Results and discussion

3.1 Hydrographic parameters of water and sediment texture

The hydrographic parameters of water were presented in table 1. Water depth of the sample sites ranged from 1.7 - 7.8 m; the deepest site was located in an artificial channel of a commercial pier. The natural depth did normally not exceed 3 m at 1 km distance away from the coast line. The sandy sediment made the sea water highly transparent. The salinity was less than what was normally observed for sea water (35 ppt) due to dilution by fresh water from a water discharge close to the sea. Dissolved oxygen (DO) was slightly lower than the standard values (6.1 - 6.6 ppm at 20 - 30 °C, 1 atm, [12]) except at Si Racha pier (site 6), Si Racha Shipping area (site 7) and Leam Chabang pier (site 8), where the deep sea piers were located causing high water turbulence and a lower organic content (data not shown). The lowest DO was found at Si Racha Fisheries Research Station (site 5) and Bang Lamung (site 9). The former was a mussel farm area and the latter consisted of local fisheries and the beach area, where there were plenty of organic matters from fisheries and anthropogenic activities along the beach. The temperature showed less variation in the tropical area and along the coastal line.

The marine sediments at all sampling sites were characterized as sand using the soil texture triangle [8] since they contained more than 60% sand particles. Their sand particles content varied from 64.20 ± 12.33 - 94.73 ± 2.34%. The sediment samples collected nearby the beach (sites 2, 3, 4 and 10) contained more than 80% sand particles with small amounts of silt and clay. The sand and silt contained fewer amounts of fine particles and organic matters which allowed only small amounts of deposition and adsorption of radionuclides and heavy metals.

Table 1. Hydrographic parameters of water along the studied areas.

Site	Depth (m)	Temp. (°C)	Transp. (m)	Salinity (ppt)	EC (mS/m)	DO (mg/L)
1	1.7	29.4	1.3	21.7	22.2	4.96
2	3.5	29.4	2.7	20.0	34.3	4.83
3	1.7	29.8	0.7	20.8	35.2	4.84
4	1.5	26.7	1.2	17.4	28.9	5.01
5	4.6	30.1	2.1	28.4	44.3	2.84
6	2.3	30.1	1.9	27.5	44.7	6.13
7	7.3	28.7	2.0	24.6	45.4	6.42
8	7.8	29.5	3.0	28.9	46.5	6.71
9	2.5	29.4	1.1	29.1	46.7	3.31
10	3.8	29.4	1.5	19.6	47.5	4.05
Mean±SD	3.67 ± 2.28	29.25 ± 0.98	1.75 ± 0.73	23.8 ± 4.42	39.57 ± 8.86	4.91 ± 1.27

Temp. = temperature; Transp. = transparency; EC = electrical conductivity; DO = dissolved oxygen.

3.2 Spatial and temporal variation of radionuclides in marine sediments

Naturally-occurring radionuclides including U-238, Th-238 and K-40 were found in marine sediment samples (table 2). The specific activity of U-238 was calculated from mean specific activities of Pb-214 (351.92 keV) and Bi-214 (609.31 keV); similarly, the specific activity of Th-232 was calculated from Tl-208 (583.00 keV) and Ac-228 (911.07 keV). The specific activity of K-40 was calculated from the photopeak at 1,460.81 keV. However, the photopeak of Cs-137 was not found in the marine sediments where its minimum detectable activity (MDA) was 2.46 Bq/kg.

UNSCEAR [13] reported the global mean specific activities of U-238, Th-232 and K-40 not exceeding 35, 30 and 400 Bq/kg, respectively. Mean specific activities of U-238, Th-232 and K-40 in sediments collected from Si Racha Fisheries Research Station (site 5) and Si Racha Shipping area (site 7) were less than the global mean values. Low specific activities of U-238 and Th-232 and high specific activity of K-40 were found at Wornapa beach (site 2) in three seasons. The mean specific activities of U-238, Th-232 and K-40 were highest in the rainy season and were also higher than the global mean values. In winter and summer, specific activities of U-238 and Th-232 were lower, while the specific activity of K-40 was much higher than the global mean values.

Due to the difference in precipitation of rainwater and the direction of Monsoon [14], the mean specific activities of the three selected radionuclides in summer and winter were higher than in rainy season except at Wastewater discharge, Bangpra (site 3). The accumulation of radionuclides in marine sediments was affected by the other factors such as geography, sand or clay beach, presence of piers, wastewater discharge sites, mussel farms and so on. The accumulation of radionuclides at Si Racha Fisheries Research Station (site 5) was lower than the other sampling sites because this area was covered with mussel farms. The high amount of calcium carbonate and the porosity of mussel shells [15] led to a high potential of adsorption resulting in low accumulation of radionuclides in the marine sediment.

The mean specific activities of U-238, Th-232 and K-40 from 10 sampling sites were provided in table 2. The values were higher than the global mean values reported by UNSCEAR (35, 30 and 400 Bq/kg, respectively).

The total activities of individual sampling sites in figure 2 indicated the lowest activity at Si Racha Fisheries Research Station (site 5) followed by Si Racha Shipping area (site 7). The total activities were highest at Wastewater discharge, Bangpra (site 3) in the rainy season. This area was located closed to the water discharge site where the strong water stream flowed through the sea causing the transportation of radionuclides from inland to the marine sediment. The radionuclides may be transferred together with the sediment by water runoff in the rainy season. The grains of clay and silt particles were very small with an enormous surface area and a net negative charge; hence, cations of radionuclides would easily be adsorbed to the surface of these particles with an electrostatic force. The specific activity of K-40 was approximately ten times the magnitude of U-238 and Th-232. Additionally, the ionic radius of K-40 (220 pm [16-17]) was approximately similar to the interlayer distance of non-expanding 2:1 silicate clay [18]; therefore, it was easily adsorbed to the sediment. The ionic radius of U-238 (175 pm) and Th-232 (180 pm [16-17]) were approximately 40 pm smaller than potassium; they could not fit into the interlayer distance of silicate clay. The expected adsorption or ion exchange at the outer surface of silicate clay, which was actually dominant to electrostatic interaction, was the only mechanism. The highest activity in the summer was found at Si Racha pier (site 6). In winter, the highest activities were found at Wornapa beach (site 2) and Bang Lamung (site 9). In addition to the large fraction of sand in these three sampling sites, the less adsorption should be found. Organic matters played an important role to enhance the adsorption by chemical bonding at their active sites (carboxylic, amino, phenolic and hydroxyl sites of organic matters). The sediment contained a small amount of organic matters ($< 0.5\%$); as a result, it did not strongly affect the adsorption mechanism of radionuclides to the surface of the sediment. The large accumulation of radionuclides could be originated from a large amount of wastewater discharge to the sea.

Table 2. Spatial and temporal variation in specific activities of radionuclides in the marine sediments.

Site	Nuclide	Specific activity (Bq/kg)			Mean ± SD
		Rainy season 2010	Winter 2010	Summer 2011	
1	^{238}U	38	56	66	53±7
	^{232}Th	63	86	92	80±7
	^{40}K	524	572	589	562±16
2	^{238}U	14	19	27	20±3
	^{232}Th	14	19	23	19±2
	^{40}K	977	954	223	718±202
3	^{238}U	128	22	41	63. ±27
	^{232}Th	208	25	23	85±50
	^{40}K	986	649	628	754±95
4	^{238}U	66	73	30	56±11
	^{232}Th	95	110	39	81±18
	^{40}K	443	556	481	493±27
5	^{238}U	18	20	20	19±0.5
	^{232}Th	18	22	24	22±1
	^{40}K	99	120	132	117±8
6	^{238}U	24	26	33	29±6
	^{232}Th	28	30	37	27±6
	^{40}K	151	153	184	494±14
7	^{238}U	24	26	33	27±2
	^{232}Th	28	30	37	31±2
	^{40}K	151	153	184	163±9
8	^{238}U	37	45	41	41±2
	^{232}Th	41	53	49	48±3
	^{40}K	287	381	366	345±24
9	^{238}U	37	69	42	49±8
	^{232}Th	48	78	50	58±8
	^{40}K	547	817	700	688±64
10	^{238}U	45	40	75	54±9
	^{232}Th	40	38	62	47±6
	^{40}K	409	379	401	396±7
Mean ± SD	^{238}U	44±10	41±6	39±6	
	^{232}Th	59±17	50±19	41±7	
	^{40}K	463±94	484±83	472±81	

Radiation hazards associated with radioactivity in the sediments were assessed from the measured specific activity as shown in table 3. The absorbed dose rate in air ranged from 27-112 (mean: 69 ± 8) nGy/h with some sites having the value higher than Thailand's mean value (77 nGy/h [13]). Radium equivalent activity (Ra_{eq}) ranged from 59 – 243 (mean: 149 ± 19) Bq/kg, which was less than the limit value (370 Bq/kg [19]). External hazard index (H_{ex}) ranged from 0.2 – 0.7 (mean: 0.4 ± 0.05), which was less than 1 [20]. Finally, annual external effective dose rate (AED_{out}) ranged from 0.03 – 0.14 mSv/y (mean: 0.08 ± 0.01 mSv/y) which was less than the global mean (0.48 mSv/y; [13]). Furthermore, the average annual external effective dose rate was lower than the maximum permissible level (MPL) [13].

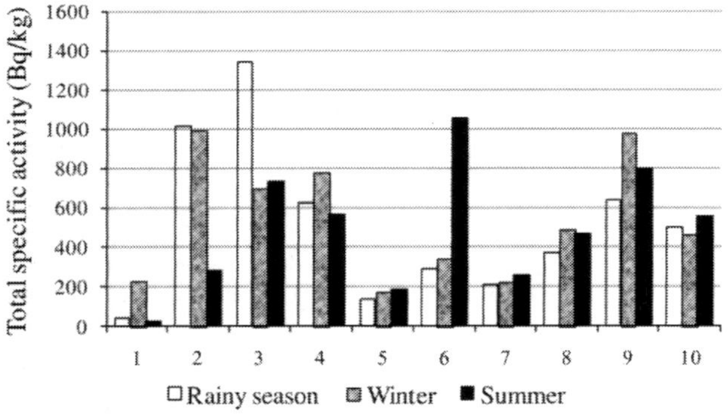

Figure 2. Temporal variation of total specific activities in the marine sediments.

Table 3. Risk index of natural radionuclides.

Site	D (nGy/h)	Ra_{eq} (Bq/kg)	H_{ex}	AED_{out} (mSv/y)
1	97	211	0.6	0.12
2	51	102	0.3	0.06
3	112	243	0.7	0.14
4	96	210	0.6	0.12
5	27	59	0.2	0.03
6	51	106	0.3	0.06
7	38	85	0.2	0.05
8	62	136	0.4	0.08
9	87	186	0.5	0.11
10	69	151	0.4	0.09
Mean ± SD	69 ± 8	149 ± 19	0.4 ± 0.05	0.08 ± 0.01

3.3 Concentrations of heavy metals in marine sediments

The high amounts of Fe (4,175 ± 183 and 8,307 ± 868 mg/kg) were found in rainy and winter seasons, respectively. The important sources of Fe were transportation and import/export goods containing iron. Zheng reported that a high concentration of metals would normally be found near the source [21]; for example, the high concentration of Zn (19,786 mg/kg) was found near the Huludao Zinc Plant; the largest zinc plants in Asia. In contrast, radionuclides could be found far away from their sources becase they could accompany various kinds of media. Besides the atmospheric phase, they could accompany fallouts or gaseous-phase radionuclides such as tritium in titrated water vapor and iodine vapor. Wind speed and direction were the most critical factors to bring them far away from the source. Radionuclides could also mobilize in the aquatic phase; for example, Cs-137 was always found in marine ecology being far away from the source. Finally, both heavy metals and radionuclides would settle down and accumulate in the marine sediments. Manganese was used as an anti-corrosion coating for columns of docks and jetties [22]. A high concentration of Mn could be harmful to embryos of marine animals [23]. Lead was the third most abundant metal to be found in marine sediments particularly at the Sanctuary of Truth (site 10), Bang Lamung (site 9) and Si Racha pier (site 8) (260.69±8.42, 233.21±7.13 and 205.74±8.20 mg/kg, respectively) with average value of 154.28±26.68 mg/kg. It was probably carried by wind or water stream from the Si Racha shipping area. Spraying of primer paints containing lead oxide may be the source of Pb in marine sediment.

In winter, Cr was the second most abundant metal at Si Racha Shipping area, Leam Chabang pier and the water discharge site, Bangpra (1,168.10 ± 126.44, 934.11 ± 32.48 and 692.23 ± 15.47 mg/kg, respectively). This was likely due to the fact that Cr was used as a metal coating to prevent oxide formation, a mixture of vanish oil and a primer in ship repairing process. Mn was the third most abundant at Fish dock and Si Racha pier (593.82±13.02 and 356.86±1.35 mg/kg).

The concentration of each metal in winter was lower than in rainy season except Pb which was probably originated from the shipyard and transported by the southwest monsoon (a seasonal rainy wind). However, the accumulation of metals in marine sediments was slightly low in the rainy season due to the dilution by rainwater.

Since the industrial estate and shipping activities spanned the area from Leam Chabang pier to the Sanctuary of Truth (site 10), contamination of heavy metals in the marine sediments was higher than other areas from Fish dock, Chonburi (site 1), to Si Racha Fishery Research Station (site 5). Fine particles with a large surface area and plenty of exchangeable sites were found leading to high accumulation of metals to their surface. From Fish dock, Chonburi, to Si Racha Fishery Research Station, sand particles dominated and the area was used for habitats and tourism resulting in the low concentration of metals.

In winter, the concentration of Pb reached its maximum followed by Hg, Cu, Cd, Ni, Zn and Cr. Similar to radionuclides, the highest level was found at Si Racha pier (site 6) followed by Wastewater discharge, Bangpra. The former might have originated from the source of contamination itself. The latter might have been originated from the source with a large amount of organic matters from water discharge, which enhanced the adsorption of metals, and also a large amount of clay and silt. Similar to Wastewater discharge, Bangpra (site 3), the coastal sediment consisted of a high percentage of silt and clay; the large amount of Pb originated from Bang Lamung (site 9) was found in coastal sediments collected from the Sanctuary of Truth (site 10).

3.4 Relation between benthos and heavy metals

Populations of benthos varied from 45 – 4,447 individuals/m^2 and the diversity indices ranged from 0 - 2.2. The diversity index reflected the relationship between species and number of individuals in each aquatic system. The highest diversity of benthos (15 species) was found at Wastewater discharge, Bangpra (site 3) followed by Wornapa beach (site 2); the Sanctuary of Truth (site 10); Fish dock, Chonburi (site 1), which comprised of 14, 13 and 12 species, respectively. The lowest diversity was found at Bang Lamung (site 9) where only 2 species were found. Annelida and Mollusca were found in each sampling site; these groups of organisms were more tolerant to high pollutant levels than other phyla. The others ware Arthropoda and Echinodermata. Some economic species including *Portunus pelagicus* and *Penaeus merguiensis* were also found. Fe, Mn, Pb and Cr in marine sediments were not significantly transported to benthos along the study sites, although these metals were highly accumulated to marine sediments.

Four phyla of benthos consisted of Annelida (16 species), Arthropoda (7 species), Mollusca (19 species) and Echinodermata (2 species); totally, 44 species were found among the ten sites. Polychaetes, shrimps, crabs, gastropod bivalves, sea cucumber and echinoderms were dominated at each station. The highly tolerant species were polychaetes (Annelida) and gastropods (Mollusca) with higher numbers of individuals (data not shown).

In this study, the radionuclide level in benthos was not measured since the numbers of benthos were not enough for radioactivity counting.

4. Conclusion

Besides the assessment of radionuclide accumulation in marine sediment along the Upper Gulf of Thailand by HPGe-gamma spectrometry, the natural radionuclides including U-238, Th-232 and K-40 were found with mean specific activities of 39 ± 6, 41 ± 7 and 472 ± 81 Bq/kg in summer; 44 ± 10, 59

± 17 and 463 ± 94 Bq/kg in rainy season, and 41 ± 6, 50 ± 9 and 484 ± 83 Bq/kg in winter, respectively. No contamination of the artificial radionuclide (Cs-137) was found. Accumulation of radionuclides in marine sediments depended on chemical properties of radionuclides, physical and chemical properties of the sediments: type of sediments, grain size and organic content, and geography: season, stream direction and flow rate. The radionuclide levels in marine sediments were higher than the global mean values reported by UNSCEAR. However, the radiation hazards associated with the radioactivity in the sediments were not expected.

Although no contamination of fission product including Cs-137 was found, the radionuclide level was very important since the sampling time was a few days before the Fukushima Daiichi disaster in 2011. The results under this study would be useful as a baseline data supporting the environmental radiation surveillance program of the country. Continuous monitoring after Fukushima disaster should be performed to tract any changes in the radioactivity levels and to ensure the radiation safety of the marine environment.

Metal contaminants in the marine sediments could be divided into two groups: less toxic (Fe and Mn) and highly toxic metals (Cd, Cu, Cr, Hg, Ni, Pb and Zn). For less toxic metals, Fe reached its highest concentration in both rainy and winter seasons, followed by Mn in the rainy season. Among toxic elements, Pb and Cr reached the highest in the rainy season (154.28±26.68 mg/kg) and in winter (578.16±126.74 mg/kg), respectively. The amount of metals was slightly higher in winter than in the rainy season due to water dilution, strong wind and waves in the rainy season.. However, the amount of heavy metals did not exceed the standard metals contamination in marine sediments. A high contamination was found at the industrial and shipping areas.

Assessment of radionuclides and heavy metals along the Upper Gulf of Thailand was very important to address the concern about human risk to exposure. The benthic communities could be used as a biological index of a tolerance level guideline for primary producers and consumers. Continuous monitoring of toxic materials not only radionuclides and heavy metals but organic compounds (PAHs, VOCs, pesticides and others) was very important in marine surveillance and could be used as baseline data for management of marine ecology. The amount of toxic substances should be evaluated together with soil parameters: pH and organic matters, to determine the kinetic and thermodynamic parameters.

Acknowledgement
We would like to express our sincere thanks to Office of Atoms for Peace, Thailand, for the financial support. We thanked the Faculty of Science @ Si Racha and Si Racha Fisheries Research Station, Faculty of Fisheries, Kasetsart University, for laboratory facilities and ships as well as sampling facilities, respectively.

References

[1] Reddy M S, Bfsha S and Joshi H V 2005 *India Chemosphere* **61** 1587-93
[2] Wang J and Chen C 2006 *Biotechnol. Adv.* **24** 427-51
[3] Croudace I W and Cundy A B 1995 *Environ. Sci. Technol.* **29** 1288-92
[4] Bothner M H, Buchholtz B M and Manheim F T 1998 *Mar. Environ. Res.* **45** 125-55
[5] Lee S V and Cundy A B 2001 *Estuar. Coast Shelf S.* **53** 619-36
[6] Finny B P and Huh C A 1989 *Environ. Sci. Technol.* **23** 294-303
[7] Tessier A, Campbell P G C and Bisson M 1979 *Anal. Chem.* **51** 844-51
[8] Boyd C E 1995 *Bottom Soils, Sediment and Pond Aquaculture* (New York: Chapman and Hall)
[9] IAEA 1989 *Technical Report Series No. 295 Measurement of Radionuclides in Food and the Environment: a Guidebook* (Vienna: IAEA)
[10] Veiga R, Sanches N, Anjos R M, Macario K, Iguatemy M, Aguiar J G, Santos A M A, Mosquera B, Bastos J, Carvalho C, Fiho M B and Umisedo N K 2006 *Radiat. Meas.* **41** 189-96
[11] Edgell K 1989 *USEPA Method Study 37 SW-846 Method 3050 Acid Digestion of Sediments, Sludges and Soils* (Cincinnati: USEPA)

[12] USEPA 2012 *Method 3050B Acid Digestion of Sediments, Sludges, and Soils* (Washington, DC: USEPA)
[13] UNSCEAR 2000 *Sources and Effects of Ionizing Radiation 2000 Report to the General Assembly with Annex B: Exposure from Natural Source of Radiation* (New York: UNSCEAR)
[14] Wongsanit S 2010 *Gross Alpha and Beta Activities in Sea Water, Sediment and Plankton in the Coastal Ecosystem: the Upper Gulf of Thailand* (Bangkok: Kasetsart University)
[15] Tubtong C and Juijunjerm R 2005 *Removal of Heavy Metals from Wastewater by Mussel Shell* (Bangkok: Srinakarinwirot University)
[16] Shannon R D and Prewit C T 1969 *Acta. Cryst.* **B25** 925-46
[17] Shannon R D 1976 *Acta. Cryst.* **A23** 751-61
[18] Essington M E 2003 *Soil and Water Chemistry, an Integrative Approach* (Boca Raton: CRP Press)
[19] OECD 1979 *Exposure to Radiation from the Natural Radioactivity in Building Materials Organization for economic Cooperation* (Paris: OECD Nuclear Energy Agency)
[20] El-Taher A and Madkour H A 2011 *Appl. Radiat. Isotopes* **69** 550-658
[21] Zheng Y F, Wang Y J, Hou C L and Guan C M 2008 *Mar. Environ. Sci* 02 (in Chinese)
[22] Parthiban G T, Palaniswamy N and Sivan V 2009 *Anti-Corros. Methods Mater.* **56** 78-83
[23] Columina M T, Domingo J L and Llobel J M 1996 *Veterinary and Human Toxicol.* **38** 7-9

Journal of Physics: Conference Series **611** (2015) 012024 doi:10.1088/1742-6596/611/1/012024

Variations of ^{210}Po activity in mussel *(Perna viridis)* of Samut Sakhon and its contribution to dose assessment

B Porntepkasemsan[1], **K Srisuksawad and W Kulsawat**
Thailand Institute of Nuclear Technology (Public Organization),
9/9 Moo 7 Ongkharak, Nakhon Nayok 26120, Thailand

E-mail: boonsom@tint.or.th

Abstract. The activities of ^{210}Po and its effective dose in green mussel *(Perna viridis)* collected from a mussel farming area in Samut Sakhon province during the period of 2012-2013 are presented. Several parameters including maximum shell length and the physiological performance of mussels using condition index and physical properties of seawater (pH, salinity, conductivity, TDS, DO and cation-anion elements) were measured. Each individual mussel was measured for its maximum shell length which was adopted as size class. The activity concentration of ^{210}Po was determined spectroscopically through its 5.30 MeV alpha particle emission, using ^{209}Po as an internal tracer. The ^{210}Po activity concentration in mussels was found to vary between 1.044 and 6.951 Bq/kg wet weight. The ^{210}Po concentration was higher in smaller-sized (≤35 mm) and lower in larger ones (40-70 mm). This confirmed that larger mussels have lower ^{210}Po activities on a weight basis. The ^{210}Po body burden (activity per mussel) ranged from 1.035 to 17.183 mBq. Contrary to the ^{210}Po concentrations, results of the body burden revealed the lower activities in smaller-sized mussels (≤35 mm) and the higher in larger-sized ones (40-70 mm). The type of fluctuations observed with ^{210}Po concentrations were interpreted as a seasonal effect. Total annual effective ^{210}Po dose due to mussel consumption was calculated to be in the range of 3.081 to 16.401 μSv. Based on the international guideline, the average dose calculated due to ^{210}Po in mussels of Samut Sakhon would not pose any significant radiological impact on human health and the mussels are considered to be safe for consumption.

1. Introduction

Among the radionuclides, naturally radioactive series of ^{238}U are the main source of radioactivity in the marine environment [1]. Basically, ^{238}U and their decay products are not required for the metabolism of the organisms. However, some of them are bioconcentrated in both hard and soft tissues of marine organisms for they can act as chemical analogs of metabolically essential elements [1, 2]. ^{210}Po is a 99.99% alpha emitter ($T_{1/2}$ = 138.4 d) with alpha particle energy of 5.3 MeV, ^{210}Po occurs in the marine environment as the decay products of ^{238}U radioactive series and it attains great significance in marine biota due to its radiotoxic properties. The toxicity of ^{210}Po is connected with its relatively high energy and that it is concentrated in the soft tissues, such as muscle, liver, kidney, and hemoglobin. The main sources of ^{210}Po entering the marine environment are atmospheric precipitation on the surface and coastal waters [3]. ^{210}Po has a tendency to get accumulated in the edible portions of

[1] To whom any correspondence should be addressed.

marine organisms, and is considered to be the most important contributors of radiation dose received by humans via fish and shellfish consumption [4, 5]

Marine molluscs such as the green mussel, *Perna virisdis* form an important component of the marine food web. They act as sentinels and considered one of the abundant primary consumers in biogenic habitats. In general, molluscs have the potential to accumulate significant amounts of radionuclides which may be biomagnified in the food chain to higher trophic levels including human consumers. They also serve as a potential bioindicator species in coastal areas by reflecting the effects of natural and anthropogenic stressors [6, 7]. In addition, mussels are considered to be an important protein source, a nutritious and delicious food among the Thai population.

As levels of ^{210}Po activity concentrations in these organisms vary throughout the year due to environmental conditions, variations in ^{210}Po input to the marine environment, and biological cycles of the species [8]. The purposes of this study were (i) to determine the seasonal variation in ^{210}Po activity concentrations in mussel *(Perna viridis)* collected from Samut Sakhon coast in 2 years' time period (2012-2013) as an extension to our previous research [9] and (ii) to determine the dose assessment of ^{210}Po in this organism.

2.Materials and methods

2.1. Sampling

Samples of mussel, *Perna viridis*, were sampling from green mussel farming area at Phanthai Norasing subdistrict, Samut Sakhon province. This region is located in the upper Gulf of Thailand where green mussel is cultured by local fishermen using bamboo poles along the coastal region.

Seawater was collected from 8 stations where 3 kilometers apart from each station and surrounding the mussel sampling poles. Seawater sample was kept in 250 mL polyethylene bottles for measurements of physical-chemical parameters such as pH, conductivity, salinity, dissolved oxygen (DO), and total dissolved solids (TDS) including cation-anion elements.

Three to four whole clusters of live green mussel samples were collected every month at about 5 meters depth and transport in an ice box to the laboratory where they were separated by size.

2.2. Sample preparation and condition index

Upon return to the laboratory, the mussel samples were rinsed with seawater and were immediately frozen at -30^{0}C until grouping. Based on shell length, several groups of mussels were made using Vernier caliper. Each size class was separated and then cleaned to remove fouling organisms such as algae, barnacles and spats. The mussels were dissected to remove the whole soft tissue from the shell (excluding byssus thread). The soft tissues were pooled from each size groups and their fresh weights were recorded. The soft tissue samples were freeze-dried to complete dryness and further homogenized and powdered after recording the dry weight.

The mussel condition index (CI), a measure of the mussel's physiological status, was calculated according to the following equation [10].

$$CI = W \cdot SC^{-1} \times 100$$

where, W is the average soft tissues dry weight (g) and SC is the shell cavity calculated as SC = total wet weight of soft tissue - dry shell weight.

2.3. Analysis of ^{210}Po

^{210}Po radionuclide analysis was performed using 20 g of homogenized mussel samples together with 0.12 Bq ^{209}Po as internal isotopic tracer. The sample was digested with 200 mL concentrated HNO_3 and then slowly heated on hotplate to reach incipient dryness. The sample residue was subsequently digested with concentrated $HClO_4$ and then HCl. The dry residue was finally dissolved in 6 M HCl and brought up to the volume of 100 mL in 0.3 M HCl. The residue was separated from supernatant by centrifugation and the Po contained in the solution plated onto a silver disc and measured by alpha-spectrometer. Measurements of polonium alpha particles emitted by the Ag disc was performed with low background 450 mm^2 ion implant detector from ORTEC EG&G connected to an Octete Plus alpha spectrometer.

Quality assurance of analytical results was ensured by analysis of IAEA certified reference materials (IAEA-414 fish muscle) and participation in inter-comparison exercises with good results.

2.4. The annual effective ingestion dose

The estimated effective ingestion dose for individuals as a result of radionuclide intake was derived from measured concentrations in sample using the appropriate ingestion dose coefficient factors (DCC) for adults recommended by IAEA [11]. This dose was calculated as follows:

$$E_{dPo\text{-}210} = A_i \times D_F \times M_F \times AI$$

where $E_{dPo\text{-}210}$ is the annual effective ingestion dose in μSv.y $^{-1}$, A_i is the activity intake (Bq/kg wet weight (ww)), D_F is the dose coefficient factor for adults (1.2 μSv/y) M_F is the modifying factor due to decay of ^{210}Po between catch and consumption (0.6) and AI is annual intake of mussel estimation. The Southeast Asia consumption rate is 4.1 kg/y dry weight [12].

3. Results and Discussion

3.1. Seawater properties and condition index

During this study, the coastal site seawater temperature varied between 21.5 °C in winter and 33.4 °C in summer. Seawater temperature generally followed the seasonal fluctuation of average air temperature of the surface seawater, which showed a minimum in December and a maximum in April.

The salinity of sea water ranged from 8.9 to 27.2. Lower salinity was found in September, October and November (figure 1). A decrease in salinity during the 24-month study period was due to the influence of high rainfall during raining season or wet season (with Monsoon or typhoon events) and, possibly, also some regional effect of Tha-Chin River and Chao-Phraya River freshwater discharges into the coastal sea. In this region, the rainfall displays a clear seasonal pattern with heavier rains in raining season and nearly zero rainfall in summer.

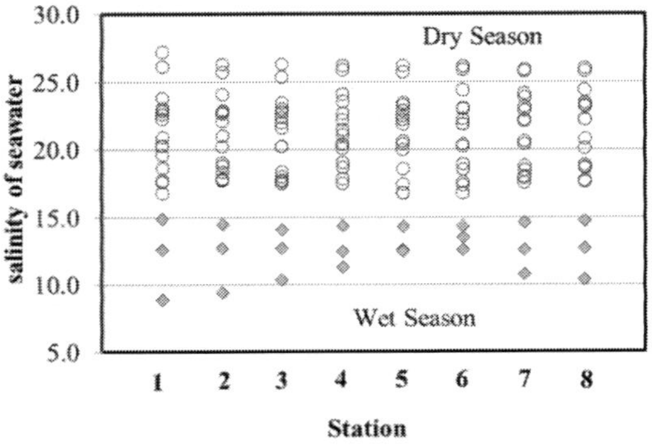

Figure 1. Salinity of seawater at 8 collecting stations.
Round symbols represent the dry season and diamond symbols represent the wet season.

Measurements of pH indicated peaks in March, April and May which was summer season or dry season. The rest of the study period were slightly constant (figure 2). The observed DO, TDS, conductivity and cation-anion elements fluctuated monthly and did not show any obvious pattern (data not shown).

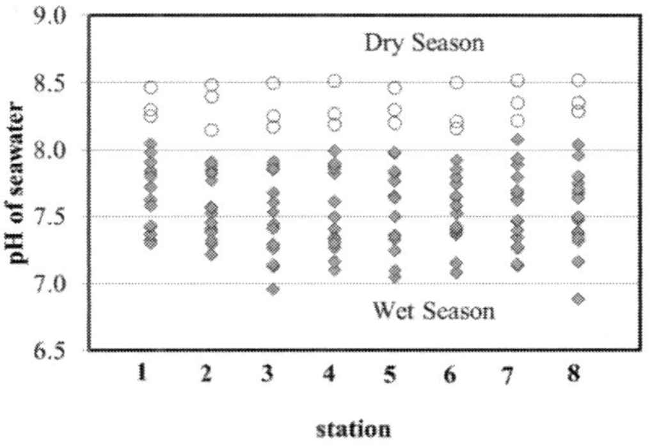

Figure 2. pH of seawater at 8 collecting stations
Round symbols represent the dry season and diamond symbols represent the wet season.

Figure 3 indicates that the mussel condition index changed from month to month. The value of condition index indicates physiological status of bivalves rather than its shell length. The higher value of condition index indicates the higher meat yield. The mussel condition index varied among size classes but was on average higher in August than in the other month. It was obviously separated into two groups: the higher was in the wet season (June – August) while the lower was in the dry season (September – December). This can be attributed to the phytoplankton bloom and spawning time. It is known that water temperature, food availability, cycle of lipid storage, and reproduction cycle (ripening of gonads and release of gametes) may influence the meat yield and biochemical compositions of mussels [13, 14]. The study area, as already mentioned, received significant drainage from the land. These abundant fresh water discharges provided more nutrients and generally higher availability of food in the study area (algae and phytoplankton). However, the continuous freshwater discharge during the wet season observed a dramatic mortality rate among the mussels and eventually deteriorated the mussel farming.

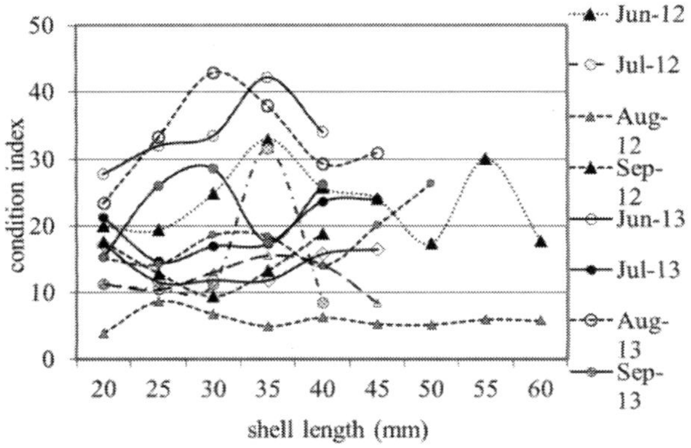

Figure 3. Condition index vs. shell length for mussels of all periods.

3.2. ^{210}Po activity concentrations

The activity of ^{210}Po in the soft tissue of green mussels based on size class or shell length are shown in figure 4. ^{210}Po activity concentration was markedly higher in the smaller sized groups (≤35 mm) and lower in the larger ones (40-70 mm). The graph showed a linear decrease when plotted against shell length (figure 4) and also showed the same trend when plotted against soft tissue dry weight (data not shown). This allometric relationship between ^{210}Po concentrations and mussel size is striking and certainly has implication to the interpretation of environmental monitoring data. The activity concentration of ^{210}Po in mussels was reported on the wet weight basis and ranged from 1.044 to 6.951 Bq/kg.

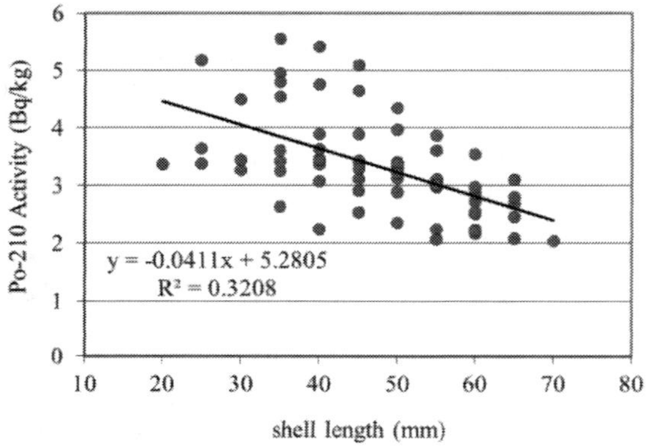

Figure 4. Relationship between Po-210 activity and shell length (size) *of Perna viridis*.

The decrease of ^{210}Po activity with increasing mussel size represented by linear functions of the type $Y = aX + b$ with $R^2 = 0.320$. The low value of R^2 found in this study was similar to the study of Calvaho et al [15] reporting $R^2 = 0.386$. The high ^{210}Po activity concentrations were found in June (3.261 – 5.178 Bq/kg), July (3.305 – 5.556 Bq/kg) and August (3.314 – 6.951 Bq/kg) which was raining season. The atmospheric total deposition of ^{210}Po is also seasonal and positively correlated with rainfall [13]. The possible source for enhanced ^{210}Po in the Upper Gulf of Thailand was the discharge from Tha Chin River and Chao Phraya River. The rivers flowed into the sampling stations carrying the drainage from agricultural land together with contaminants from industrial factories.

It is interesting to note that ^{210}Po activities in mussel soft tissues (radionuclide body burden) increased with increasing shell length (figure 5). This means that the activity of ^{210}Po significantly increased in proportion to the amount of soft tissues and thus to body size or age. However, in all size classes no meaningful variation was found between ^{210}Po body burden and different months of the year. The data on mussels of the same size class covering a 2-year period indicated that whole-body activity concentration remained constant ($R^2 = 0$) regardless of the mussel condition index (figure 6). This suggested that radionuclide body burden did not change significantly over the seasons of the year, despite changes in body weight in relation to the accumulation of lipids and the ripening of the gonad.

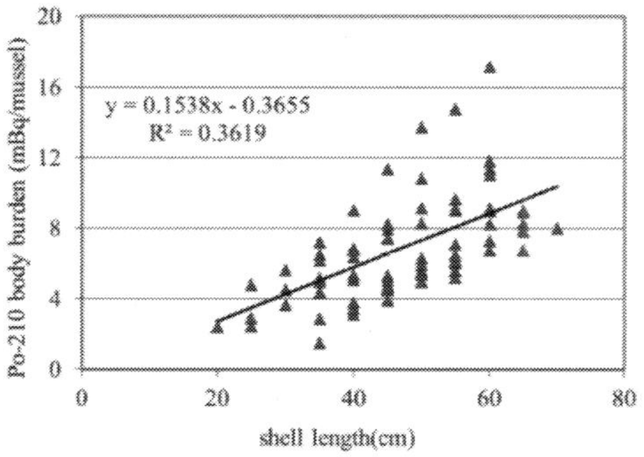

Figure 5. Relationship between ^{210}Po body burden and shell length.

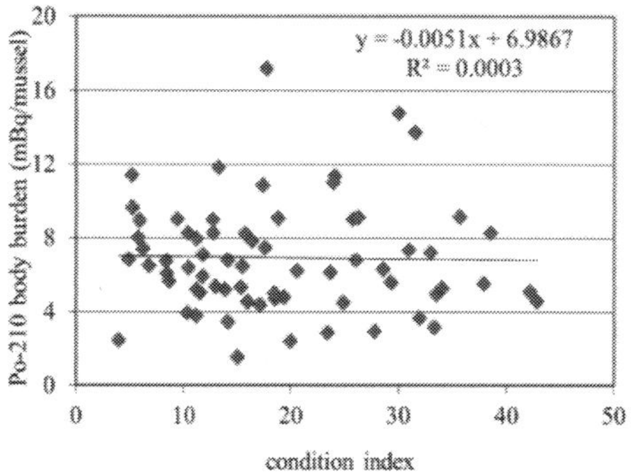

Figure 6. Relationship between ^{210}Po body burden and condition index.

3.3. The annual effective ingestion dose

The diet of the population living along the coast consisted primarily of fish and seafood, which could lead to a higher radiation exposure and subsequent relatively-high radiation doses due to ^{210}Po intake. Annual effective ingestion doses received by an individual were evaluated and found to be in the range of 3.081 to 16.401 μSv. The calculated dose values were compared with those published elsewhere. Aarkrog *et al* [16] estimated the individual dose to be between 51 and 91 μSv/y, with a value of 160 μSv/y for a critical group consuming large quantities of seafood products. Connan et al [17] reported a range of 78 – 129 μSv/y for the adult population from western English Channel, France. Pollard *et al* [18] calculated a dose of 19 μSv/y. Nielsen et al [19] estimated a dose of 700 μSv/y and Jia *et al* [20] gave exposure values between 50 and 200 μSv/y. In the southern coast of India, the annual ingestion dose was estimated to range from 5.1 and 34.9 μSv/y [21]. In this study, the annual effective doses were much lower than in the literatures which can be explained that the annual intake of mussel for Thai people are less than the other countries [12], the different species i.e., *Perna viridis*, *Perna perna*, *Mytilus galloprovincialis* and *Mytilus edulis* have the unique radionuclide accumulation capacities [13, 15] and the different regions have their own environmental conditions.

The International Commission on Radiological Protection [ICRP] has reported a maximum dose of 1000 μSv/y from practices in addition to natural radioactivity to the public, which was estimated to be 2400 μSv/y on average [22, 23]. The calculated doses for exposure to ^{210}Po from consuming green mussels collected from Samut Sakhon province were much less when compared to the ICRP and the global levels. Therefore, it could be concluded that in the mussel farming area of Samut Sakhon, the annual effective dose due to ^{210}Po via green mussel consumption would not cause any health hazards to the general public.

4. Conclusion

The rapid population growth coupled with a rising seafood demand has led aquaculture to become the fastest growing food sector in Thailand [12]. Phanthai Norasing subdistrict, Samut Sakhon province, is one of the promoted shellfish farming areas, and shellfish is one of the most important mariculture product within the country. ^{210}Po concentrations (Bq/kg wet weight) and ^{210}Po body burden (Bq/mussel) observed in the bio-indicator mussel samples from green mussel *(Perna viridis)* farming area in Phanthai Norasing subdistrict, Samut Sakhon province, was presented. The following conclusions could be derived:

1. Seawater properties including pH, conductivity, salinity, DO, TDS and cation-anion elements were measured. The values of pH and salinity were higher in dry season and lower in wet season. However, the conductivity, DO, TDS and cation-anion elements did not show any obvious patterns.

2. The ^{210}Po activity concentrations were found to vary between 1.044 and 6.951 Bq/kg ww and the fluctuation could be interpreted as a seasonal effect which peaked in the wet season and lower in the dry season. Also, ^{210}Po concentrations decreased with increasing mussel size.
3. The ^{210}Po activities in mussel soft tissues (radionuclide body burden) were 1.035 – 17.183 mBq/mussel. The total amount of radionuclide in mussel soft tissues did not fluctuate significantly through the year and did not correlate with the physiological condition (condition index) of mussels. The ^{210}Po body burden revealed the lower activities in smaller-sized mussels and the higher in the larger-sized group.
4. Green mussels *(Perna viridis)* from the study site displayed seasonal fluctuation of ^{210}Po concentrations which were caused by mussel body size fluctuation and were not due to changes in ^{210}Po body burden.
5. The estimated consequent annual effective ^{210}Po doses due to mussel consumption were calculated. They were found to be in the range of 3.081– 16.401 μSv. The obtained effective dose levels of mussels in the studied area were considered radiologically safe for human intake.

Acknowledgments
The authors are grateful to anonymous reviewers and the editorial staff of the International Nuclear Science and Technology Conference 2014 (INST2014) for assistance in the revision and production of this manuscript. This research work was financially supported by Ministry of Science and Technology, Government of Thailand.

References

[1] Kilic O, Belivermis M, Cotuk Y and Topcuoglu S 2014 *Mar. Pol. Bull* **80** 325-29
[2] Sunith Shine S R, Feroz Khan M and Godwin Wesley S 2013 *Mar. Pol. Bull* **75** 276-82
[3] Fowler S W 2011 *J. Environ. Radioact.* **102** 448-61
[4] Mishra S, Bhalke S, Pandit G G and Puranik V D 2009 *Chemosphere* **76** 402-6
[5] Strok M and Smodis B 2011 *Chemosphere* **82** 970-76
[6] Rozmaric M, Rogic M, Benedik L, Strok M and Barisic D 2013 *Chemosphere* **93** 2063-68
[7] Carroll M L, Johnson B J, Henkes G A, McMahon K W, Voronkov A, Ambrose Jr. W G and Denisenko S.G 2009 *Mar. Pol. Bull* **59** 193-206.
[8] Ugur A, Ozden B and Filizok I 2011 *Chemosphere* **83** 1102-07
[9] Porntepkasemsan B, Srisuksawad K and Permnamtip V 2014 *Advanced Materials Research* **931-932** 624-8
[10] Crosby M P and Gale L D 1990 *J. of Shellfish Res* **9(1)** 233-7
[11] IAEA 2003 *International basic safety standards for protection against ionizing radiation and for the safety of radiation sources IAEA Safety series No. 115* (Vienna: IAEA)
[12] FAO 2012 *Report of World Review of Fisheries and Aquaculture, FAO Fisheries Circular No.1074* (Rome: FAO) pp 165
[13] Calvalho F, Oliveira J M and Alberto G 2011 *J of Environ Radioact* **102** 128-137
[14] Orban E, Di Lena G, Nevigato T, Casini I, Marzetti A and Caproni R 2002 *Food Chemistry* **77** 57-65
[15] Calvaho F, Oliveira J M, Alberto G and Vives i Batlle J 2010 *Mar. Pollut. Bull.* **60** (10) 1734-42
[16] Aarkrog A, Baxter M S, Bettencourt A O, Bojanowski R, Bologa A, Charmasson S, Cunha I, Delfanti R, Duran E, Holm E, Jeffree Livington H D, Mahapanyawong S, Nies H, Osvath I, Pingyu I, Povinec P P, Sanchez A Smith J N, and Swift D 1997, *J. Environ. Radioact.* **34(1)** 69-90
[17] Connan O, Germain P, Solier L, and Gouret G 2007, *J. Environ. Radioact.* **97(2-3)** 168-88
[18] Pollard D, Ryan T P, and Dowdall A 1998, *Radiat. Prot. Dosim.* **75(1-4)** 139-42
[19] Nielsen S P, Bengston P, Bojanowski R, Hagel P, Herrmann J, Iius E, Jakobson E, Motiejunas S, Panteleev Y, Skujina A, and Suplinska M 1999, *Total Environ.* **237** 133-41

[20] Jia G, Belli M, Sansone U, Rosamilia S, and Blasi M 2003, *J. Radioanal. Nucl. Chem.* **256 (3)** 513-28

[21] Feroz Khan M., Godwin Wesley S. and Rajan M.P 2014 *J. Environ Radiact.* **138** 410-6

[22] International Commission on Radiological Protection [ICRP] 1991 *ICRP Publication no. 60* (Oxford: Pergamon Press)

[23] United Nations Scientific Committee on the Effects of Atomic Radiation 2000 *UNSCEAR Report to the general assembly with scientific annexe* (New York: United Nations)

International Nuclear Science and Technology Conference 2014 (INST2014) IOP Publishing
Journal of Physics: Conference Series **611** (2015) 012025 doi:10.1088/1742-6596/611/1/012025

Analysis of radon in shallow-well water: a case study at Phichit subdistrict in Songkhla province, Thailand

A Charoensri[1], S Siriboonprapob and N Sastri
Bureau of Technical Support for Safety Regulation, Office of Atoms for Peace, Bangkok, Thailand

E-mail: apisara@oaep.go.th

Abstract. Radon levels were measured in shallow-well water samples collected from Phichit subdistrict in Songkhla province, Thailand. A total of 35 water samples from shallow-wells were collected and measured for the radon concentration. The measurements were performed using a RAD7 portable radon detector. The radon concentrations varied from 0.18 ± 0.07 to 98.1 ± 5.92 Bq/L with a mean value of 16.76 ± 2.33 Bq/L. These recorded values were compared with the safe limit values recommended for drinking water by various health and environmental protection agencies. Thirty-four percent of the recorded values were within the safe limit of 11 Bq/L recommended by the US Environmental Protection Agency. The annual effective dose from ingestion and inhalation of radon was also evaluated. The estimated total effective dose varied from 0.48 to 262.91 μSv/year. The total effective dose in most of the samples (~90%) in this study was within the safe limit (0.1 mSv/year) recommended by the World Health Organization (WHO) and the European Council.

1. Introduction

Radon gas and its radioactive isotopes receive special attention among all other naturally occurring radioactive materials because they contribute the largest amount of total annual effective dose to humans [1]. The most important aspect of high radon concentrations is its hazard on human health as it is a cause of lung cancer [2-4]. Moreover, a very high radon concentration in drinking water can also lead to a significant risk of the stomach and gastrointestinal cancer [5, 6]. Knowledge of the levels of radon in each source including household water, particularly water from ground sources is necessary to protect the public from consequences of excessive exposure to radiations. Radon in water has been measured in many parts of the world due to the risk of radiation exposure from drinking water [7-11].

Most of radon in buildings comes directly from soil that is in contact with or beneath the basement or foundation. Radon is also found in groundwater and will enter a home whenever this water is used. In many situations such as showering, washing clothes and flushing toilets, radon can be released from the water and then mixed with the indoor air. Thus, radon from water contributes to the total inhalation risk associated with radon in indoor air. In addition to this, drinking water contains dissolved radon, and the radiation emitted by radon and its radioactive decay products can expose sensitive cells in the stomach as well as other organs to radiation-induced damage once they are absorbed into the bloodstream.

[1] To whom any correspondence should be addressed.

Even though, radon is often being discussed as a threat to human health, it shall also be mentioned here that it can be used as an ideal tracer for a considerable variety of applications in the fields of environmental geology and hydrogeology. For instance, radon, which is a noble gas, has a strong affinity to non-aqueous phase liquids (NAPLs). This property makes it applicable as naturally occurring partitioning tracer for assessing residual NAPL contamination of aquifers [12]. Due to the distinct water/air partitioning behavior of radon and its straightforward on-site detectability, the radon distribution pattern in the groundwater can be used as an appropriate measurement for assessing the progression of an *in situ* air sparging (IAS) measurement as a function of space and time [13].

From the result of a survey at Moo 3, Phichit subdistrict, Na Mom district, Songkhla province, people in this area still used the shallow-well water for consumption. Hence, it was reasonable to investigate the radon levels in the shallow-well water being used for drinking. The purpose of this study was to investigate the radon levels in shallow-well water from the chosen villages and towns in Phichit subdistrict, which was used for drinking purpose and to determine its health hazards.

2. Materials and methods

2.1. Description of the area

Shallow wells in Moo 3, Phichit subdistrict, Na Mom district, Songkhla province, were chosen for this research. This study area waslocated on the eastern side of the southern part of Thailand (latitude 6.992476 north and longitude 100′01″-101′06″ east). The altitude was approximately 4 meters above sea level. Its area was approximately 40.5 square kilometers or 25 - 312.5 Rai. The study area is surrounded by steep and tall mountains with the center flat plain used for planting rice, rubber and fruit trees. The geology of the study area consisted of granite with its conversion as shown in figure 1 [14]. The influence of gas and hot solution was determined by detecting feldspar's conversion to mineral kaolin granite; the metal elements in rocks would be eroded and dissolved in the soil and groundwater easily [14].

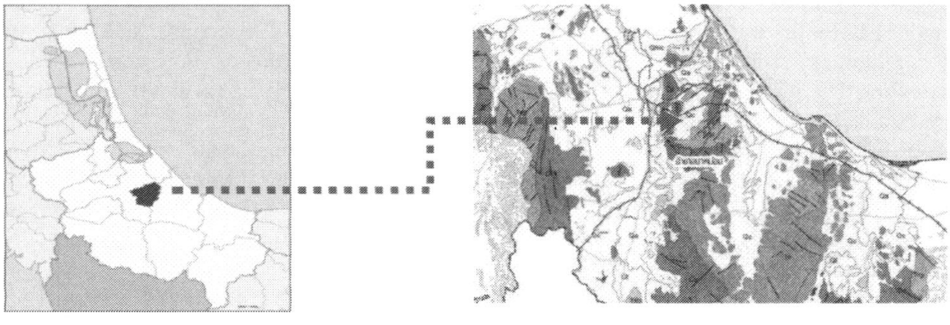

Figure 3. Map of the study area in Phichit subdistrict, Songkhla province [14].

2.2. Sample collection and measurements of radon activity concentration

A total of 35 water samples from Moo 3, Phichit subdistrict, were collected and measured for radon concentration. Shallow-well water was used by the residents in this area for consumption without prior treatment. The RAD7 portable radon detector (Durridge Company Inc.) was used in this study for the radon concentration measurement. The schematic diagram of RAD7 H_2O assembly is shown in figure 2. [15] The water samples were taken in 250-ml vials designed for the RAD7 device and provided by the manufacturer. During the sampling period the weather condition should be fairly stable. The water was hand pumped for 5-10 min prior to sample collection. The water sampling was quite complicated because the radon gas is easily escaped from water. The sampling has to be done without any aeration which might lead to outgassing. Hence, the water samples should be collected in such a way that no bubbling. In this research, water sample was measured simultaneously at the sampling site after collection. The time difference between taking the sample and measuring was only few minutes; hence no decay of radon occurred in the water samples. For accurate readings, the RAD7 has been dried out thoroughly to reduce the relative humidity to below 10% before making each measurement.

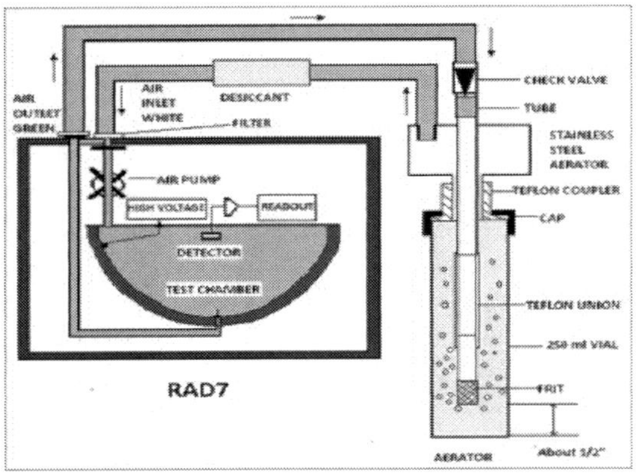

Figure 2. The schematic diagram of RAD7-H_2O assembly [15].

RAD7-H_2O is an addition to RAD7 that enables measurements of radon in water over a concentration range from less than 0.37 Bq/L up to 0.15×10^5 Bq/L. The operation of this instrument is based on the following principles: (1) radon is expelled from water sample by using a bubbling kit, (2) expelled radon enters a hemisphere chamber by air circulation, (3) polonium decayed from radon is collected onto a silicon solid state detector by an electric field, and (4) radon concentration is estimated from the count rate of polonium. RAD7-H_2O gives results within 30 min after measuring with a sensitivity that matches or exceeds that of the liquid scintillation method. The RAD7-H_2O method employs a closed loop aeration scheme whereby the air volume and water volume are constant and independent of the flow rate. The air that recirculates through the water continuously extracts the dissolved radon until a state of equilibrium develops. The RAD7-H_2O system reaches this state of equilibrium within about 5 min, after which no more radon can be extracted from the water.

The extraction efficiency or percentage of radon removed from the water to the air loop is very high about 94% for a 250 mL sample. The exact value of the extraction efficiency depends somewhat on ambient temperature, but it is almost always well above 90%. The RAD7 detector converts alpha radiation directly to an electric signal. RAD7 has the ability to tell the difference between the new radon daughters and the old radon daughters left from previous tests [15].

2.3 Evaluation of mean annual effective dose

Radon enters human body through ingestion and through inhalation as radon is released from water to indoor air. Therefore, radon in water is a source of radiation dose to both stomach and lungs. The annual effective dose for ingestion and inhalation were calculated according to parameters introduced by United Nations Scientific Committee on the Effects of Atomic Radiation (UNSCEAR) report.[16]

For ingestion, the following parameters were used:

- The effective dose coefficient from ingestion of radon is 3.5 nSv/Bq [16];
- Annual intakes by infants, children and adults are 100, 75 and 50 L, respectively;
- The annual effective doses from ingestion corresponding to 1 Bq/L are 0.35 μSv/year for infants, 0.26 μSv/year for children, and 0.18 μSv/year for adults.

For inhalation, the following parameters were used:

- Ratio of radon in air to radon in tap water supply is 10^{-4} Bq/m^3;
- Average indoor occupancy time per person is 7000 h/year;
- Equilibrium factor between radon and its progeny is 0.4;
- Dose conversion factor for radon exposure is 9 nSv/Bq/h/m^{-3}, and

- The annual effective dose from inhalation corresponding to the concentration of 1 Bq/L in tap water is 2.5 μSv/year.

The World Health Organization (WHO) [17] and European (EU) Council [18] recommended 0.1 mSv/year annual effective dose from drinking water as the safe limit of the three radioisotopes: ^{222}Rn, ^{3}H and ^{40}K.

3. Results and discussion

The results of radon measurements in shallow-well water from Phichit subdistrict are shown in table 1. The radon concentration in the samples ranged from 0.18± 0.07 to 98.1± 5.92 Bq/L with a mean value of 16.7 ± 2.33 Bq/L. Many health and environmental protection agencies recommended a safe limit of radon in drinking water for human beings. The US Environment Protection Agency proposed the allowed maximum contamination level for radon concentration in water to be 11 Bq/L [19]. UNSCEAR suggested a value of radon concentration in water for human consumption between 4 and 40 Bq/L [1]. These levels represented the concentrations that would not pose any significant risk to human health from drinking water over a lifetime. The recorded values of radon concentration in the shallow-well water were within the safe limit recommended by US Environmental Protection Agency [19] and UNSCEAR [1]. All measured radon activity concentrations in samples were well below the European Commission recommended reference level for radon in drinking water of 100 Bq/L [20].

From Table 1, the annual effective doses from ingestion and inhalation of radon were calculated according to the parameters introduced by UNSCEAR (2000) report [16]. The range of annual effective doses from ingestion and inhalation of radon were 0.03 - 17.66 and 0.45 - 245.25 μSv/year, respectively. The estimated total annual effective dose due to ingestion and inhalation ranged from 0.48 to 262.91 μSv/year. The World Health Organization [17] and EU Council [17] recommended 0.1 mSv/year annual effective dose from drinking water as the safe limit for three radioisotopes: ^{222}Rn, ^{3}H and ^{40}K. Hence, the total annual effective dose in most water samples (~90%) of this study was within the safe limit.

4. Conclusions

In this paper, the results of the radon measurements in 35 shallow-well water samples collected from Phichit district in Songkhla province were presented. The measurements were performed by using a RAD7 electronic radon detector. The majority of the shallow-well water samples were within the internationally recommended safe limit and therefore it was safe for drinking purposes. The total effective dose in most samples in this study were also within the safe limit (0.1 mSv/year) recommended by WHO and EU Council. There was no significant radiological risk due to radon ingestion for the residents in this area.

Table 1. Results of radon measurements in Shallow-well water

Sample No.	Radon Concentration in shallow-well water (Bq/L)				Temperature (^{0}C)	Annual effective dose (μSv/year)		
	Minimum	Maximum	Mean	SD		Ingestion	Inhalation	Total
1	3.93	6.73	5.52	1.17	25.5	0.99	13.80	14.79
2	6.47	8.13	7.35	0.70	25.5	1.32	18.37	19.69
3	8.82	9.70	9.19	0.39	25.2	1.65	22.97	24.62
4	23.30	26.90	24.70	1.73	24.3	4.45	61.75	66.20
5	0.29	0.73	0.51	0.19	23.4	0.09	1.27	1.36
6	77.80	96.30	89.00	8.11	24.0	16.02	222.5	238.52
7	90.10	104.00	98.10	5.92	24.3	17.66	245.25	262.91
8	23.60	32.30	29.00	3.78	24.5	5.22	72.50	77.72
9	5.91	8.28	7.02	1.08	24.6	1.26	17.55	18.81
10	9.41	11.50	10.40	0.88	24.3	1.87	26.00	27.87
11	11.50	16.60	13.60	2.13	24.3	2.44	34.00	36.44
12	6.43	8.09	7.27	0.68	24.3	1.30	18.17	19.47
13	5.59	7.94	6.65	1.18	26.4	1.20	16.62	17.82
14	55.90	68.20	59.90	5.61	25.6	10.78	149.75	160.53
15	3.86	5.65	4.56	0.78	24.0	0.82	11.40	12.22
16	10.60	15.20	12.60	1.95	24.0	2.27	31.50	33.77
17	86.20	93.00	90.70	3.03	24.2	16.32	226.75	243.07
18	27.20	30.10	28.70	1.25	24.8	5.17	71.75	76.92
19	13.80	18.80	16.70	2.33	27.4	3.00	41.75	44.75
20	2.42	2.86	2.61	0.18	33.8	0.47	6.52	6.99
21	14.70	16.70	15.60	0.83	33.5	2.80	39.00	41.80
22	1.14	1.57	1.39	0.18	32.8	0.25	3.47	3.72
23	2.00	3.37	2.75	0.82	36.2	0.50	6.87	7.37
24	1.29	1.86	1.61	0.24	34.1	0.29	4.02	4.31
25	7.87	11.30	9.82	1.43	35.9	1.77	24.55	26.32
26	1.86	3.13	2.64	0.55	33.2	0.47	6.60	7.07
27	1.14	2.29	1.57	0.55	37.4	0.28	3.92	4.20
28	0.86	2.43	1.50	0.68	32.8	0.27	3.75	4.02
29	3.15	5.43	4.32	0.98	34.1	0.78	10.80	11.58
30	0.43	1.29	0.79	0.38	33.5	0.14	1.97	2.11
31	0.14	0.29	0.18	0.07	34.7	0.03	0.45	0.48
32	0.57	2.00	1.11	0.62	36.2	0.20	2.77	2.97
33	1.00	3.29	1.93	1.07	34.4	0.34	4.82	5.16
34	4.43	5.58	5.21	0.53	33.2	0.93	13.02	13.95
35	11.20	12.8	12.10	0.74	37.1	2.18	30.25	32.43

SD-Standard deviation

References

[1] UNSCEAR 2008 *Sources and Effects of Ionizing Radiation 2008* (New York: UNSCEAR)
[2] Folger P F, Nyberg P, Wanty R B and Poeter E 1994 *Health Phys.* **67** 245-53
[3] Khan A J 2000 A study of indoor radon levels in Indian dwellings influencing factors and lung cancer risks *Radiat. Meas.* **32** 87-92
[4] Duggal V, Rani A, Mehra R 2014 Measurement of indoor radon concentration and assessment of doses in different districts of Northern Rajasthan *Indoor Built Environ.* **23** 1142-50
[5] Kendall G M and Smith T J 2002 Dose to organs and tissues from radon and its decay products *J. Radiol. Prot.* **22** 389-406

[6] Zhuo W, Iida T and Yang X 2001 Occurrence of ^{222}Rn, ^{226}Ra, ^{228}Ra and U in groundwater in Fujian province *J. Environ. Radioact.* **53** 111-200

[7] Rani A, Mehra R, Duggal V 2013 Radon monitoring in groundwater samples from some areas of Northern Rajasthan *Radiat. Protect Dosimetry* **153** 496-501

[8] Duggal V, Mehra R and Rani A 2013 Determination of ^{222}Rn level in groundwater using a RAD7 detector in the Bathinda district of Punjab *Radiat. Prot. Dosimetry* **156** 239-45

[9] Ramola R C, Rawat R B, Kandari M S and Choubey V M 1997 Measurement of radon in drinking water and indoor air *Radiat. Prot Dosimetry* **74** 103-5

[10] Sonkawade R G, Ram R, Kanjilal D and Ramola R C 2004 Radon in tube well drinking water and indoor air *Indoor Built Environ.* **13** 383-5

[11] Ali N, Khan E U, Akhter P, Khan F, Waheed A 2010 Estimation of mean annual effective dose through radon concentration in the water and indoor air of Islamabad and Murree *Radiat. Prot. Dosimetry* **141** 183-91

[12] Schubert M, Paschke A, Lau S, Geyer W and Knoller K 2007 Radon as a naturally occurring tracer for the assessment of residual NAPL contamination of aquifers *Environ. Pollut.* **145** 920-7

[13] Schubert M, Schmidt A, Muller K and Weiss H 2011 Using radon-222 as indicator for the evaluation of the efficiency of groundwater remediation by in situ air sparging *J. Environ. Radioact.* **102** 193-9

[14] Pungrassami T 1984 Tin mineralization of the Thung Pho - Thung Khamin mining district Changwat Songkhla Proceedings of the *Conference of Applications of Geology and the National Development* (Bangkok: Chulalongkorn University) pp 207-14

[15] Durridge Company, Inc. 2012 *RAD7, RAD H_2O Accessory Owner's Manual* (Billerica, MA: Durridge Company, Inc.)

[16] United Nations Scientific Committee on the Effects of Atomic Radiation 2000 *Sources and Effects of Ionizing Radiation* vol. 1 (New York: United Nations)

[17] World Health Organization 2004 *Guidelines for Drinking water Quality* vol. 1 (Geneva: World Health Organization)

[18] European Union Commission 1998 Council Directive 98/83/EC of 3 November 1998 on the Quality of Water Intended for Human Consumption *Official J. E.U. Comm.* **L330** 32-54

[19] United States Environmental Protection Agency 1991 *National Primary Drinking Water Regulations* (Washington, DC: US Government Printing Office)

[20] European Commission 2001 Commission recommendation of 20 December 2001 on the protection of the public against exposure to radon in drinking water supplies *Official J. E.U. Comm.* **L344** 85-8

International Nuclear Science and Technology Conference 2014 (INST2014) IOP Publishing
Journal of Physics: Conference Series **611** (2015) 012026 doi:10.1088/1742-6596/611/1/012026

Spatial distribution of ^{137}Cs in surface soil under different land uses in Chao Phraya watershed: Potential used as sediment source tracing

K Srisuksawad[1,3], B porntepkasemsan[1], N Noipow[1], A Omanee[1], W Wiriyakitnateekul[2], R Chouybudha[2], P Srimawong[2]
[1]Thailand Institute of Nuclear Technology, Nakhon Nayok, Thailand 26120
[2] Land Development Department, Bangkok, Thailand 10900

E-mail: kanitha.srisuksawat@gmail.com

Abstract. Sediment fingerprinting techniques involves the discrimination of sediment sources based on differences in source material properties and quantification of the relative contributions from these sources to sediment delivered downstream to the river catchments. Results of the previous study indicated that fallout radionuclides (FRNs); ^{137}Cs and excess ^{210}Pb ($^{210}Pb_{ex}$) are the most suitable radionuclides to be used as sediments sources tracers. This study investigated the spatial distribution of ^{137}Cs in soil under different land uses in Chao Phraya watershed; the most significant watershed in Thailand. Emphasis was placed on discriminating among potential sediment sources including the cultivated (upland crops), pasture field, uncultivated (swamp, forest, and grass field), and channel erosion (stream and river bank). One hundred and twenty four soil samples were collected from all sources and determining for ^{137}Cs. The ^{137}Cs mass activities in pasture areas varied from the limit of detection (LLD) to 1.22±0.05 with the average of 0.64±0.14 Bq kg^{-1}. In cultivated areas the ^{137}Cs mass activities varied from LLD to 1.41±0.04 with the average of 0.38±0.04 Bq kg^{-1}.The ^{137}Cs mass activities were higher in uncultivated areas varied from 0.12±0.03 to 1.73±0.05 with the average of 0.76±0.15 Bq kg^{-1}. The ^{137}Cs mass activities in channel bank varied from LLD to 1.16±0.04 with the average of 0.39±0.05 Bq kg^{-1}.GIS and geospatial interpolations revealed pattern in the spatial concentrations of ^{137}Cs and indicated important differences in its distributions showing the differences behaviour of ^{137}Cs in different land uses.

1. Introduction

Information on sediment source or provenance is difficult to obtain using conventional monitoring techniques, but in most environments it can be provided by sediment source fingerprinting techniques. The sediment source fingerprinting techniques involve the discrimination of sediment sources based on differences in source material properties and quantification of the relative contributions from these sources to sediment delivered downstream to the river catchments [1]. The fingerprinting procedures either employs statistical testing of a range of source material tracer properties to select a subset that discriminate sources [2] or the other approaches based exclusively on selected properties such as fallout radionuclides [3,4] or mineral magnetic measurements [5,6]. The key difference between these early approaches was the priori selection of specific tracers properties based on well established source

[3] To whom any correspondence should be addressed.

differences, in contrast to multi-parameter source fingerprinting which relies on statistical selection of subset of properties to discriminate sources. Currently, there are many reports on the successful use of environmental radionuclides in tracing sediment sources for many different environment and locations around the world [4, 7-18].The potential for using environmental radionuclides for sediment source tracing has been recognized particularly a group of environmental radionuclides that reach the land surface from the atmosphere, namely, fallout radionuclides (FRNs) . The FRNs most widely used are ^{137}Cs and excess ^{210}Pb ($^{210}Pb_{ex}$) [4, 7, 8, 10-19].

Caesium-137 (half-life 30.17 yr) is produced during nuclear fission. It worldwide presents in the environment during the period from 1952 to the mid 1980s. During the nuclear weapon tests, ^{137}Cs was injected into the stratosphere where it circulated globally [20, 21]. Caesium-137 moved back to the troposphere with deposition on earth thus being strongly related to local precipitation patterns and rates [22]. Caesium-137 down to earth was strongly and rapidly adsorbed to cation exchange sites on clay and organic soil particles, subsequently showing limited mobility by chemical processes and being essentially non-exchangeable [23]. These are making the ^{137}Cs the valuable fingerprints for distinguishing surface and subsurface sediment sources and surface sources influenced by mixing and disturbance.

In this study we investigated the spatial distribution of ^{137}Cs in soil under different land uses in Chao Phraya watershed. The objective was to study the differences in the ^{137}Cs activity concentrations among potential sediment sources in order to use it as sediment sources tracing in the catchment. Emphasis was placed on discriminating among potential sediment sources included the cultivated (upland crops), pasture field, uncultivated (swamp, forest, and grass field), and channel erosion (stream and river bank). Geostatistics and Geographical Information Systems (GIS) techniques were applied together for mapping of the ^{137}Cs spatial distribution in surface soil.

2. Study area

Chao Phraya watershed occupies the area of 20,523 km^2 and covers 16 provinces in Central Thailand. It is the most important watershed in Thailand. Seventy percent of its catchment is the agricultural land comprising of 55% rice, 18% upland crops (maize, cassava and sugarcane), and 23% fruit trees and vegetables. Chao Phraya watershed encounters flooding in rainy season every year. In 2011 the watershed experienced the severe flooding which affected millions of farmland and area wide. Runoff from upland agricultural catchments to lowland flood plain has damaged 2.01 M ha of farmland including 1.5 M ha rice fields, 0.3 M ha upland crops and 0.12 M ha fruit trees. Soil under flooding has lost their water holding capacity and aggregate stability made them sensitive to soil erosion. Determination of the extent and severity of soil erosion in agricultural landscape is required to provide information on erosion, sedimentation and associated loss of nutrients resulting from flood events. In addition identification and apportion hot spots of land degradation are crucial information to be used in land inventory data highlighting main land management factors that may contribute to soil instability and erosion during an heavy rainfall events.This information is valuable in targeting appropriate soil and water conservation practices or strategies where they are the best deployed to the flood preparedness. With the well flooded preparedness, the severity of the flood impacts should be reduced.

Two small sub catchment with differences in their geologic origin and one flat plain were selected for this study : Khlong Dong sub catchment (15°35′3.23″-16°5′10.57″N , 99°35′3.19″-100°5′6.65″E; 80% younger terrace deposit, 15% alluvial deposit, 5% erosion surface and foot slope of culluvium of metamorphic rock) with 5-15% mean slope and located in Nakhon Sawan and Kamphaeng Phet province; Huai Lumpayan- Huai Phu Kirio sub catchment (15°8′12.37″-15°22′1.78″N, 100° 29′24.61″-100°52′40.3″N; 50% younger terrace deposit, 20% alluvial deposit, 5% erosion surface and foot slope, and 25% residuum and colluvium of sedimentary rock) with 5-35% mean slope and located in Lopburi province; and lower Chao Phraya flat plain (14°22′15.08″-15°32′51.27″N, 100°6′26.36″-100°32′39.82″E; 100% alluvial flood plain) and located Chai-Nat province.

Within each catchment, three potential sediment sources were defined: pasture (paddy field), cultivated (mixed cultivated) and channel bank.

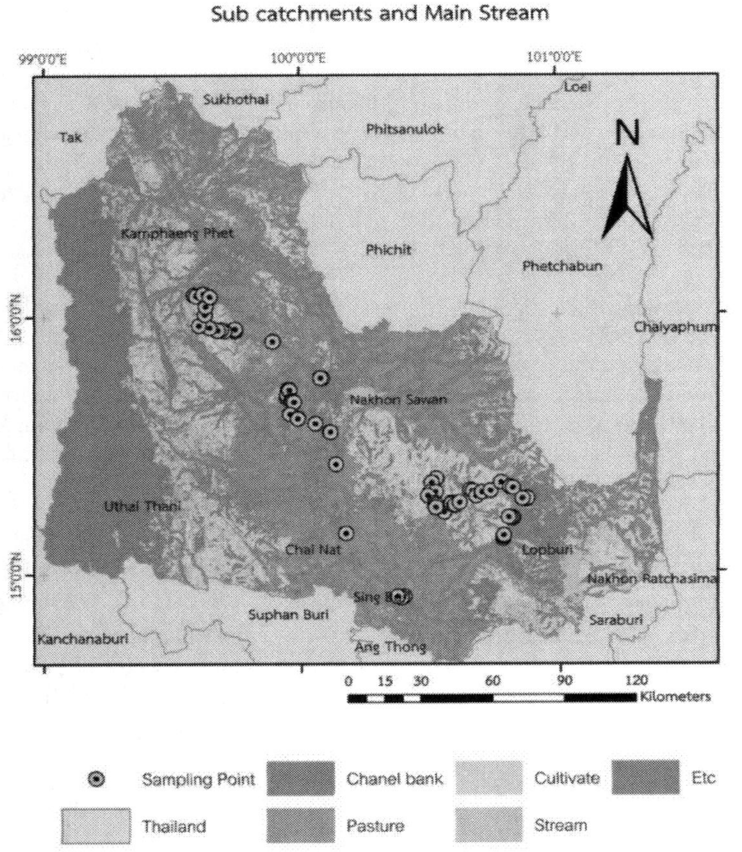

Figure 1. Land use scheme and soil sampling position

3. Materials and methods

One hundred and twenty four samples of soil and sediment samples were collected at sites of different land uses (Fig. 1). Soil and sediment were collected at the surface (0-2 cm depth) of sources material from areas with good connectivity to the stream. For the channel bank, the individual samples comprised composite samples of material from the full vertical extent of the bank profile. The source material samples retrieved from each sampling point were typically 0.5-1 kg in order to provide sufficient mass for subsequent laboratory analyses.

The target sediment and source material collected from the catchments were air-drying, disaggregation and sieving through a 2 mm sieve. The < 2 mm fraction of the potential sources materials and target sediments were used for subsequent radionuclide analyses.

An aliquot of about 500 g of each sample was placed in a Marinelli beaker in preparation for gamma spectrometer analysis. The mass activity concentration of ^{137}Cs were simultaneously determined by gamma –ray spectrometry, using an ORTEC high resolution Ge detector of 60% relative efficiency, coupled to a PC based digital analyzer system employing ORTEC Gamma Vision software. Count time was in excess of 80,000 sec per sample, providing results with analytical precision of around 1 % at the 95% level of confidence. The efficiency of the detector system was established using mixed radionuclides standard provided by AMALE INTERNATIONAL type EG-ML 1443-67-5.

4. Results and discussion

Within the study area the ^{137} Cs mass activity concentrations according to land uses have been calculated and the range and mean value ± SE (1σ) are shown in Table 1. The detailed information of the data is as the followings:

Table 1. Mean and range values for ^{137}Cs of different land uses. Means are reported with standard error (SE).The units are Bqkg^{-1}.

Radionuclide	Pasture	Cultivated	Uncultivated	Channel erosion
^{137}Cs				
Mean ± SE	0.64 ± 0.14	0.38 ± 0.04	0.76 ± 0.15	0.39 ± 0.05
Range	LLD-1.22	LLD-1.41	0.12-1.73	LLD-1.16
No. of sample	12	63	16	33

The mean value of ^{137}Cs activity concentrations in hill slope forest topsoil, hill slope cropland topsoil, and river sediments in Shan State, Myanmar were reported of 3.80±3.20 (n=11), 1.33±0.64 (n=11) and 0.67±0.40 (n=48) Bqkg^{-1}, respectively. At Belaga in Sarawak, Malaysia, an unseived (bulk) topsoil sample from the forest ridge had the ^{137}Cs activity concentration of 1.57±0.69 Bqkg^{-1}, and in addition six samples out of twelve of river sediment had the mean value of 0.60±0.10 Bqkg^{-1}.The ^{137}Cs activity concentrations in hill slope and river sediments (Cimanuk River catchment) of 0.88±0.19 Bqkg^{-1} (n=3) and 0.69±0.23 Bqkg^{-1} (n=4), were found in West Java, Indonesia. In East Java the ^{137}Cs activity was not found in bulk samples from hill slope cropland soils in Brantus River catchment. In Timor Leste, the weighted average of ^{137}Cs activity for hill slope topsoils was 2.11± 0.26 Bqkg^{-1}(n=20), however, the ^{137}Cs were not detected in river sediments [24].The ^{137}Cs activity concentrations in this study are therefore in the range of ^{137}Cs activity concentrations found at the same altitude in neighbouring countries.

The coefficients of variation (CV) of our study are also comparable with others. Mabit et al. [25] stated that the CV of reference samples in forested sites was large (0.19-0.47), but low in pasture and grassland (0.051-0.41). The CV of our results is within the range (0.13-0.22). It is found that the ^{137}Cs activity concentrations for all land uses in Huai Lampayan – Huai Phu Kirio sub catchment are higher than those in Khlong Dong sub catchment. The scattered plan of sampling, variation in slope gradient, the differences in parent rock and rate of soil mixing during planting has resulted in the variation of ^{137}Cs activity in samples. The variation of data between two sub catchments highlighted the important role of parent rocks and rate of soil disturbance in the study area.

Geostatistics together with geographic information system (GIS) was used for mapping of soil redistribution based on ^{137}Cs activity [26, 27]. This technique is applicable to small catchments with similar land use and topographic conditions [28]. In this present study, ^{137}Cs activity distribution was mapping based on land uses and similarities in each sub catchments. ^{137}Cs activity values determined for each land use within catchment was extended to other areas of the same land use within the same catchment. The ^{137}Cs distribution map is shown in Fig. 2. It revealed patterned in the spatial concentrations of ^{137}Cs and indicated important differences in its distributions showing the differences behaviour of ^{137}Cs in different land uses.

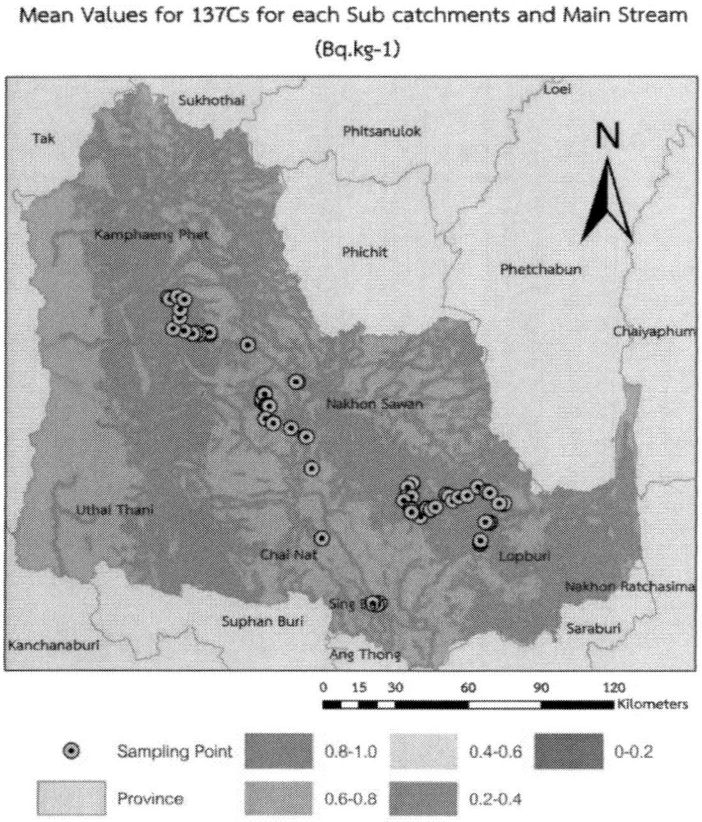

Figure 2. Caesium-137 distribution map of the Chao Phraya watershed.

5. Conclusion

Mass activity concentrations of ^{137}Cs in surface soil (0-2 cm depth) of the potential sediment sources in two sub catchments and one flood plain of the Chao Phraya watershed were determined. The activity concentrations were in the range of those in the neighbouring countries. The activity distribution reflected the important role of parent rock and rate of soil disturbance to the variation of the ^{137}Cs activity. The ^{137}Cs activity distribution map was carried out applying the Geostatistics together with geographic information system (GIS).

Acknowledgments

The authors gratefully acknowledge the financial assistance provided by Thailand Institute of Nuclear Technology (TINT- project no. 5701212). We would like to thank the International Atomic Energy Agency (IAEA) for supporting the mission of the expert, Prof. Yong Li of the Chinese Academy of Agricultural Science under the project RAS 5055 and THA 5050. The technical advice of Prof. Yong Li based on the application of the FRNs and his guidance on the soil sampling technique is deep gratitude.

References

[1] Smith H G and Blake W H 2014 *Geomorphology* **204** 177
[2] Collins A L and Walling D E 2002 *J Hydrol* **261** 218
[3] Wallbrink P and Murray A 1993 *Hydrol Process* **7** 297
[4] Wallbrink P, Murray A, Olley J and Olive L 1998 *Water Resour Res* **34** 879
[5] Oldfield F, Rummery T A, Thompson R and Walling D E 1979 *Water Resour Res* **15(8)** W0808
[6] Yu L and Oldfield F 1989 *Quaternary Res* **32** 168
[7] Olley J M, Murray A S, Mackenzie D H and Edwards K 1993 *Water Resour Res* **29** 1037
[8] Wallbrink P J, Olley J M, Murray A S and Olive L J 1996 *Erosion and Sediment Yield: Global and Regional Perspectives* (Wallingford: IAHS Press) p 347
[9] Krishnappen B G, Chambers P A, Benoy G and Culp J 2009 *Can J Civ Eng* **36** 1622
[10] Mizugaki S, Onda Y, Fukuyama T, Koga S, Asai H and Hiramatsu S 2008 *Hydrol Process* **22** 4519
[11] Froelich W and Walling D E 1997 *Human Impact on Erosion and Sedimentation* (Wallingford: IAHS Press) p 159
[12] Walling D E 2003 *Proc. Oslo workshop, Erosion and Sediment Transport Management in River: Technological and Methodologies Advances* (IAHS Press) p 57
[13] Walling D E 2005 *Sci Total Environ* **344** 150
[14] Walling D E 2006 *Soil Erosion and Sediment Redistribution in river Catchments: Measurement, Modeling and Management* (CABI UK) p 13
[15] Nagel G N and Ritchie J 1999 *Phys Geogr* **20** 348
[16] Nagel G N and Ritchie J 2004 *Land Degrad Dev* **14** 1
[17] Matisoff G, Bonniwell E C, and Whiting P J 2002 *J Environ Qual* **31** 62
[18] Mukundan R, Radcliff D E, Ritchie J C, Rissie L M and Mckinley R A 2010 *J Environ Qual* **39** 348
[19] Sculler P, Walling D E, Iroume A, Quilodran C, Castillo A and Navas A 2013 *J Environ Radioac* **124** 147
[20] UNSCEAR 1982 *Ionizing Radiations: Sources and Biological Effects. Report to the General Assembly (New York : United Nations)*
[21] UNSCEAR 2000 *Sources and Effects of Ionizing Radiation, Report to the General Assembly (New York : United Nations)*
[22] Ritchie J C and McHenry J R 1990 *J Environ Qual* **19** 215
[23] Schultz R K 1965 *Health Phys* **11** 1317
[24] Furuichi T and Wasson R J 2013 *J Asian Earth Sci* **77** 108
[25] Mabit L, Benmansour M and Walling D W 2008 *J Environ Radioac* **99** 1799
[26] Mabit L and Bernard C 2007 *J Environ Radioac* **97** 206
[27] Mabit L, Bernard C, Makhlouf M and Laverdiere M R 2008 *Geoderma* **145** 245
[28] Gharibreza M, Raj J K, Yusoff I, Othman Z, Tahir W Z W M and Ashraf M A 2013 *Soil Till Res* **131** 1

International Nuclear Science and Technology Conference 2014 (INST2014) IOP Publishing
Journal of Physics: Conference Series **611** (2015) 012027 doi:10.1088/1742-6596/611/1/012027

Measurements of indoor radon concentrations in Chaiya and Tha Chana districts, Surat Thani province, Thailand

K Titipornpun[1,5], A Titipornpun[2], P Sola[3] and T Bhongsuwan[4]
[1] Department of Physics, Faculty of Science and Technology, Suratthani Rajabhat University, Surat Thani 84100, Thailand
[2] Suratpittaya School, Muang district, Surat Thani 84000, Thailand
[3]Thailand Institute of Nuclear Technology (Public organization), Nakhon Nayok 26120, Thailand
[4] Department of Physics, Faculty of Science, Prince of Songkla University, Songkla 90110, Thailand

E-mail: kmaswiwat@hotmail.com

Abstract. Chaiya and Tha Chana districts of Surat Thani province are located in the areas with high levels of equivalent uranium at the ground surface, which have been identified as sources of radon. A survey measurement of indoor radon concentrations was carried out in 248 houses, using CR-39 detectors in closed cups. The geometric mean of indoor radon concentrations in Chaiya and Tha Chana districts were 26 ± 2 Bq•m^{-3} and 30 ± 2 Bq•m^{-3}, respectively. Although the minimum radon concentration was 4 Bq•m^{-3} in both locations, the maximum radon concentration was found to be 159 Bq•m^{-3} in Tha Chana district, while it was 88 Bq•m^{-3} in Chaiya district. The level of radon concentrations above the action level (148 Bq•m^{-3}) recommended by the United States Environmental Protection Agency was only found in two houses, which accounted for 1% of the total buildings surveyed in this present study. The majority of houses, which accounted for 94% of the total buildings surveyed, showed the radon concentration below the action level. As these houses had access to air flow during the daytime through open doors and windows, it is likely that such ventilation was sufficient to keep radon at a low concentration.

1. Introduction

Radon gas (^{222}Rn) is well known to pose public health risk as it has been reported that indoor radon is the second leading cause of lung cancer after tobacco smoking [1]. In addition, radon is also an important source of the natural radiation. The effective dose of radon gas from inhalation is more than half of all natural radiation sources [2]. Because radon is an inert gas, it can easily enter into houses through cracks in walls and solid floors, gaps in suspended floors and around service pipes, construction joints, cavities inside walls and water supply [3]. The United States Environmental Protection Agency (U.S. EPA) has recommended the "action level" of 148 Bq•m^{-3} to reduce lung cancer risk due to radon exposure in homes [4]. While the World Health Organization (WHO) proposed a reference level of indoor radon concentrations at 100 Bq•m^{-3} for minimization of health

[5] To whom any correspondence should be addressed.

hazards [5]. Therefore, the measurement of the indoor radon concentration is important and has been investigated around the world. The worldwide arithmetic mean of the indoor radon levels has been estimated at 39 Bq•m^{-3} [5]. In Thailand, some measurements of indoor radon concentrations have been studied. The indoor radon concentrations were measured at Phu Wiang district of Khon Kaen province, Saraphi district of Chiang Mai province and Na Mom district of Songkhla province which are located in the northeastern, northern and southern regions of Thailand, respectively. In these areas, no excess level of radon above the action level of EPA was found [6]. A feasibility of geogenic indoor radon mapping from airborne radiometric survey had been studied on three basins of northern Thailand. The estimated indoor radon zone map showed that there was no alluvial and terrace basins of northern Thailand found in the risk zone [7]. In southern Thailand, the indoor radon concentrations were measured in Songkhla province. The result showed that the highest average radon concentration above the action level of EPA [8] was found in Na Mom district. However, the risk zone due to indoor radon concentration cannot be identified without measurement.

The main objective of this study was to measure the indoor radon concentrations in Chaiya and Tha Chana districts of Surat Thani province, southern Thailand. In addition, variation in indoor radon concentrations due to building materials was investigated.

2. Survey location

Chaiya and Tha Chana districts, which are located in the northern part of Surat Thani province in the south of Thailand, were selected as the survey locations to collect data on radon concentration (figure 1). Chaiya and Tha Chana districts were subdivided into 9 subdistricts (54 villages) and 6 subdistricts (81 villages), respectively. In addition to our data collection, radiometric surveys of airborne particles were performed by the Department of Mineral Resources of Thailand. The unit of uranium concentration at the isotope equilibrium for uranium-238 series is shown in part per million, equivalent uranium (ppm eU). The equivalent uranium at the ground surface in more than half of the investigated locations were higher than 3 ppm eU, the level that has been identified as the high level of indoor radon [9]. A total of 248 buildings were randomly selected for radon measurement. These included 131 buildings from 9 subdistricts in Chaiya district and 117 buildings from 6 subdistricts in Tha Chana district. All buildings were separated to be dwellings (7 %), retail shops (79%), private offices (8 %), and public offices (6 %). Most buildings (97%) included in the survey area had one story, while the rest (3%) had two stories. The foundations of buildings were made from slab-on-ground (98%) and crawl space (2%), respectively.

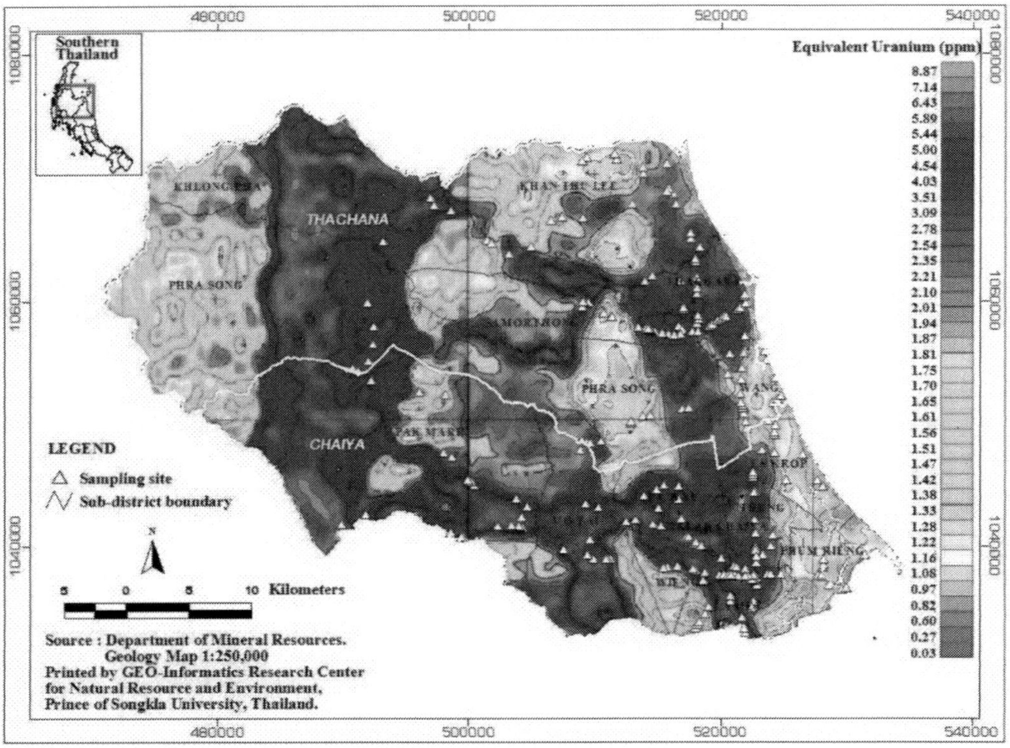

Figure 1. Map of Chaiya and Tha Chana districts, Surat Thani province, southern Thailand showing the survey locations (triangle symbol) and the equivalent uranium at the ground surface. The white line divides the territory into two districts.

3. Materials and Methods

A total number of 248 buildings were surveyed for measurements of indoor radon concentrations using CR-39 nuclear track detectors. The large rectangular sheets of CR-39 (29 cm x 32 cm, 1 mm thickness) were supplied by the Track Analysis Systems Ltd, UK. The sheets were then cut into 1.5 cm x 1.5 cm chips and each was numbered on the right corner for identification purpose. Control chips were stored in sealed bags at -15°C to minimize aging and gas leakage. For indoor radon measurement, each CR-39 chip was fixed by a small piece of the adhesive tape to the bottom centre of the plastic cup. Each cup has an 8.5 cm diameter orifice, a 5 cm diameter base and a depth of 9.5 cm. The orifice of each cup was closed with cling film to allow only ^{222}Rn gas to pass through the filter and to exclude the nongaseous radon daughters from entering the dosimeter [10, 11]. In each building, a detector was mainly placed in the living room for dwellings, and in the working room for retail shops, private and public offices. The height of each detector was placed about 1.5 m above the floor as representative of breath height inside the rooms and away from the windows and doors. All detectors were installed on the first floor for 40 days, the time period that was within the range of established protocol that allowed for the quick and reliable assessment of radon measurement [3, 4]. After exposure to radon in air, all measured detectors including background chips were chemically etched in 6.25 M of NaOH solution at 85°C for 100 minutes [12]. After chemical etching, each detector was washed with distilled water and then dried out. The alpha track density caused by radon occurred in each detector was counted manually under an optical microscope. To relate track density with indoor radon concentration, the dosimeters were evaluated at Thailand Institute of Nuclear Technology (Public organization), Nakhon Nayok province, Thailand, following the method of Sola *et al.* [13]. The track densities were converted into radon concentrations using the calibration factor of 0.052 track•cm^{-2}•d^{-1} per Bq•m^{-3}.

4. Results and discussion

4.1. Variation in indoor radon concentrations.

Figure 2 shows the frequency distribution histograms of the number of rooms as a function of indoor radon concentrations in the buildings in Chaiya and Tha Chana districts. The histogram of Chaiya district collected from 131 buildings (figure 2a) appeared to be slightly skewed to the right, whereas the histogram for Tha Chana district collected from 117 buildings (figure 2b) appeared to be strongly skewed. The majority of rooms had indoor radon concentrations in the range of 21 to 40 Bq m^{-3} at both locations. Because of the skew in the data distribution of Chaiya and Tha Chana districts, the data were transformed by applying the natural logarithm shown in figure 2c (Chaiya district) and figure 2d (Tha Chana district). The frequency distributions of the indoor radon concentrations at both locations were well described by the log-normal model. Therefore, the geometric mean (GM) and geometric standard deviation (GSD) as well as minimum (Min) and maximum (Max) values were used to describe the results and for statistical comparisons.

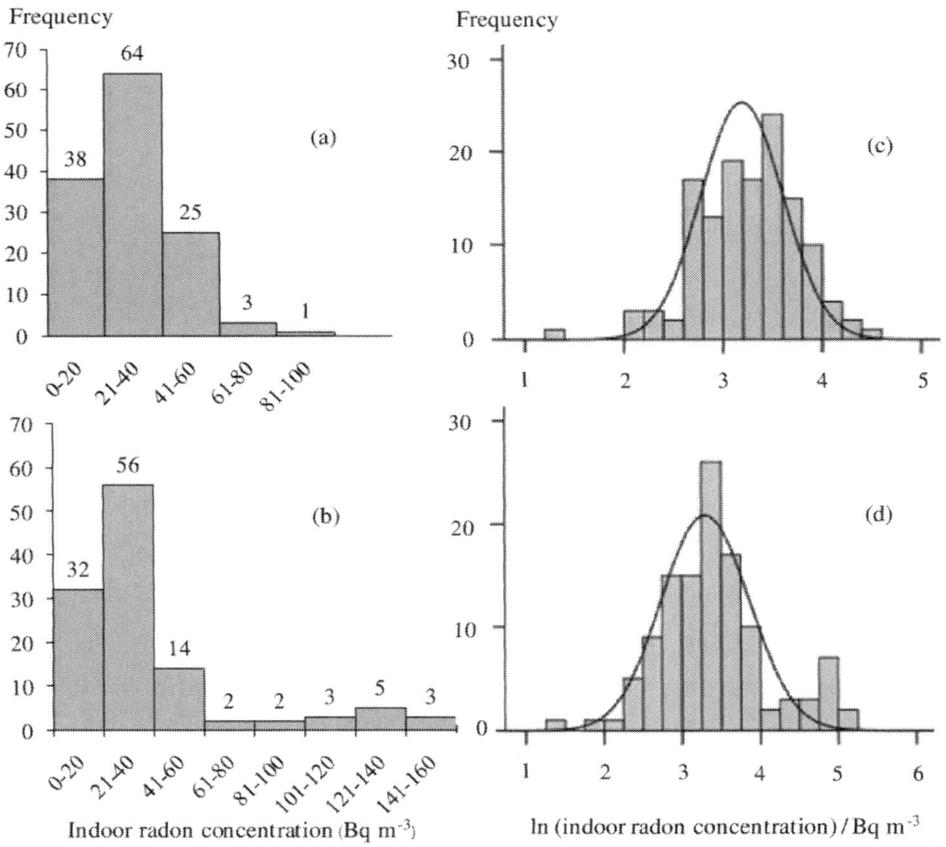

Figure 2. Frequency distributions of the indoor radon concentrations under normal plot and logarithmic transformation in Chaiya (a, c) and Tha Chana (b, d) districts.

Table 1 summarizes the statistics of indoor radon concentrations in Chaiya and Tha Chana districts. In Chaiya district, the indoor radon concentrations were found to vary from 4 - 88 Bq m^{-3}, whereas the indoor radon concentrations in Tha Chana district varied from 4-159 Bq•m^{-3}. The minimum indoor radon concentration was 4 Bq•m^{-3} in both Phum Rieng and Wang subdistricts located in Chaiya and Tha Chana districts, respectively. The maximum radon concentration was 88 Bq•m^{-3} in Pak Mark subdistrict of Chaiya district, whereas it was 159 Bq•m^{-3} in Khlong Pha subdistrict of Tha Chana district. The minimum value was found in the coastal areas, whereas the maximum values were found in the mountainous areas.

Table 1. Indoor radon concentrations in Chaiya and Tha Chana districts.

District/ subdistrict	Number of buildings	Indoor radon concentration (Bq•m^{-3})					
		Min	Max	GM[a]	GSD[b]	% (No.) >100 Bq•m^{-3}	% (No.) >148 Bq•m^{-3}
Chaiya district	**131**	**4**	**88**	**26**	**2**	**-**	**-**
1. Talad Chaiya	12	11	49	24	2	-	-
2. Ta Krop	14	16	39	24	1	-	-
3. Thung	15	16	49	24	1	-	-
4. Pak Mark	20	19	88	39	2	-	-
5. Pa Way	18	9	71	30	2	-	-
6. Phum Rieng	14	4	51	19	2	-	-
7. Mo Tai	9	16	58	33	2	-	-
8. La Med	17	11	57	24	2	-	-
9. Wieng	12	8	47	20	2	-	-
Tha Chana district	**117**	**4**	**159**	**30**	**2**	**9% (11)**	**2% (2)**
1. Khlong Pha	12	7	159	36	2	8% (1)	8% (1)
2. Khan Thu Lee	21	12	151	48	2	29% (6)	5% (1)
3. Tha Chana	23	13	136	30	2	13% (3)	-
4. Phra Song	20	7	72	27	2	-	-
5. Wang	21	4	48	20	2	-	-
6. Samor Thong	20	16	130	30	2	5% (1)	-
Total	**248**	**4**	**159**	**28**	**2**	**4% (11)**	**1% (2)**

[a]GM = geometric mean
[b]GSD = geometric standard deviation.

Comparison of the geometric mean indoor radon concentrations among the subdistricts showed statistically significant differences ($p < 0.05$, one-way ANOVA) in both districts. In Chaiya district, the lowest geometric mean of radon concentration of 19 ± 2 Bq•m^{-3} was found in Phum Rieng subdistrict, whereas the highest geometric mean of 39 ± 2 Bq•m^{-3} was found in Pak Mark subdistrict ($p < 0.05$, Tukey's HSD). In Tha Chana district, the lowest geometric mean of indoor radon concentration (20 ± 2 Bq•m^{-3}) was found in Wang subdistrict, while the highest value (48 ± 2 Bq•m^{-3}) was found in Khan Thu Lee subdistrict, which were significantly different ($p < 0.05$, Tukey's HSD). These results showed that the lowest and the highest geometric means of indoor radon concentrations were found in the areas located in low and high equivalent uranium at the ground surface, respectively (see figure 1).

The geometric mean indoor radon concentration for the entirely investigated areas of Chaiya and Tha Chana districts were 26 ± 2 Bq•m^{-3} and 30 ± 2 Bq•m^{-3}, respectively. The geometric mean of radon concentrations in Tha Chana district was significantly higher than in Chaiya district ($p < 0.05$, independent sample t-test). However, the geometric mean of radon concentration for the entire data (28 ± 2 Bq•m^{-3}) was lower than the worldwide average (39 Bq•m^{-3}) of indoor radon levels [5].

Comparing indoor radon concentrations with the WHO and U.S. EPA action levels, the results showed that 9% of 117 buildings in Tha Chana district exceeded the 100 Bq•m^{-3} action level recommended by WHO [5] while only 2% exceeded the 148 Bq•m^{-3} action level recommended by U.S. EPA [4]. In Chaiya district, no excess level of radon concentration above the action levels was found. With respect to the action levels, the indoor radon concentrations were found to exceed 100 Bq•m^{-3} in 4% of the buildings in the two districts together, while they exceeded 148 Bq•m^{-3} in only 1% of the buildings. We supposed that the reason for this was that most of the buildings in both districts (94%) were well ventilated because their windows and doors were left open throughout the day.

4.2. Variation in indoor radon concentrations due to building materials.
All of the building materials surveyed in this study were categorized into two different groups; wood-brick-mortar and brick-mortar. The wood-brick-mortar combination was accounted for 42% of total buildings, while the other 58% of total buildings were built from a combination of brick and mortar. The minimum and maximum indoor radon values measured from the buildings made of wood, brick and mortar were 7 Bq•m^{-3} and 130 Bq•m^{-3}, respectively, while the minimum and maximum values measured from the buildings made of brick and mortar were 4 Bq•m^{-3} and 159 Bq•m^{-3}, respectively (table 2). The geometric mean of indoor radon concentrations for the brick-mortar buildings was significantly higher than the wood-brick-mortar buildings ($p < 0.05$). This result may be due to the fact that wood is not a source of radon, while radon exhalation in bricks has been reported [14, 15, 16]. Most buildings in southern Thailand are nowadays made of brick and mortar, therefore the risk of indoor radon concentration due to building materials should be emphasized.

Table 2. Variation in indoor radon concentrations due to different building materials in Chaiya and Tha Chana districts.

Types of building materials	Number of Buildings	% of buildings	Indoor radon concentration (Bq•m^{-3})			
			Min	Max	GM[a]	GSD[b]
Wood, brick and mortar	104	42	7	130	24	2
Brick and mortar	144	58	4	159	31	2

[a]GM = geometric mean
[b]GSD = geometric standard deviation.

5. Conclusion
The geometric mean indoor radon concentrations in Chaiya and Tha Chana districts were 26 ± 2 and 30 ± 2 Bq•m^{-3}, respectively. The geometric mean indoor concentration for entire surveyed areas was 28 ± 2 Bq•m^{-3}. Only 1% of 248 buildings surveyed were found to have radon concentration level that exceeded the action level recommended by the U.S. EPA. Moreover, the geometric mean indoor radon concentration of the buildings made of brick and mortar was higher than the buildings made of wood, brick and mortar.

Acknowledgments
This research was sponsored by Suratthani Rajabhat University, Thailand. The authors would like to thank all participants in the investigated areas for the assessment of radon concentrations. We also gratefully thanked the Department of Mineral Resources of Thailand for providing the map of equivalent uranium at ground surface and the Geographic Informatics Research Centre for Natural Resource and Environment, Prince of Songkla University, Thailand, for the help on constructing a survey map corresponding to equivalent uranium at ground surface.

References

[1] IARC (International Agency for Research on Cancer) 1988 *IARC Monographs on the evaluation of carcinogenic risks to humans* **43** (Lyon: IARC)
[2] United Nations Scientific Committee on the Effects of Atomic Radiation 2000 *Report to the general assembly* (New York: United Nations)
[3] United States Environmental Protection Agency 2012 *EPA 402/K-12/002* (Washington D.C.: EPA)
[4] United States Environmental Protection Agency 1992 *EPA/402/R-62/011* (Washington, D.C.: EPA)
[5] World Health Organization 2009 *WHO Handbook on Indoor Radon: A Public Health Perspective* (Geneva: WHO press)
[6] Wanabongse P, Tokonami S and Bovornkitti S 2005 *International Congress Series* **1276** 208-9
[7] Wattananikorn K, Emharruthai S and Wanaphongse P 2008 *Radiat. Meas.* **43** 85-90
[8] Titipornpun A 1999 *Master of Science Thesis in Physics graduate school: in Thai* (Songkla: Prince of Songkla University)
[9] Duval J S 1988 *EOS, Transections American Geophysical Union* **70** 496-502
[10] Abu-Jarad F, Fremlin J H and Bull R A 1982 *Phys. Med. Biol.* **25** 683-94
[11] Abel-Ghany H A 2008 *Iran. J. Radiat. Res.* **6** 113-20
[12] Bhongsuwan T, Pisapak P and Duerrast H 2011 *Songklanakarin J. Sci. Technol.* **33** 609-16
[13] Sola B, Wanabongse P and Chankaw N 2009 *Siam Physics Congress* (Phetchaburi, Thailand)
[14] Malanca A, Pessina V, Dallara G, Luce C N and Gaidolfi L 1995 *Appl. Radiat. Isot.* **46** 1387-92
[15] Rizzo S, Brai M, Basile S, Bellia S and Hauser S 2001 *Appl. Radiat. Isot.* **55** 259-65
[16] Faheem M and Matiullah 2008 *Radiat. Meas.* **43** 1458-62

A simple laboratory system for diffusive radon flux measurements

C Kranrod[1,3], S Chanyotha[1], S Tonlublao[1] and W C Burnett[2,4]

[1] Department of Nuclear Engineering, Faculty of Engineering, Chulalongkorn University, Bangkok, Thailand

[2] Department of Earth, Ocean and Atmospheric Sciences, Florida State University, Tallahassee, Florida, USA

E-mail: kranrodc@gmail.com

Abstract. This study designed a simple, custom-made system to estimate the diffusive radon flux from solid materials (e.g., sediments, soils, building materials). Determination of the radon flux is based on the measurement of the radon activity in the air over time inside a closed loop system. For sediments, the system consists of wet sediment and water inside a gas-tight flask connected in a closed loop to a drying system and a radon analyzer (Durridge RAD7). The flux is determined based on an initial slope method in which the slope of radon activities vs. time plot during the first 12 h is evaluated. The slope is then multiplied by the total air volume and divided by the exposed sediment area to obtain the radon flux. The minimal thickness or mass of wet sediment should be about 4 cm. or (equivalent to approximately 150 g of wet sediment) to obtain a reliable radon diffusive flux in this study.

1. Introduction

The magnitude of terrestrial groundwater discharge to coastal water bodies has been reported in recent literature, with global fluxes estimated to range from 0.01 to 10% of total riverine discharge [1]. Therefore, groundwater discharge may also be an important pathway for the diffuse pollution to the coastal zone where coastal aquifers have become contaminated by septic systems [2]. These include nutrient elements, nitrogen, phosphorus and silicon [3-6] and dissolved gases such as methane (CH_4) and radon [7-9]. ^{222}Rn has seen an increasing use as a tracer for studying the exchange rate across both air-sea and sediment-water interfaces. In addition, ^{222}Rn has also been used in a wide range of environments for tracing submarine groundwater discharged (SGD) into the coastal zone and rivers. ^{222}Rn, which has a half-life of 3.82 days, is a naturally occurring radionuclide that originates from the ^{238}U decay series through the alpha decay of ^{226}Ra. Because it is a noble gas, it behaves conservatively in most chemical and biological reactions in nature. Its distribution will be controlled by balancing between the production from its parent (^{226}Ra), radioactive decay, and the physical transport process, including the biological mixing of sediment [10].

In the mass balance models that have been used for the estimation of sediment-water transport processes or SGD, information is needed regarding the flux of ^{222}Rn via the diffusion from sediments and the ^{222}Rn activity of the discharging water. The diffusive term can be assessed in a number of

[3] To whom any correspondence should be addressed.

[4] Visiting Professor in the Nuclear Engineering/Nuclear Technology Program at Chulalongkorn University, Bangkok, Thailand.

ways including *in situ* benthic flux chambers and laboratory incubations of sediment cores [11, 12]. All of these approaches are time consuming and may present numerous logistical difficulties. In some cases there is a considerable amount of spatial variability and an alternative approach is needed. One useful method is to conduct sediment equilibration in the laboratory where representative sediment sampled from the field site are reacted with water over sufficient period (e.g. 20 days) to ensure equilibrium between the ^{222}Rn in the water and the ^{222}Rn produced by the sediment-bound ^{226}Ra. Since typical SGD flow rates are on the order of centimetres per day, ^{222}Rn in the groundwater should have obtained the same level of equilibrium as measured in the laboratory.

In this paper, we present a simple, custom-made setup using an automated radon-in-air analyzer that is capable of generating data necessary for the evaluation of radon flux from sediments in a matter of hours to a day. Our results have demonstrated the feasibility and reliability of the setup and its potential use to estimate radon flux from all types of solid materials.

2. Experimental

2.1 Sediment samples

The sediments used in this study were collected on July 2013 from sampling station No.32 in the northern part of Tonle Sap Lake, Cambodia. Tonle Sap, the largest freshwater lake in Southeast Asia, is a classic flood-pulse system and hosts one of the most productive inland fisheries in the world, accounting for more than 75% of Cambodia's inland fish caught and about 60% of the country's protein needs [13].

2.2 Laboratory Setup

A 700-mL wide-mouth glass bottle with tight lid connected to plastic tubes was used as a reaction flask. The minimal mass/thickness of the sediment was evaluated by running the experiment with the same sediment under different mass/thickness conditions. We tested 25, 50, 100, 150, 200, 250 and 300g of wet sediment. All wet sediment-weight samples were added to a separate flask and then the sediment was covered with 300 ml of radon free water. After the sample had settled, the height of wet sediment in the flasks was measured. The wet sediment was added to the reaction flask without any modification since the sampling. The sediment was kept in a closed container the entire time from the sampling until the start of experiment. The flask was connected in a closed loop with a Tygon tube to a laboratory drying tube filled with Drierite desiccant and then to a Durridge RAD7 radon-in-air analyzer (figure 1).

The RAD7 radon detector is a high performance instrument that can determine the activity of ^{222}Rn continuously by means of electrostatic collection and measurement of the alpha-emitting daughters, ^{218}Po and ^{214}Po. The internal air pump of the RAD7 was set to the "auto" mode, which allowed the pump to operate continuously until the relative humidity reached 10% or less. Once that level is obtained, the pump operates for 1 min in very 5-min interval. The air is directed to the inlet tube of the reaction flask, which is submerged in the water with a gas diffusion stone on the end. The air is then channelled through the plastic outlet to the dryer and the radon analyzer. For most of the experiments the program on the RAD7 was set to run on a 2-hour cycles up to48 h, although the results reported here represent just the first 12 hours. For prolonged experiments, two drying tubes were set in series to ensure the low level of humidity. The temperature in the room was monitored and held steady at 24.5±0.5 °C.

2.3 Leak tests

It is known that the internal air pump of the RAD7 radon analyzer has a slow steady air leak [14,15]. This occurs because there is a pressure drop across the inlet filter and a pressure rise across the internal air pump and the pump itself is not 100% airtight. The extent of this leakage is very low and can be ignored for most radon analysis routines for up to several hours. However, as shown here, the leakage must be accounted for to get accurate results for a closed loop application with runs up to a few days.

Since the experimental system is a closed air loop, decay and leakage would lead to lower activities over time once radon has entered the system. If there was no leakage at all, observed drops in ^{222}Rn activities would be equal to the expected losses by decay. However, after an elevated activity of ^{222}Rn was introduced into the system, we observed slow drops in the activity above the decay losses virtually every time when the system operated longer than about 12 h. The extent of these losses could be evaluated by plotting the observed activities over time compared to those based on theoretical decay. This experiment was done by introducing a relatively high activity (~20,000 -30,000 Bq/m^3) of radon into the system from a Pylon ^{226}Ra standard source (Model RN-1025) that contains 98.138 kBq ^{226}Ra and monitored the activities for over about 40 h. The difference between the observed and the theoretical activity at each time step (1 h) could be evaluated by assessing a leak factor (LF) in the following manner:

$$LF(\% / hr) = 100 \bullet \left(\frac{A_{decay} - A_{obs}}{A_{decay}} \right) / t \tag{1}$$

where A_{decay} represents the theoretical activity (Bq/m^3) at each time step; A_{obs} is the observed activity (Bq/m^3) at the same time; and t is the cumulative time (h) from the start of the experiment. Our final leak factor term is based on the mean of the last five experimental results.

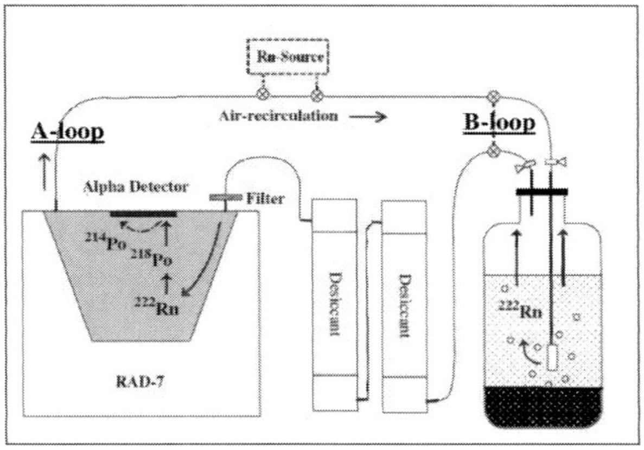

Figure 1. Schematic sketch of the experimental setup for evaluating ^{222}Rn released from sediment over time. The dashed lines with 3-way (⊕) are bypasses for the ^{222}Rn source to separate two parts of the system: the A-loop and the B-loop. See details in text.

3. Results and discussion

3.1 Leakage rates

We repeated leak tests 3 times on the same RAD7 (serial #1444) that was used for all the sediment equilibration experiments. Two of the tests were run with the pump in "auto" mode (result of one test is shown in figure 2a) and the third test was performed without the pump running (figure 2b). The results show the activities of ^{222}Rn steadily differ from the theoretical decay line as a function of time when the pump is run longer than about 12 h. To evaluate the extent of leakage, a graph of observed activities time was plotted in comparison to the theoretical decay. After 40 h, the observed trend is about 20% lower than the decay line. When the test ran with the pump turned off, there was very little difference between the observed and theoretical decay.

The loss rate of RAD7 was calculated to be 0.510±0.006 % per h based on the data in figure 2a. Since the estimated total air volume in our system was 1850 cm^3, the average rate of exchange between ambient air and the air inside the system was 9.3 cm^3 per hour. Xu [15] reported a RAD7 air loss of 5% in 20 h, which was equivalent to 0.25% per hour. This reported rate level was only half of the rate presented in this study. However, their experimental setup had four RAD7s in line together with a 12-liter reservoir for a total estimated air volume of 15,350 cm^3. This equates to an average air exchange rate of about 9.6 cm^3 per hour for each RAD7 pump, which was close to the value presented here. Therefore, it should be evaluated for each RAD7 used when running a closed loop for more than a few h.

Determination of the leakage rate, it has to correct the observed values by applying the following equation:

$$A_{cor} = A_{obs} + \left(\frac{LF}{100} \bullet A_{obs} \bullet t \right) \tag{2}$$

where A_{cor} is the activity of ^{222}Rn corrected for leakage. These are the activities that will be used to evaluate the radon diffusive flux.

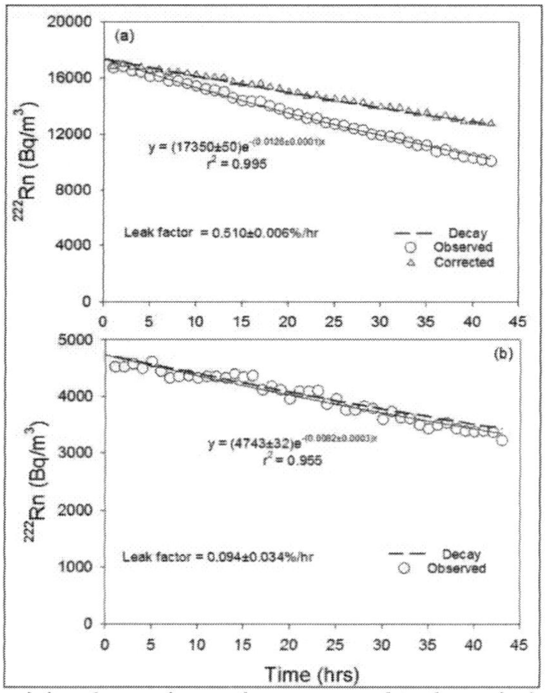

Figure 2. (a) Radon activity observed over time compared to theoretical decay. (b) A leakage test performed in exactly the same manner as in "a" but without the RAD7 air pump. In this case, the observed radon activities are in good agreement with the decay curve.

3.2 Radon diffusion from sediments

Radon activity in the air observed from the closed loop experiments are plotted as a function of time. Figure 3a shows the results of a 96 h run and the first 12 h are shown separately (Figure 3b). The ^{222}Rn activity was shown to steadily increase. As more time passes, it can observe that the activity trend started to bend as a consequence of the leakage and decay. Therefore, we correct for the leakage as described above (equation 2) and then base on our analysis the leakage-corrected results.

In order to determine the radon diffusion rate from the sediment in this study, one could rely on the nearly linear gradient during the first several hours in the closed system. The diffusive flux of radon (J_D) can be derived from the initial slope approach and is calculated as follows [14]:

$$J_D(Bq/m^2day) = \frac{slope\,(Bq/m^3hr) \bullet V_{air}(m^3)}{Area\,(m^2)} \bullet \frac{24hr}{day} \tag{3}$$

where V_{air} is the total air volume (1.850 x 10^{-3}m^3) and the area is the exposed area of the sediment in the reaction flask (0.0067 m^2). Therefore, based on the data shown in figure 3b, the linear slope of 2.79±0.28 Bq/m^3 hour was evaluated after the first 12 hours and the flux was calculated using equation 3 at 18.5±1.9 Bq/m^2 day.

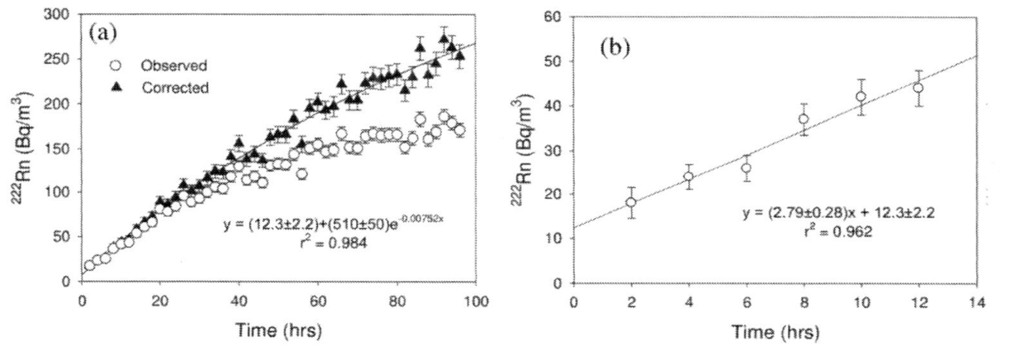

Figure 3. (a) A full 96-hour run on the 150 g sediment showing the best fit to an exponentially increasing function rising to a maximum. (b) First 12 h of the same run showing the linear gradient of 2.79±0.28 Bq/m^3.h and y-intercept regression of 12.3±2.2 Bq/m^3[14].

In the case of evaluating the radon diffusion from sediment by running the experiment for a few days to estimate the equilibrium ^{222}Rn activity, longer-term results can be used for curve fitting to evaluate the equilibrium level for the estimation of the radon flux. In our study, the same samples were run for 96 h to perform a best-fit analysis of the leakage-corrected results (figure 3a) using an exponentially increasing function rising to a maximum in the form [14]:

$$y = y_o + A_{eq}(1 - e^{-\lambda_{222}t}) \tag{4}$$

where the y-intercept (Bq/m^3), y_o, is taken from the linear regression based on results from the first several hours (Figure 3b); A_{eq} is the calculated equilibrium activity (Bq/m^3); and λ_{222} is the decay constant of ^{222}Rn (0.00752/hr). The equilibrium value can be determined by varying the A_{eq} coefficient until the best match is obtained for the trend of the leakage-corrected ^{222}Rn results over time. Therefore, the solid line through the leak-corrected points (figure 3a) represents the best-fit with an equilibrium value of 510±50 Bq/m^3. The radon flux can be calculated from the following equation [14]:

$$J_D\,(\text{Bq/m}^2\text{ day}) = \left(\frac{A_{eq}\,(\text{Bq/m}^3) \bullet V_{air}\,(\text{m}^3) + A_{water}\,(\text{Bq/m}^3) \bullet V_{water}\,(\text{m}^3)}{Area\,(\text{m}^2)}\right) \bullet \lambda_{222}\,(\text{1/day}) \tag{5}$$

where A_{water} is the ^{222}Rn activity (Bq/m^3) in the water; V_{air} and V_{water} represent the volumes of the air and water respectively; and λ_{222} is the decay constant of ^{222}Rn (0.181/day). Figure 3b shows the y-intercept of 12.3 Bq/m^3, which is close to the average radon activity in the room air (5 - 20 Bq/m^3) used to circulate through the closed system. A_{eq} is calculated based on the curve fitting equation (equation 4) and with a water volume of 3.0 x 10^{-4} m^3. The total air volume for our system was estimated to be 1.85 x 10^{-3} m^3. A_{water} is calculated based on the derived equilibrium ^{222}Rn activity in the air multiplied by the temperature dependent solubility coefficient, k [16]:

$$k = 0.105 + 0.405e^{-0.0502T} \tag{6}$$

where T *is* the temperature in °C (varied from 24.5 to 25.5 °C during all the experiments).

A comparison of the two approaches used in estimating the radon diffusive flux from sediments shows that the equilibrium approach generally provides better precision with lower uncertainties (table 1). The initial slope method had propagated uncertainties estimated to be between 10 to 30% while the equilibrium approach displayed about 10% of uncertainty for all the runs reported here. The initial slope method depends on assessing the radon-in-air activities that are quite low (e.g., < 40 Bq/m^3 for measurement 2: table 1) while the equilibrium approach is based on activities that 4-5 times higher. Thus, there is less uncertainty in individual measurements and less scattering of data in the curve fitting.

Table 1. Calculated radon diffusion from Tonle Sap sediment sample ran in triplicate based on: (1) a linear gradient over the first 12 hours of the experiment; and (2) an equilibrium model.

Measurement Number	(1) Initial Slope (Bq/m^{3}hour)	(1)Flux Bq/m^2 day	(2) Equilibrium Bq/m^3	(2) Flux Bq/m^2 day
1	3.8±1.1	25.4±7.3	580±20	30.2±2.4
2	2.8±0.3	18.5±1.9	510±30	26.5±2.4
3	2.7±0.3	17.6±2.0	480±30	25.0±2.6

Table 2. Calculated radon diffusion flux from 7 different mass/thickness conditions of wet sediment.

ID	Sediment Weight (g)	Sediment Thickness (cm)	Initial Slope (Bq/m^{3}hour)	Flux Bq/m^2 day
1	25	1.3	0.5±0.1	3.4±0.9
2	50	1.8	1.6±0.1	10.3±0.9
3	100	2.5	2.7±0.3	17.7±2.0
4	150	3.8	2.8±0.3	18.4±1.9
5	200	5.3	2.9±0.6	19.0±4.2
6	250	6.0	2.5±0.3	16.8±2.3
7	300	6.8	2.6±0.8	17.5±5.4

To evaluate the minimal thickness/mass of the wet sediment for a reliable radon diffusive flux in this study, the linear gradient was used during the first 12 h for each different mass/thickness condition. Table 2 shows the calculated radon diffusion flux from 7 different mass/thickness conditions. These results (figure 4) show that radon flux increases with the sediment weight from 25 g (0.5 cm) to 100 g (2.5 cm) whereas the radon flux values at sediment weight of 100g (2.5 cm) to 300 g (7.6 cm) seem to be constant within the experiment uncertainty. Corbett [17] reported that most radon diffusion in muddy lake sediments does occur in the upper 4-5 cm, which is consistent with our results. At some point, the sample weight of wet sediments should reach "infinite thickness", i.e., additional mass in same area has no effect. Based on our results, one reaches an infinite thickness at about 4 cm. (wet sediment weight of 150g in this case).

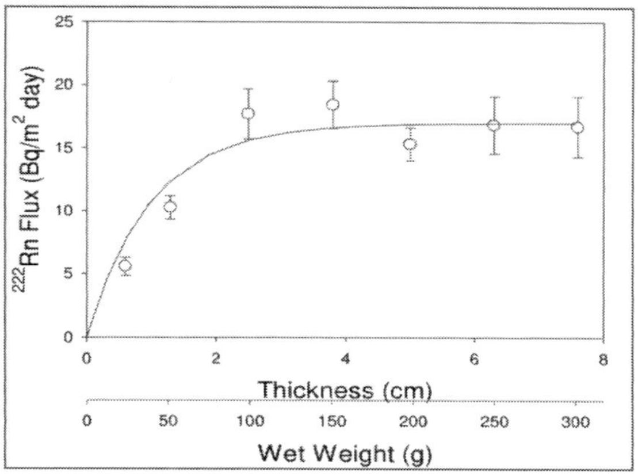

Figure 4. Radon diffusion flux from different mass/thickness conditions.

4. Conclusions

To calculate the reliable and accurate flux, an initial slope approach could be used during the first 12 h. During this period, the small leakage of RAD7 could be ignored. In addition, the thickness of wet sediment of about 4 cm. (wet weight of 150g) was required for a reliable estimation of radon diffusive flux from muddy sediments.

References

[1] Taniguchi M, Burnett W C, Cable J E, and Turner J V 2002 Investigation of submarine groundwater discharge *Hydrol. Process.* **16** 2115

[2] Buddemeier R W 1996 Groundwater discharge in the coastal zone *Proc. of an international symposium*: *LOICZ/R&S/96-8 (Texel)*

[3] Rowe G, Clifford C H, Smith K L Jr. and Hamilton P 1975 Benthic nutrient regeneration and its coupling to primary productivity in coastal waters *Nature* **255** 215

[4] Howarth R W 1988 Nutrient limitation of net primary production in marine ecosystems. *Annu. Rev. Ecol. Evol. Syst.* **19** 89

[5] Rocha C, Galvão H and Barbosa A 2002 Role of transient silicon limitation in the development of cyanobacteria blooms in the Guadiana estuary, south-western Iberia. *Mar. Ecol. Prog. Ser.* **228** 35

[6] Gilbert P M, Heil C A, O'Neil J M, Dennison W C and O'Donohue M J H 2006. Nitrogen, phosphorus, silica, and carbon in Moreton Bay, Queensland, Australia: Differential limitation of phytoplankton biomass and production. *Estura. Coast.* **29** 209

[7] Hammond D, Simpson H J, and Mathieu G. 1975 Methane and Radon-222 as Tracers for Mechanisms of Exchange across the Sediment-Water Interface in the Hudson River Estuary In Marine Chemistry in the Coastal Environment *American Chemical Society Symposium Series 18* ed Church T M (Washington D C: American Chemical Society)

[8] Kameyama S, Tsunogai U, Gamo T, Gamo T, Zhang J, Suzuki M A I and Koyama Y 2005 Geochemical studies on submarine groundwater discharges in Toyama Bay using methane as a tracer *Chikyu Kagaku* **39** 131

[9] Cable J E, Burnett W C, Chanton J P and Weatherly G 1996 Modeling groundwater flow into the ocean based on ^{222}Rn *Earth Planet Sci. Lett.* **144** 591

[10] Cothern C R 1987 *Environmental Radon* (*Environmental Science Research* Vol 35) ed Cothern C R and Smith J E Jr (New York: Plenum) chapter Properties pp 1-29
[11] Martens C S, Kipphut G W and Klump J V 1980 Sediment-water chemical exchange in the coastal zone traced by *in situ* radon-222 flux measurements. *Science* **208** 285
[12] Corbett D R, Burnett W C and Cable P H 1997 Tracing of groundwater input into Par Pond Savannah River Site by Rn-222. *J. Hydrol.* **203** 209
[13] Lamberts D 2006 The Tonle Sap Lake as a productive ecosystem Int. *J. Water Resur. D.* **22** 481
[14] Chanyotha S, Kranrod C and Burnett W C 2014 Assessing diffusive fluxes and pore water radon activities via a single automated experiment. *J. Radioanal. Nucl. Chem.* **301** 581
[15] Xu B, Burnett W C, Lane-Smith D and Yu Z G 2010 A simple laboratory-based radon calibration system *J. Radioanal. Nucl.Chem.* doi: 10.1007/s10967-009-0427-6
[16] Schubert M, Paschke A, Lieberman E and Burnett W C 2012 Air-water partitioning of ^{222}Rn and its dependence on water temperature and salinity *Environ. Sci. Tech.* **46** 3905
[17] Corbett D R, Burnett W C, Cable P H and Clark S B 1998 A multiple approach to the determination of radon fluxes from sediments *J. Radioanal. Nucl. Chem.* **236** 247

^{99m}Tc-ciprofloxacin for diagnosis of bacterial infection

A Aungurarat[1], T Ngamprayad, M Dangprasert, S Phumkem and B Jowanaridhi
Radioisotope Center, Thailand Institute of Nuclear Technology, Bangkok, Thailand

E-mail: jangkanan@yahoo.com

Abstract. Preparation of ^{99m}Tc-ciprofloxacin for diagnosis of bacterial infection was investigated by varying factors which affected this compound. The optimum conditions for preparation of ^{99m}Tc-ciprofloxacin and a lyophilized kit for Tc-99m labelling were studied. The results from biodistribution study showed that the percentages of the injected dose per gram tissues of infected area at 1 and 3 hours after injection were around 0.25– 0.56. ^{99m}Tc-ciprofloxacin was found sterile, pyrogen-free and non-toxic. Radiochemical purity was greater than 90% with greater than 6 hours of stability.

1. Introduction

At present, infectious diseases remain a major health problem and cause of death worldwide, particularly in developing countries. Diagnosis, therapy and control of the infection are challenges faced by medical personnel. Nuclear medicine techniques, because of their sensitivity, are often used in the diagnosis of fever of unknown origins or suspected bacterial infection especially deep-seated infection such as intra-abdominal abscesses, endocarditis and osteomyelitis.

The radiopharmaceuticals routinely used for infection imaging include ^{67}Ga-citrate, ^{99m}Tc- or ^{111}In-labelled leukocytes and ^{99m}Tc- or ^{111}In-labelled human immunoglobulin (HIG). However, each of them has limitations such as a low blood clearance, the time-consuming procedure, difficulties in blood handling or a viral infection risk [1-2]. Therefore, a novel approach using ^{99m}Tc-labelled antibiotics has gained much attention owing to its antibacterial activity.

The first antibiotic that was commercialized for infection imaging was ciprofloxacin [3-4]. Ciprofloxacin, a fluoroquinolone broad spectrum antibiotic, binds to and inhibits DNA gyrase of bacteria. It is metabolized in the liver and excreted via the kidneys. The molecular structure of ciprofloxacin is shown in figure 1. The preparation of ^{99m}Tc-ciprofloxacin is much easier than that of ^{99m}Tc- labelled leukocytes. These features make ^{99m}Tc-ciprofloxacin potentially useful for infection imaging. As previously reported, ^{99m}Tc-ciprofloxacin has been extensively evaluated with good sensitivity for diagnosis and localization to a wide varieties of bacterial infection lesions [5-11]. Moreover, ^{99m}Tc-ciprofloxacin showed high sensitivity that could aid the physicians in detecting the lesions in their early stages and would be useful for monitoring clinical response.

[1] To whom any correspondence should be addressed.

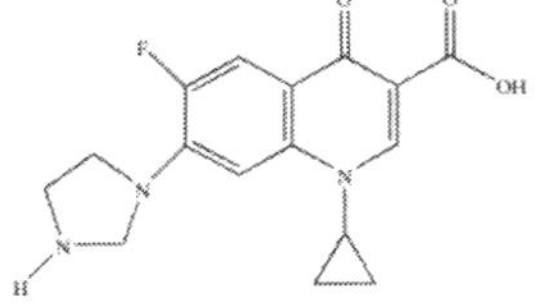

Figure 1. Molecular structure of ciprofloxacin.

The objectives of this study were to optimize the condition for radiolabeling of ^{99m}Tc-ciprofloxacin and to develop a lyophilized kit for ^{99m}Tc labelling. On the other hand, the capability of ^{99m}Tc-ciprofloxacin to distinguish between bacterial infection and sterile inflammation is still controversial [12-15]. Therefore, the biodistribution of ^{99m}Tc-ciprofloxacin in septic and sterile inflammation-induced mice was also evaluated.

2. Materials and methods

2.1 Materials

The chemicals used in this study were obtained from commercial sources (Sigma, Aldrich, Fluka and Merck). The ^{99}Mo/^{99m}Tc generator used in this study was the product of Sam Young Unitech Co., Ltd., Korea.

2.2 Optimization of the labelling condition

$SnCl_2.2H_2O$ was added into ciprofloxacin solution. Then, sodium pertechnetate was added into the solution after the pH was adjusted. After incubation, the mixture was filtered through 0.22 µm cellulose acetate membrane filter and then the quality control was performed.

Various labelling conditions were studied including the amount of both $SnCl_2.2H_2O$ (10 – 400 µg) and ciprofloxacin (1 – 10 mg), pH 2 – 10, activity of Na^{99m}TcO$_4$ (10 – 50 mCi), the incubation temperature (25, 60 and 100°C) and the total volume of the reaction mixture (1 – 5 mL) [5,8].

The radiochemical purity (RCP) was analysed by ITLC-SG or Whatman#31ET in acetone (R_f of free TcO_4^- = 0.9 – 1.0) and ITLC-SG in a mixture of ethanol : H_2O : NH_4OH (2:5:1) (R_f of H.R. and tin colloid = 0 – 0.1). The radiochemical purity of ^{99m}Tc-ciprofloxacin should be not less than 90.0%. Whereas labelling efficiency was obtained by the same procedure using the pre-filtered solution.

2.3 Stannous kit formulation

A mixture of $SnCl_2.2H_2O$ and sodium tartrate was formulated as a lyophilized kit. The mole ratio of $SnCl_2.2H_2O$ and sodium tartrate at 1:1, 1:2, 1:3, 1:4 and 1:5 when $SnCl_2.2H_2O$ were 400 and 500 µg were studied.

2.4 Stability test

The stability of ^{99m}Tc-ciprofloxacin was analyzed by ITLC at 0.5, 2, 4 and 6 h after labelling.

2.5 Biodistribution in mice

^{99m}Tc-ciprofloxacin biodistribution in male ICR mice, weighted 25 – 30 g, was performed in 5 groups (6 mice per group). Each group was pre-treated at the right thigh muscle 1 day prior to the test with *Staphylococcus aureus* (*S. aureus*), *Pseudomonas aerugenosa* (*P. aerugenosa*), heat-killed *S. aureus*, heat-killed *P. aerugenosa* and turpentine.

When the inflammation of the inoculated muscle was apparent, the mice were injected intravenously with 20 – 25 µCi / 0.1 mL ^{99m}Tc-ciprofloxacin. The urine of each mouse was collected and counted in the gamma counter. Three mice of each group were killed at 1 and 3 h after injection. Tissue samples were dissected and measured in the gamma counter. The results were expressed as percentages of the injected dose per gram (%ID/g).

2.6 Sterility, pyrogen and toxicity test

The sterility test was performed according to the compendial method stated in the United State Pharmacopeia (USP). Briefly, the test samples were incubated with both soybean casein digest medium (SCDM) and fluid thioglycollate medium (FTM) for 14 days at 22.5±2.5°C and 32.5±2.5°C respectively.

The pyrogen test is a test for bacterial endotoxin using the gel-clot technique. The LAL test kits from Lonza USA were used in this study.

The toxicity was tested in 5 healthy male ICR mice, weighing 25 - 30 g. One mg of ciprofloxacin solution was intravenously injected into each mouse. The mice were observed over 24 hours following injection. All mice must survive and show no symptom of abnormal reaction.

2.7 High performance liquid chromatography (HPLC) analysis

The compounds were analysed by HPLC using 4 x 250 μ-Bondapack connected to an ultraviolet-visible (UV-VIS) spectrometer PDA at 254 nm and a flow cell YtSi detector for gamma energy less than 70 keV, eluted with 1.5 mL/min of a mixture of acetonitrile (eluent A) and 0.025 M H_3PO_4 pH 3 (eluent B). The gradient was as follows: 0 – 3 min of 100% eluent A, 4 – 5 min of 13% eluent A and 6 – 20 min of 100% eluent A [16].

3. Experimental results

3.1 Results

The optimum condition for wet labelling was 2 mg ciprofloxacin, 40 μg $SnCl_2.2H_2O$, pH 4 – 5 labelled with 10 – 30 mCi $Na^{99m}TcO_4$ in a total volume of 1 – 4 mL. The solution was incubated in boiling water for 10 min and let stand at room temperature for 20 min. Finally, the reaction mixture was filtered through 0.22 μm cellulose acetate membrane filter. The RCP of ^{99m}Tc-ciprofloxacin was not less than 90% up to 6 h at room temperature.

The effect of $SnCl_2.2H_2O$ quantity was shown in figure 2. The labelling efficiency was greater than 90% when the quantity of $SnCl_2.2H_2O$ was 30 - 60 μg.

The effect of ciprofloxacin quantity was shown in figure 3. The labelling efficiency was greater than 90% when the quantity of ciprofloxacin was 1 - 6 mg.

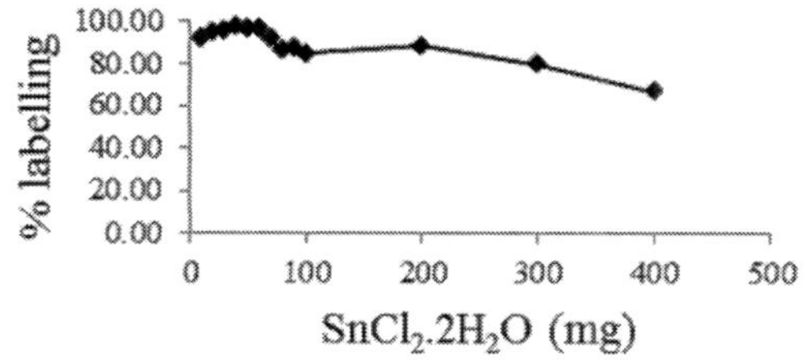

Figure 2. Effect of $SnCl_2.2H_2O$ quantity on labelling efficiency.

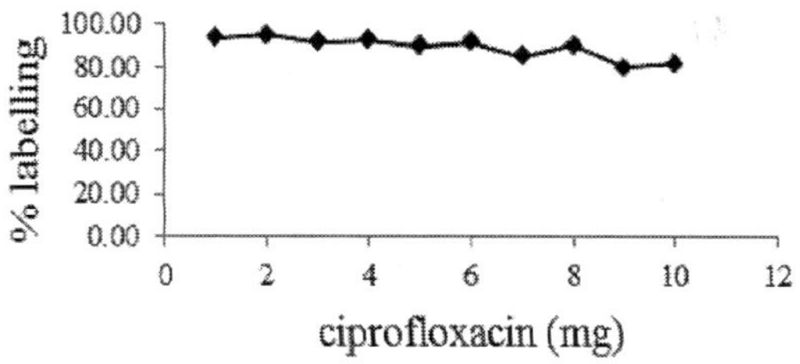

Figure 3. Effect of ciprofloxacin quantity on labelling efficiency.

The effect of pH was shown in figure 4. The labelling efficiency was greater than 90% when pH of ^{99m}Tc-ciprofloxacin was 4 – 5.

The effect of total volume of ^{99m}Tc-ciprofloxacin was shown in figure 5. The RCP was greater than 90% when total volume of ^{99m}Tc-ciprofloxacin was 1 – 4 mL.

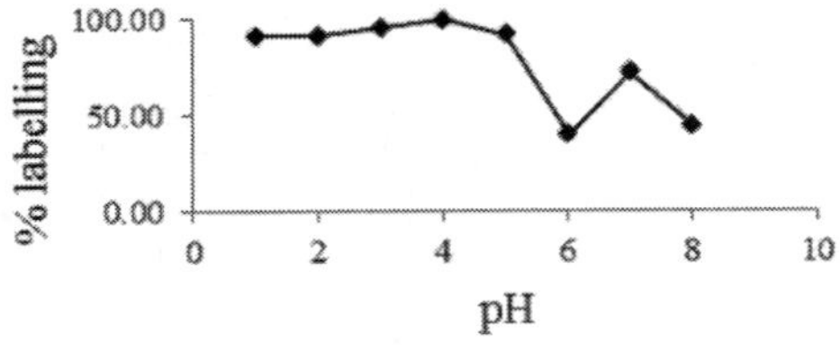

Figure 4. Effect of pH on labelling efficiency.

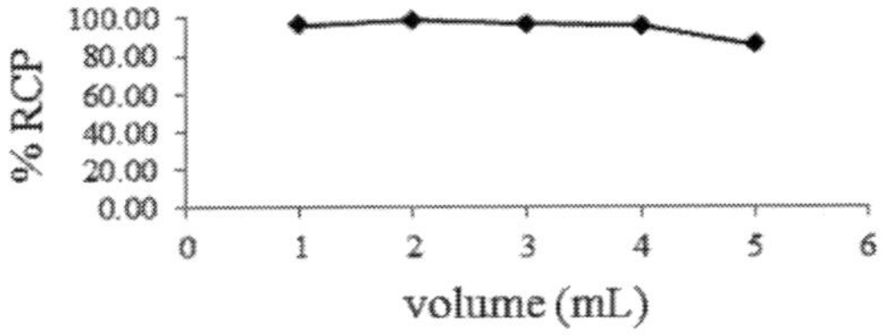

Figure 5. Effect of volume on RCP.

The effect of activity of $Na^{99m}TcO_4$ was shown in figure 6. The RCP was greater than 90% when the radioactivity of ^{99m}Tc-ciprofloxacin was 10 – 30 mCi.

The effect of incubation temperature was shown in figure 7 - 8. The RCP was greater than 90% when incubation was performed at 60°C and 100°C for 10 min.

For lyophilized kit, the mixtures of 1:1, 1:2, 1:3, 1:4 and 1:5 mole ratio of 400 µg and 500 µg $SnCl_2.2H_2O$: sodium tartrate were lyophilized and labelled with $NaTcO_4$ by the same condition as wet labelling. ^{99m}Tc-ciprofloxacin from lyophilized kit was sterile, pyrogen-free, non-toxic and having not less than 90% RCP when left at room temperature for 6 h (figure 9).

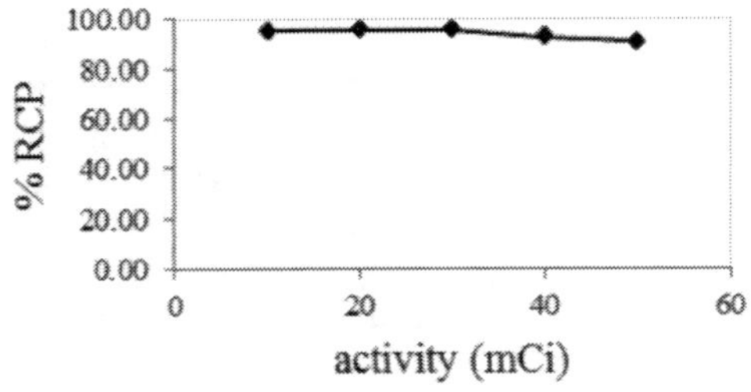

Figure 6. Effect of raidoactivity on RCP.

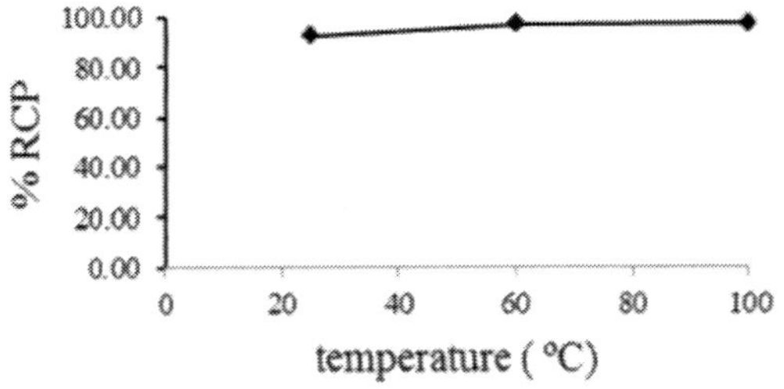

Figure 7. Effect of temperature on RCP.

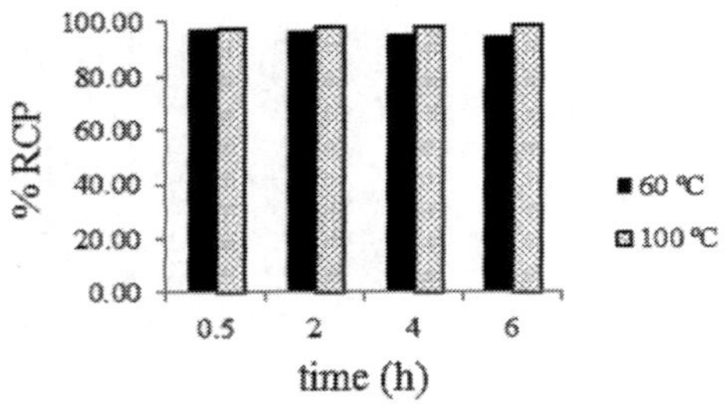

Figure 8. Stability of ^{99m}Tc-ciprofloxacin.

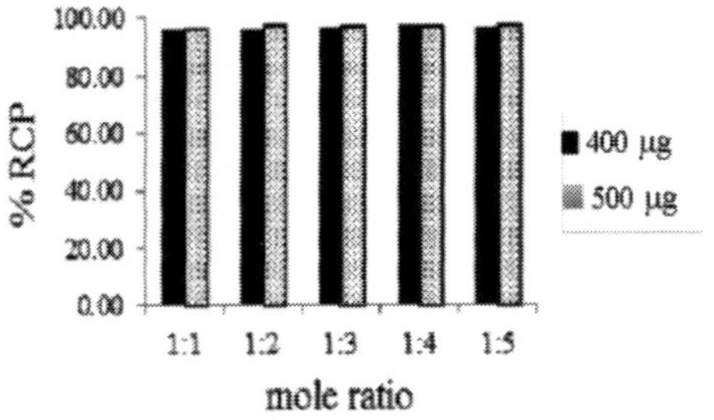

Figure 9. Effect of mole ratio on RCP.

From HPLC analysis, the UV chromatogram of ciprofloxacin showed a peak with the retention time of 8.8 min (figure 10a) and the same peak position was found on ^{99m}Tc-ciprofloxacin radiochromatogram (Figure 10b). The peak at 1.3 – 1.6 min on radiochromatogram represented free TcO_4^- (figure 10c).

The biodistribution of ^{99m}Tc-ciprofloxacin in mice was shown in figures 11 and 12. The infected:uninfected muscle ratio showed no differences between septic and turpentine-induced inflammation.

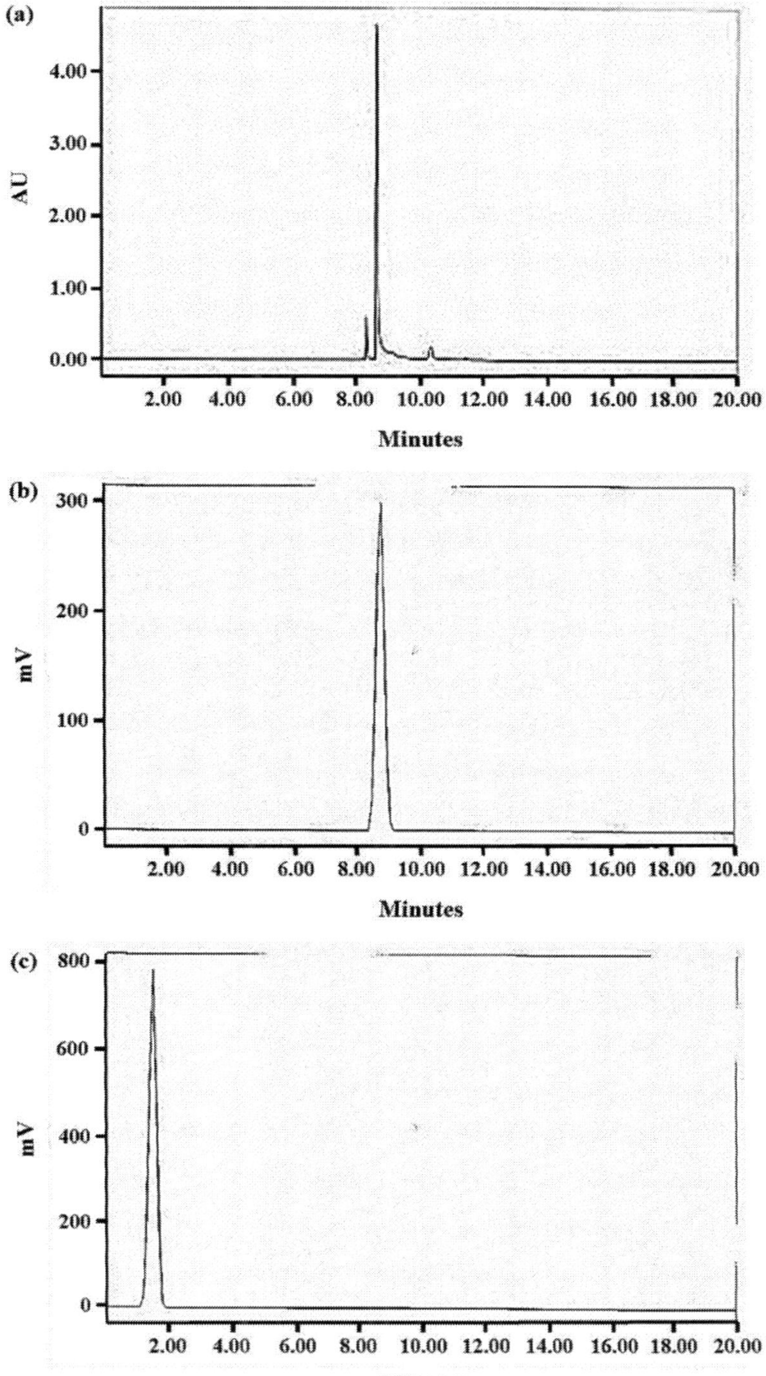

Figure 10. HPLC chromatogram (a) ciprofloxacin (b)^{99m}Tc-ciprofloxacin (c) $Na^{99m}TcO_4$.

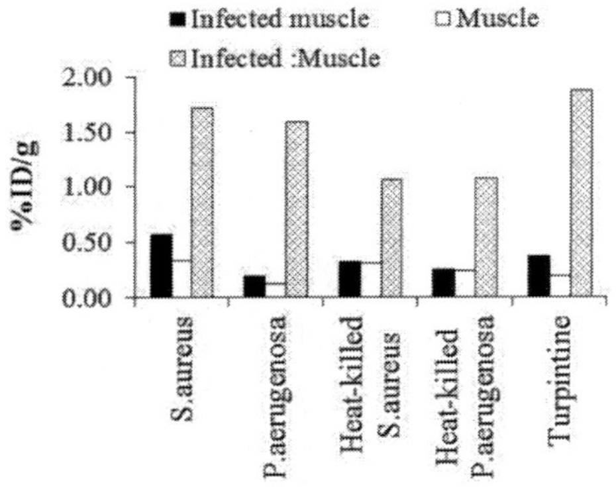

Figure 11. Biodistribution of ^{99m}Tc-ciprofloxacin in mice 1 hr after injection.

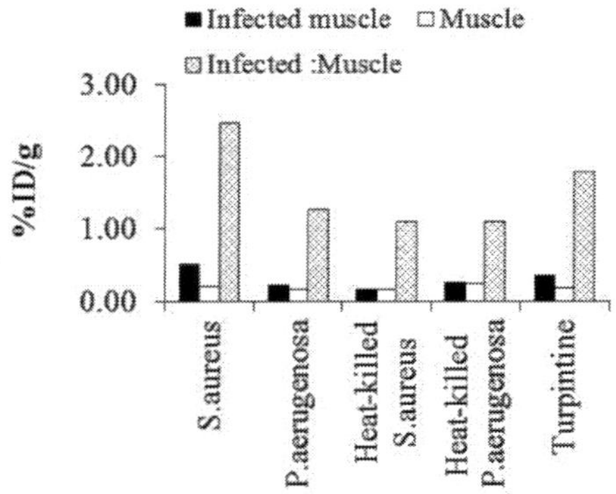

Figure 12. Biodistribution of ^{99m}Tc-ciprofloxacin in mice 3 hr after injection.

4. Discussion and conclusion

Chemical and biological quality control results of ^{99m}Tc-ciprofloxacin preparation in this study were satisfactory. The optimization of wet labelling condition was preliminary to the preparation of lyophilized kit. The $SnCl_2.2H_2O$: sodium tartrate lyophilized kit was developed to ease the labelling procedure and reduce radiation exposure to workers.

In this study, ^{99m}Tc-ciprofloxacin was a clear colorless solution with pH 4 – 5, having not less than 90% RCP up to 6 h at room temperature, sterile, pyrogen-free and non-toxic. It is promising for detection of septic and aseptic inflammation.

Acknowledgments

The researchers thank Mrs. Suchada Pongpat, a nuclear science specialist, Research and Development Division, Thailand Institute of Nuclear Technology, for providing the bacteria.

References

[1] Rennen H J J M, Boerman O C, Oyen W J G and Corsten F H M 2001 Imaging infection/inflammation in the new millennium *Eur. J. Nucl. Med.* **28** 241–52

[2] Chianelli M, Mather S J, Martin-Comin J and Signore A 1997 Radiopharmaceuticals for the study of inflammatory processes: a review *Nucl. Med. Commun.* **18** 437–55

[3] Hooper D C and Wolfson J S 1991 Fluoroquinolone antimicrobial agents *N. Engl. J. Med.* **20** 384–93

[4] Chu T W and Fernandes P B 1989 Structure-activity relationships of the fluoroquinolones *Antimicrob. Agents Chemother.* **30** 131–5

[5] Rien H S et al 2004 Synthesis and comparison of ^{99m}Tc-enrofloxacin and ^{99m}Tc-ciprofloxacin *J. Nucl. Med.* **45** 2088-94

[6] Vinjamuri S et al 1996 Comparison of ^{99m}Tc Infecton imaging with radiolabelled white-cell imaging in the evaluation of bacterial infection *Lancet* **347** 233-5

[7] Britton K E and Vinjurami S 1997 Clinical evaluation of technetium-99m Infecton for the localisation of bacterial infection *Eur. J. Nucl. Med.* **24** 553–6

[8] Larikka M J, Ahonen A K and Britton K E 2002 ^{99m}Tc-ciprofloxacin (Infecton) imaging in the diagnostic of knee prosthesis infections *Nucl. Med. Commun.* **23** 167–70

[9] Malamitsi J et al 2003 Infecton : a ^{99m}Tc-ciprofloxacin radiopharmaceutical for the detection of bone infection *Clin. Microbiol. Infect.* **9** 101–9

[10] Hall A V, Solanki K K, Vinjurami S, Britton K E and Das S 1998 Evaluation of the efficacy of ^{99m}Tc-Infecton, a novel agent for detecting sites of infection *J. Clin. Pathol.* **51** 215–9

[11] Britton K E et al 2002 Imaging bacterial infection with ^{99m}Tc-ciprofloxacin (Infecton) *J. Clin. Pathol.* **55** 817–23
[12] Kleisner I, Komarek P, Komarkova I and Konopkova M 2002 A new technique of ^{99m}Tc-ciprofloxacin kit preparation *Nuklearmedizin* **41** 224–9
[13] Dumarey N, Blocklet D, Appelboom T, Tant L and Schoutens A 2002 Infecton is not specific for bacterial osteo-articular infective pathology *Eur. J. Nucl. Med. Mol. Imaging* **29** 530-5
[14] Sarda L et al 2003 Inability of ^{99m}Tc-ciprofloxacin scintigraphy to discriminate between septic and sterile osteoarticular diseases *J. Nucl. Med.* **44** 920–6
[15] Sarda L, Saleh-Mghir A, Peker C, Meulemans A, Crémieux A C and Le Guludec D 2002 Evaluation of ^{99m}Tc-ciprofloxacin scintigraphy in a rabbit model of *Staphylococcus aureus* prosthetic joint infection *J. Nucl. Med.* **43** 239–45
[16] The United States Pharmacopeial Convention 2007 *The United States Pharmacopeia 30-National Formulary 25* (Rockville: The United States Pharmacopeial Convention) pp 1755-6

Gamma irradiation induces DNA double-strand breaks in fibroblasts: a model study for the development of biodosimetry

P Uttayarat[1], T Tangtong[1], K Sukapirom[2] and K Boonsirichai[1]
[1] Thailand Institute of Nuclear Technology (Public Organization), Ongkharak, Nakhon Nayok 26120 Thailand
[2] Division of Instruments for Research, Office for Research and Development, Faculty of Medicine, Siriraj Hospital, Bangkok 10700 Thailand

E-mail: puttayar@alumni.upenn.edu

Abstract. Double-strand breaks (DSBs) of DNAs induced by ionizing radiation can pose detrimental damages on organisms which include genetic instability and cell death. It is necessary to be able to assess health risks associated with irradiation from both accidental and therapeutic exposures in a timely manner for proper medical treatments. This present study showed the first attempt to develop a biodosimetric measure in Thailand based on the quantification of phosphorylated histone H2AX (γ-H2AX) formed at DSB sites with an aim to establish a dose response curve using a two-dimensional (2D) cell culture model. Human dermal fibroblasts were grown into monolayers before irradiated by gamma rays from a Co-60 source in a custom-made lead chamber at doses 0, 0.2, 1, 2 and 4 Gy and a dose rate of 0.21 Gy/min. After 30 min post exposure, γ-H2AX proteins were immunofluorescently labelled for evaluation by confocal microscopy and flow cytometry. The accumulation of phosphorylated γ-H2AX proteins at DSBs appeared as nuclear foci with the most prominent intensity at 4 Gy. Linear regression analysis of flow cytometric data showed a linear response (R^2 = 0.9862) of foci intensity in proportion to irradiation dose. In addition, the fraction of cell viability was shown to decrease at higher doses. This technique can be further developed as a quick assessment tool to identify individuals subjected to accidental radiation in parallel to other established biodosimetric measures.

1. Introduction

The structure of DNA which stores the cells' genetic materials is constantly subject to lesions generated by both internal and external agents [1–3]. Internally, products released from normal metabolic activities inside the cells including reactive oxygen species can induce base damage, sugar damage and breaks on one of the DNA strands [1,3]. Unlike the single-strand break (SSB) damage, lesions that induce simultaneous breaks on both strands of the DNA helices at sites that are located sufficiently close to one another are known as double-strand breaks (DSBs). The primary source of external agents responsible for DNA DSBs is ionizing radiation. These DSB damages are much more lethal to the cells as they lead to genetic instability if incorrectly repaired or cell death when the damage remains unrepaired [1,3].

After sensing the damage, the cells rapidly respond to DSBs through various mechanisms including DNA repair, cell cycle checkpoints and apoptosis. The latter serves to eliminate severely damaged cells that have lost the ability to regulate their normal growth [4]. The repair of DNA DSBs relies on two major processes, non-homologous end joining (NHEJ) and homologous recombination (HR) [1,2,4]. NHEJ is the major pathway in non-replicating cells, whereas HR functions in the late S/G2 phase [2,4]. Proteins that are involved in the early stages of cellular responses in sensing the damage and lead to proper repair processes have been the central focus of research in recent years. One of the key factors in such event is phosphorylation of the core histone variant H2AX on the serine-139 residue (γ-H2AX) [3,5,6] in the vicinity of DSBs, which in turn recruits other protein machinery such as MRN complex, ATM, 53BP1 and MDC1 proteins [4] to the damaged sites. The phosphorylation of γ-H2AX can be visualized through the formation of foci inside the cells' nuclei [3,7]. At low doses of radiation, each γ-H2AX focus approximately represents one DSB [5,6] and it has been noted that up to thousands of H2AX protein can be phosphorylated per DSB [3] at higher doses. The importance of H2AX in the DNA repair process was confirmed by previous study [3] which showed the high sensitivity and genomic instability of H2AX-deficient cells to ionizing radiation compared to control.

As a marker for DNA damage, γ-H2AX can be used as a biodosimetric tool to assess the extent of DSBs induced by accidental or therapeutic exposures to ionizing radiation. In the former case, quantification of γ-H2AX foci using blood samples has shown a promising potential [6] that supplements other standardized methods such as a dicentric assay for population triage. In therapeutic exposure, radiotherapy can potentially induce injuries to the skin [5], in which the γ-H2AX assay may be applied to estimate the resulting dose and the extent of such injuries. In this present study, we attempted to develop γ-H2AX foci as a biodosimetric measure as a part of the biodosimetry program initiated by the Office of Atoms for Peace, Thailand. Using the 2D cell culture model, we first established a dose response curve for gamma radiation in the range of 0 - 4 Gy.

2. Materials and methods

2.1 Cell culture

Normal human dermal fibroblasts (HDFBs) passages 8 - 12 were cultured in Dulbecco's modified Eagle's medium (DMEM) supplemented with 10% fetal bovine serum, 50,000 μg streptomycin and 50,000 U penicillin in 6-well tissue culture plates. The media was refreshed daily. HDFB monolayers at day 2 post confluence were used in all experiments.

2.2 Gamma irradiation

Lead containers with shelves were custom built as illustrated in figure 1 to accommodate samples during irradiation by gamma rays from the Co-60 source. Doses in the range of 0 - 5 Gy were first calibrated using optically stimulated luminescence dosimeters. HDFB monolayers were irradiated for the total doses of 0.2, 1, 2, and 4 Gy at the dose rate of 0.21 Gy/min. Non-irradiated HDFBs (0 Gy) were used as sham control. Within 30 min after exposure to gamma irradiation, HDFBs were washed with phosphate buffered saline and treated with trypsin for characterization of γ-H2AX foci by flow cytometry and laser scanning confocal microscopy. In addition, separate sets of HDFBs were subcultured and maintained until 90% confluence before harvesting for quantification of DNA fragmentation by gel electrophoresis.

Figure 1. Experimental setup inside the gamma radiation facility.

2.3 Immunofluorescence staining of γ-H2AX
All staining steps were done on ice. Following treatment with trypsin, HDFBs were fixed with 4% paraformaldehyde for 20 min, washed, centrifuged, and resuspended in 0.1% Triton-X 100 for 5 min. The cells were then incubated in 1% bovine serum albumin (BSA) solution for 30 min to block non-specific binding prior to incubation with mouse anti-human γ-H2AX clone JBW301 (EMD Millipore, USA) antibody at a dilution of 1:200 in 1% BSA solution for 1 h. After washing in phosphate buffered saline (PBS), the cells were incubated in a 1:500 goat anti-mouse Alexa Flour 488 (Invitrogen, USA) solution for 45 min, washed, and resuspended in 1% paraformaldehyde for analysis by flow cytometry. For visualization by confocal microscopy, a small drop of cells was mixed with antifade agent (ProLong Gold, Life Technologies, USA) and mounted on a glass slide.

2.4 Flow cytometry
Immunofluorescence intensity of γ-H2AX was measured using a BD FACSCaliber flow cytometer (BD Biosciences, USA). Data were collected from 3 independent experiments with at least 10,000 cells for each dose. Analysis was done using BD CellQuest and FlowJo softwares (BD Biosciences, USA; FlowJo, USA).

2.5 Visualization of γ-H2AX foci
For microscopy imaging, the cell nuclei were counterstained with Hoechst 33258 (Life Technologies, USA). Images of γ-H2AX foci were captured on a Zeiss LSM510 Meta microscope (Carl Zeiss, Jena, Germany). About 30 – 50 cells were imaged per irradiation treatment.

2.6 DNA fragmentation assay
Harvested HDFBs were incubated at 37 °C for 1 h in 10 mM Tris-HCl, pH 8.0, 20 mM EDTA, 200 mM NaCl, 0.2 % Triton X-100 and 100 µg/mL proteinase K. Genomic DNA was precipitated from the supernatant with an equal volume of isopropanol and 100 mM NaCl at -20 °C, and then dissolved in NanoPure water with RNase A. 1 µg of genomic DNA was then separated on 0.6% agarose gel in Tris-borate buffer and stained with ethidium bromide for visualization.

2.7 Cell viability assay
A portion of HDFBs harvested for DNA fragmentation assay was mixed with an equal volume of 0.4% Trypan blue. After 10 min, cells were counted on a hemocytometer. Viable cells did not uptake the dye and appeared transparent, whereas non-viable cells were permeable to the dye and appeared blue.

2.7 Data analysis
To construct a dose response curve, the collected γ-H2AX intensities at each irradiation dose were normalized to control. Data were collected from 3 independent experiments and presented as mean ± standard error of mean.

3. Results and discussion

3.1 Gamma radiation-induced DNA DSBs
As a result of radiation-induced DNA damage, the immediate repair of DNA DSBs can be detected by fluorescence imaging of γ-H2AX phosphorylation, which appeared as bright nuclear foci (figure 2). In our present study, the dose-dependent increase of γ-H2AX proteins was evident by the increase in foci intensity within 30 min after exposure to radiation. At the lowest irradiation dose of 0.2 Gy (figure 2a), a few bright foci appeared sporadically inside the cell nuclei, whereas the uniform distribution of foci with prominent intensity was observed at the highest dose of 4 Gy investigated in this study (figure 2b).

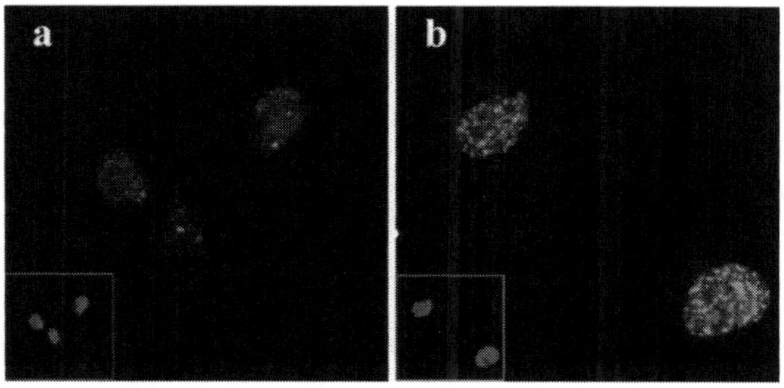

Figure 2. Confocal fluorescence images of γ-H2AX foci (green) formed inside the cells' nuclei after exposure to gamma irradiation at (a) 0.2 Gy and (b) 4 Gy. The nuclei (blue) were counterstained with Hoechst 33258 and shown on the lower left corner.

For quantitative measurement, the intensity of phosphorylated γ-H2AX foci at each irradiation treatment was analyzed by flow cytometry. By normalizing the collected flow intensity at all doses to the value at 0 Gy, our data showed a linear increase in foci intensity in proportion to irradiation dose (R^2 = 0.9862) (figure 3), indicating the extent of DSB damage at increasing doses. At the highest dose of 4 Gy, the phosphorylation of γ-H2AX showed a two-fold increase compared to sham control.

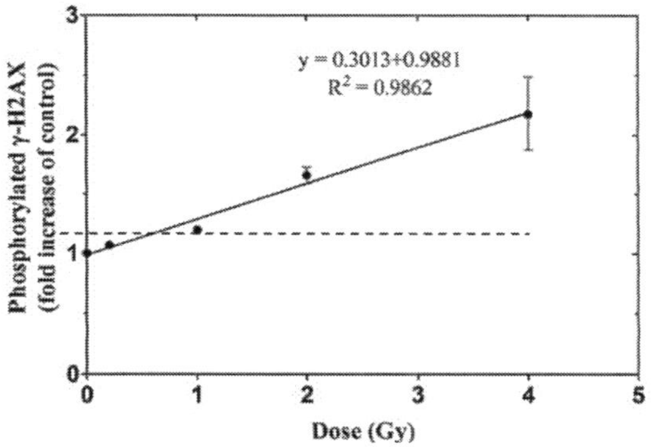

Figure 3. Graph represents the dose response relationship of γ-H2AX foci to gamma irradiation in HDFB cell model. Data were normalized to control (0 Gy) and shown as a fold increase in γ-H2AX intensity. Dashed line marks the background level.

This increase in foci intensity was in agreement with previous studies of X-ray [7,8] and gamma [5,6] irradiation-induced DNA DSBs of fetal fibroblasts [7], human lymphocytes [6,8], minipig skin tissues [5,9] and skin equivalents [6] that reported the linear relationship of γ-H2AX phosphorylation to irradiation dose. As an immediate cellular response to irradiation, the foci intensity peaked at 30 min post exposure [6,7] before attenuated over time as DNA repair proceeded [5-7,10]. However, this linear relationship of foci to irradiation dose was maintained at 48 h as reported in the case of lymphocytes, fibroblasts and keratinocytes [6] after gamma irradiation at doses 0 - 5 Gy. In addition, the residual of γ-H2AX foci was also reported in minipig skin biopsies at day 3 following 50-Gy exposure to gamma radiation [5]. This persistence in the formation of foci, which may indicate delayed or impaired repair of DSBs due to complex DNA damage, enables the use of γ-H2AX assay to supplement other biodosimetric measures for population triage during the first 48 h after accidental exposure for proper medical treatments. In addition, the assay has been shown to be implemented in the automated setup [11], which also facilitates the process of large-scale samples. Although only a 30-min time point was the focus in this present study to first establish the dose response curve, the levels of γ-H2AX phosphorylation at longer time points and the use of other biological samples such as blood will be further investigated to fully complete the time-dependent, dose response relationship.

High dose radiation from radiotherapy can pose injury to the skin as well as the long-term effects such as fibrosis and skin cancer [5]. In this present study, we have used dermal fibroblasts in a 2D culture model to quantify the extent of DNA damage as an initial step towards the development of 3D culture model to substitute for the use of animals or biopsies in future study. Such results may provide the better estimation of cutaneous radiation reaction [5] in individuals subjected to therapeutic exposure.

3.2 Apoptosis and cell viability

Radiation-induced DNA DSBs can result in cell apoptosis if the damaged DNA is not repaired and the cells lose the ability to control their normal growth. During cell apoptosis, endonucleases are stimulated to induce fragmentation of DNA, which can be detected by agarose gel electrophoresis. In our study, HDFBs were first irradiated by gamma rays at the selected doses of 0.2 and 4 Gy, subcultured and then harvested for DNA fragmentation assay on day 4. Compared to the DNA ladder in the range of 0 - 20 kbp (figure 4), we did not observe any DNA fragments from HDFB extracts after exposure to both irradiation doses. However, DNA fragments in the range of 180-900 bp have been previously reported [10] in lymphocytes after exposure to X-ray irradiation at doses > 1 Gy. Such discrepancy in the data may be attributed to the apoptosis-resistance nature of fibroblasts. In comparison to keratinocytes, fibroblasts were two times more resistant upon exposure to 2 Gy soft X-ray [12]. Metabolically, HDFBs could recover their level of glutathione, which helps protect the cells from radiation-induced oxidative damage, within 72 h after 1 Gy gamma irradiation [13]. Gamma radiation-induced apoptosis in many cell types is mediated partly through Fas/FasL interaction [14,15]. Under normal culture conditions, HDFBs express FLIP protein which confers resistance to Fas-induced apoptosis [16] and therefore would also contribute to HDFBs' resistance to radiation-induced apoptosis. In addition, Bmi-1, a polycomb group RING finger protein, has also been shown to contribute to radioresistance in HDFBs especially at a low radiation dose [17].

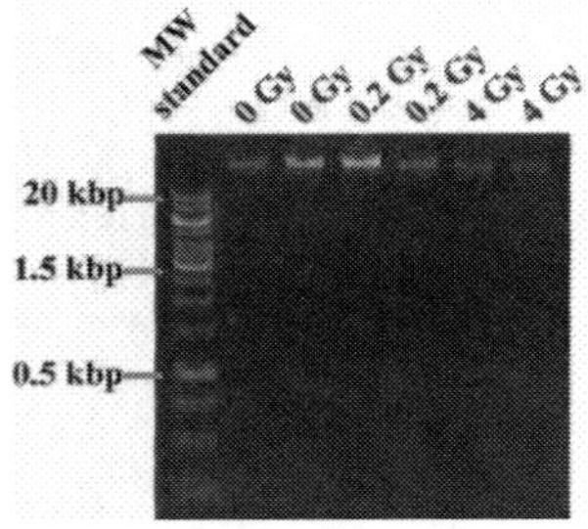

Figure 4. DNA fragmentation assay of HDFB extracts on 0.6% agarose gel

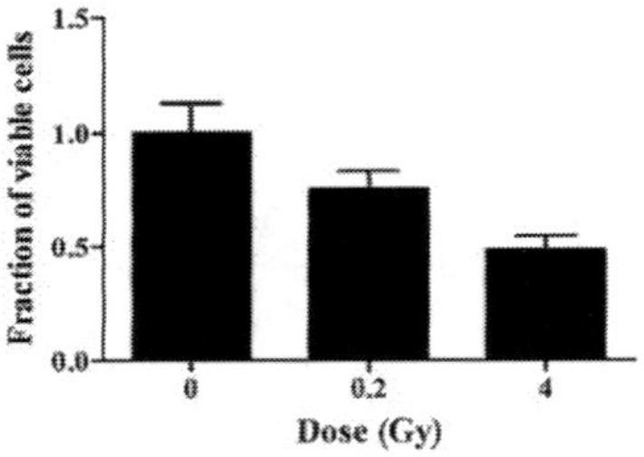

Figure 5. Fraction of cell viability analyzed by trypan blue assay

Although DNA fragments were not detected, the number of cells, represented by fraction of cell viability (figure 5), decreased from 1 to 0.5 as the irradiation dose increased from 0 – 4 Gy. This decline in the cell number suggests that some other cellular functions may be altered by irradiation and later affect the number of remaining, viable cells in subcultures. Similar observations have been reported for HDFBs [13], human pulmonary endothelial cells [18], SSC4 oral squamous cell carcinoma cells and MCF7 breast cancer cells [19] exposed to gamma radiation. In certain reports, reduction in cell viability has been shown to occur concomitantly with induction of apoptotic markers [18,20-22]. However, cell cycle arrest and reduction in clonogenic ability may also explain the reduction in cell counts upon a subculture [19,20]. To further investigate the mechanisms involved, apoptotic markers such as caspase-

3 activity, the distribution of cell population in various phases of the cell cycle and the clonogenic ability of irradiated HDFBs and will be investigated.

4. Conclusion

In this present study, we have established the dose-response relationship of DNA DSBs, measured from γ-H2AX phorphorylation, to gamma radiation using human-derived dermal fibroblasts as a cell model. The phosphorylation of γ-H2AX at DSB sites showed a linear increase with irradiation dose in the range of 0 - 4 Gy tested in this present study, supporting its application as a biodosimetric tool. Despite the absence of DNA fragmentation, the increase in irradiation dose had an effect on the number of live cells remained in subculture post irradiation.

Acknowledgments

This research is sponsored by the Thailand Institute of Nuclear Technology (Public Organization). All of the radiation processes were aided by staffs at the Gem Irradiation Center, Thailand. The author thanked Ms. Pucharee Songprakhon at the Division of Molecular Medicine, Department of Research and Development, Faculty of Medicine Siriraj Hospital for her help on confocal microscopy.

References

[1] Jeggo P A and Löbrich M 2007 DNA double-strand breaks: their cellular and clinical impact? *Oncogene* **26** 7717–9
[2] Thorslund T and West S C 2007 BRCA2: a universal recombinase regulator *Oncogene* **26** 7720–30
[3] Kuo L J and Yang L-X 2008 Gamma-H2AX - a novel biomarker for DNA double-strand breaks *In Vivo* **22** 305–9
[4] Van Gent D C and van der Burg M 2007 Non-homologous end-joining, a sticky affair *Oncogene* **26** 7731–40
[5] Ahmed E A, Agay D, Schrock G, Drouet M, Meineke V and Scherthan H 2012 Persistent DNA damage after high dose *in vivo* gamma exposure of minipig skin *PLoS One* **7** e39521
[6] Redon C E, Dickey J S, Bonner W M and Sedelnikova O A 2009 γ-H2AX as a biomarker of DNA damage induced by ionizing radiation in human peripheral blood lymphocytes and artificial skin *Adv. Space Res.* **43** 1171–8
[7] Beck M, Moreels M, Jacquet P, Van Oostveldt P, De Vos W H and Baatout S 2012 X-irradiation induces cell death in fetal fibroblasts *Int. J. Mol. Med.* **30** 114–8
[8] Brand M, Sommer M, Achenbach S, Anders K, Lell M, Löbrich M, Uder M and Kuefner M A 2012 X-ray induced DNA double-strand breaks in coronary CT angiography: comparison of sequential, low-pitch helical and high-pitch helical data acquisition *Eur. J. Radiol.* **81** e357–62
[9] Moroni M, Maeda D, Whitnall M H, Bonner W M and Redon C E 2013 Evaluation of the gamma-H2AX assay for radiation biodosimetry in a swine model *Int. J. Mol. Sci.* **14** 14119–35
[10] Ghardi M, Moreels M, Chatelain B, Chatelain C and Baatout S 2012 Radiation-induced double strand breaks and subsequent apoptotic DNA fragmentation in human peripheral blood mononuclear cells. *Int. J. Mol. Med.* **29** 769–80
[11] Garty G, Chen Y, Salerno A, Turner H, Zhang J, Lyulko O, Bertucci A, Xu Y, Wang H, Simaan N, Randers-Pehrson G, Yao Y L, Amundson S A and Brenner D J 2010 The RABIT: a rapid automated biodosimetry tool for radiological triage *Health Phys.* **98** 209–17
[12] Slonina D, Brankovic K, Panteleeva A and Dorr W 2001 Chromosomal damage and survival of keratinocytes and fibroblasts after irradiation with 200 kV or 25 kV X-rays *Rep. Pract. Oncol. Radiother.* **6** 39

[13] Kwon Y-K, Ha I J, Bae H W, Jang W G, Yun H J, Kim S R, Lee E K, Kang C-M and Hwang G S 2014 Dose-dependent metabolic alterations in human cells exposed to gamma irradiation *PLOS ONE* doi:10.1371/journal.pone.0113573

[14] Reap E A, Roof K, Maynor K, Borrero M, Booker J and Cohen P L 1997 Radiation and stress-induced apoptosis: A role for Fas/Fas ligand interactions *Proc. Natl. Acad. Sci. USA* **94** 5750–5

[15] Embree-Ku M, Venturini D and Boekelheide K 2002 Fas is involved in the p53-dependent apoptotic response to ionizing radiation in mouse testis *Biol. Reprod.* **66** 1456-61

[16] Santiago B, Galindo M, Palao G and Pablos J L 2004 Intracellular regulation of Fas-induced apoptosis in human fibroblasts by extracellular factors and cycloheximide *J. Immunol.* **172** 560-6

[17] Bae S, Kim K, Cha H J, Choi Y, Shin S H, An I, Lee J H, Song J, Yang K H, Nam S Y and An S 2014 Altered microRNA expression profiles are involved in resistance to low-dose ionizing radiation in the absence of BMI1 in human dermal fibroblasts *Int. J. Oncol.* **45** 1618-28

[18] Yu J, Piao B, Pei Y, Qi X and Hua B 2010 Protective effects of tetrahydropalmatine against γ-radiation induced damage to human endothelial cells *Life Sci.* **87** 55–63

[19] Risinger A L, Natarajan M, Thomas Jr. C.R. and Mooberry S L 2011 The taccalonolides, novel microtubule stabilizers, and c-radiation have additive effects on cellular viability *Cancer Lett.* **307** 104–11

[20] Keta O, Todorović D, Popović N, Korićanac L, Cuttone G, Petrović I and Ristić-Fira A 2014 Radiosensitivity of human ovarian carcinoma and melanoma cells to γ-rays and protons *Arch. Med. Sci.* **10** 578–86

[21] Ghavami L, Goliaei B, Taghizadeha B and Nikoofar A 2014 Effects of barley β-glucan on radiation damage in the human hepatoma cell line HepG2 *Mutat. Res. Genet. Toxicol. Environ. Mutagen.* **775–776** 1–6

[22] Ma H, Rao L, Wang H L, Mao Z W, Lei R H, Yang Z Y, Qing H and Deng Y L 2013 Transcriptome analysis of glioma cells for the dynamic response to c-irradiation and dual regulation of apoptosis genes: a new insight into radiotherapy for glioblastomas *Cell Death Dis.* **4** e895

An evaluation on the design of beam shaping assembly based on the D-T reaction for BNCT

M Asnal[1,3], T Liamsuwan[2] and T Onjun[1]
[1] School of Manufacturing Systems and Mechanical Engineering, Sirindhorn International Institute of Technology, Thammasat University, Pathum Thani, Thailand
[2] Nuclear Research and Development Division, Thailand Institute of Nuclear Technology, Ongkharak, Nakhon Nayok, Thailand

E-mail: muharaniasnal@yahoo.co.id

Abstract. Boron Neutron Capture Therapy (BNCT) can be achieved by using a compact neutron generator such as a compact D-T neutron source, in which neutron energy must be in the epithermal energy range with sufficient flux. For these requirements, a Beam Shaping Assembly (BSA) is needed. In this paper, three BSA designs based on the D-T reaction for BNCT are discussed. It is found that the BSA configuration designed by Rasouli et al. satisfies all of the International Atomic Energy Agency (IAEA) criteria. It consists of 14 cm uranium as multiplier, 23 cm TiF_3 and 36 cm Fluental as moderator, 4 cm Fe as fast neutron filter, 1 mm Li as thermal neutron filter, 2.6 cm Bi as gamma ray filter, and Pb as collimator and reflector. It is also found that use of specific filters is important for removing the fast and thermal neutrons and gamma contamination. Moreover, an appropriate neutron source plays a key role in providing a proper epithermal flux.

1. Introduction

Boron Neutron Capture Therapy (BNCT) is an indirect radiotherapy for destruction of cancer cells. To utilize BCNT, ^{10}B enriched compounds, preferentially concentrated in malignant tissues, must be first introduced into the patient's body. The tissues are then irradiated by a low-energy neutron beam, which can induce $^{10}B(n,^{4}He)^{7}Li$ reactions producing secondary charged particles with high biological effectiveness.

One of the critical parameters in BNCT is the neutron flux of a neutron source. It must be an appropriate rate for a sufficiently long enough time. The neutron source providing the required neutron beam with high neutron flux has previously seen only in a nuclear reactor. However, a suitable nuclear reactor for such purpose is expensive and difficult to be realized in or near urban areas. This leads to the development of other neutron sources for BNCT. For example, Montagnini et al. [1] proposed a compact neutron generator based on the D-T fusion reaction as a suitable neutron source for in-hospital treatments. It has greater safety, lower energy for an incident deuteron beam, lower cost, smaller size, and higher social acceptability.

Owing to the fact that the typical neutron energy emitted from D-T fusion reaction is 14.1 MeV, these neutrons cannot be used directly in BNCT. Therefore, they are moderated to the epithermal energy

[3] To whom any correspondence should be addressed.

range, between 1 eV and 10 keV. For this purpose, neutrons must be treated within a particular system, called a beam shaping assembly (BSA). By using appropriate geometry configuration and material selection in BSA design, the quality and intensity of neutron beam can be optimized. Further, as the intensity of the beam determines the treatment time, it must be considered in the BSA design as well. According to the International Atomic Energy Agency (IAEA) recommendation, epithermal neutron flux should be on the order of 10^9 n/cm^2 s. It is pivotal to note that since the neutron flux usually decreases greatly when neutrons pass through different components of BSA [1], it is necessary to increase the number of neutrons emitted from the neutron source. In this paper, the main components of BSA are compared and three BSA designs are discussed.

2. BSA Components

As mentioned earlier, a BSA is basically designed to moderate high energy neutrons to lower energies. In addition, it must remove fast and thermal neutrons and gamma contaminations. A BSA consists of the multiplier, moderator, neutron filter, reflector, gamma shielding and collimator as major components. In order to obtain a suitable neutron beam for BNCT, a proper design procedure must be employed.

2.1. Multiplier

A multiplier is used to increase the number of neutrons emitted from a neutron source, due to the fact that the neutron flux is significantly decreased when neutrons pass through different materials of the BSA. Hence, the material of the multiplier should be properly chosen. Uranium is a typical material used as a neutron multiplier to increase the number of neutrons via fission reactions.

Rasouli et al. [3] used an isotropic source of 14.1 MeV neutrons placed at the centre of a sphere made of natural uranium to determine the optimal thickness of uranium, generating the highest flux of neutrons. The number of neutrons per neutron source was found to increase with the uranium's thickness and reached a maximum value with a 14 cm radius. They also compared the geometry of uranium, namely cylindrical and spherical. It was found that the uranium geometry played a minor role on neutron multiplication. Two cylinders containing uranium at the maximum point generate approximately the same number of neutrons as generated by the spherical uranium. Thus, a sphere is certainly preferable due to its lightness.

Eskandaria and Kashian [4], employed Pb and ^{238}U as multiplier materials. The large cross section of Pb for (n, 2n) reaction compensate for the neutron losses by leakage and absorption during moderation. Moreover, the neutron flux was maximized by the use of ^{238}U as a neutron multiplier. In this design, Pb was modelled with 1, 2, 3, 4, 5, 6, 8, 10, and 12 cm thicknesses. The suitable thickness for Pb was found to be 3 cm. Further, the layer with ^{238}U was modelled with 0.5, 1, 1.5, 2, 2.5, 4, and 5 cm thicknesses, and the suitable thickness was found to be 2 cm.

Bi, U, and Pb containing acceptable (n, 2n) cross sections for 14 MeV neutrons were examined by Faghihi using MCNP5 code [5, 6]. Results showed that 5 cm Bi was considered as a suitable multiplier. Pb was also another good choice of a multiplier. However, due to gamma emissions, U was not recommended as a multiplier.

2.2. Moderator

As mentioned previously, the multiplied generated neutrons must be moderated to the epithermal range using proper moderator materials. These materials must have low scattering cross sections at desirable epithermal energies, but high scattering cross sections at higher energies. It should be noted that moderator materials with high absorption cross section absorb many neutrons before they have reached the patient, leading to high gamma ray contamination.

For the moderator, Fluental and Fe are considered to be suitable materials [7]. Fluental is a neutron moderator material developed at VTT in Finland [8] and composed of 69% AlF_3, 30% Al and 1% LiF. Fe is however preferred due to its lower cost and higher abundance. Magnesium fluorine (MgF_2) is also proven useful for producing neutrons in the epithermal energy range [9]. Although U is still used as a moderator as well, it was recently considered an improper material owing to its harmful gamma emissions [5].

Verbeke et al. [9] proposed medium materials such as Fluental, Fe, Al, Ni, AlF_3, Al+AlF_3, and LiF as moderators for BNCT. These materials were proven to approximately fulfil IAEA criteria. In fact, because of their low cost compared to Fluental, Al and AlF_3 are commonly used as moderator materials in BSA for BNCT. However, the low density of Al and AlF_3 compared to Fluental is one of their main drawbacks. Relatively thick Al and AlF_3 are required for the same neutron attenuation as achieved with Fluental.

De Boer [10] evaluated Al followed by Al_2O_3 or AlF_3 near the beam exit as a moderator. These combinations exhibited efficient performance since the O and F cross sections fill in the valleys between the energy resonances peaks of Al. Results also showed that AlF_3 was a suitable moderator for neutrons with energies less than 10 keV, whereas Al_2O_3 was a better moderator for fast neutrons with energies which are higher than 10 keV.

Miyamaru and Murata [11] evaluated D_2O, Fluental, carbon, and beryllium as moderator materials. Results showed that D_2O was suitable for epithermal neutron production with the highest flux and most decreased number of fast neutrons. D_2O, H_2O, Li/D_2O, and Fe were also evaluated as moderator materials by Mofakham [12] and D_2O was found to be a suitable material in terms of both absolute and relative thermal-epithermal fluency in the emergent neutron beam.

For finding proper moderators, Rasouli et al. [3] investigated 16 different materials, namely TiF_3, Fluental, AlF_3, Al_2O_3, MgF_2, BeO, Fe, Al, H_2O, CF_2, Mn, PbF_2, LiF, Co, Cu and PbF_4. In their work, the geometry of the moderator and reflector are cylindrical. The epithermal flux and the ratio of epithermal neutron flux over fast neutron flux ($\varphi_{\rm epi}/\varphi_{\rm Fast}$) associated with these materials were calculated using the MCNP4C code [3, 13]. Results showed that TiF_3 with 23 cm thickness was suitable material for a moderator. However, although the epithermal flux for TiF_3 reached its maximum value, the resulting $\varphi_{epi}/\varphi_{Fast}$ did not satisfy the IAEA's recommended value, which might be due to the use of uranium as a multiplier that can intensify the fast neutron flux. For increasing $\varphi_{\rm epi}/\varphi_{\rm Fast}$, a second moderator was used to reduce the number of fast neutrons. When 23 cm thick TiF_3 was selected as the first moderator, the value of $\varphi_{\rm epi}/\varphi_{\rm Fast}$ of 40 cm Fluental as the second moderator was shown to be larger than the critical value recommended by IAEA. Thus, it is suitable to use Fluental as the second moderator in this case.

According to Yani [14], materials containing F or Mg have a maximum value of fast to epithermal over macroscopic absorption cross sections ($\sum s_{f/epi}/\sum\gamma$). In terms of neutron accumulation in epithermal energy range during the slowing down from source energies, these materials are claimed to be better than other moderator materials. Due to this reason, Eskandari and Kashian [4] also used AlF_3 as a moderator. It has a large ratio of $\sum s_{f/epi}/\sum\gamma$. Based on their work, the ratio of fast neutrons to total neutron flux between 60 and 70 cm of AlF_3 was constant and therefore 60 cm AlF_3 was selected as a moderator.

In another study by Koivunoro et al. [7], the best results for moderator materials were achieved with Fe and Fluental. The material combination of Fluental is ideal to decrease the neutron flux to the desired energy range of ~10 keV without over-moderation. In this system, a 28 cm D–T layer of iron was first used to decrease the fast neutron energy in the range of 1–14.1 MeV. Then, a 34 cm D–T layer of Fluental was used to decrease the neutron flux in the range of >100 keV.

Due to the fact that a good moderator for BNCT should not induce high gamma ray contamination, a material inducing less gamma dose rate must be selected as a moderator material. Based on these criteria, Faghihi [5] selected Fe with 50 cm length and 80 cm thickness as the moderator material.

Furthermore, in order to reduce cost, successive stacks of Al, polytetrafluoroethylene (PTFE) and LiF (Al/PTFE/LiF) were used as a moderator [15]. The reason is that the neutron interaction cross section in Al and PTFE can effectively moderate fast neutrons. Results showed an acceptable behaviour of the proposed moderator and the advantage of irradiating the target with near-resonance-energy protons (2.3 MeV) because of the high-neutron yield at this energy. This leads to production of less fast neutrons and ultimately the lowest treatment times. The 34 cm Al/PTFE/LiF moderator was found to be the suitable option.

2.3. Fast and thermal neutron filter

One of the key considerations in designing a BNCT beam is reducing the fast neutron component of the incident beam. In many cases, the fast neutron flux cannot be properly decreased by the moderator due to a surplus of the fast and thermal neutrons. Because of this, a fast neutron filter material that can reduce the fast component of the flux, resulting in an increased epithermal one, is required. The material commonly used is Al or its compounds [16]. According to the recommendations of IAEA, the desirable minimum beam intensity would be 10^9 epithermal neutrons $cm^{-2}s^{-1}$. In addition, in order to increase the ratio of epithermal neutron flux over thermal neutron flux ($\varphi_{epi}/\varphi_{thermal}$), the thermal neutron flux should be reduced as much as possible. Thermal neutrons cause too high a radiation dose to the skin and soft tissue. In order to reduce these undesired doses, a material functioning as the thermal neutron filter must be used. Elements with good thermal neutron absorbers are ^3He, ^{10}B, ^6Li, ^{14}N, and Cl. Since ^3He is very rare and has a high potential as an energy source, it is not very useful as a thermal neutron absorber [10].

^6Li was considered a suitable thermal neutron filter due to smaller the release of gamma rays in the process. Its high thermal absorption cross section helps to remove thermal neutrons from clinical neutron beams. Thus, lower thermal contamination and lower gamma contamination due to thermal neutron absorption is the result. Gao [17] used ^6Li as a filter for BNCT. The filter was able to increase the average energy of the epithermal neutrons in the epithermal neutron beam. Venhuizen [18] also examined the performance of gadolinium (Gd) as the thermal neutron filter using MCNP calculations [19]. Results, however, showed that Gd is not suitable for this application owing to gamma contamination issue.

Because of its high inelastic scattering cross section, Fe was used as neutron filter material [3]. It decreased the fast neutron flux with energy up to 14.1 MeV. Fe with a thickness of 4 cm was found to satisfy the IAEA's criteria because D_{fn}/φ_{epi} was less than 2×10^{-13} Gy cm^2. Further, due to filter addition, the fast and thermal neutron contaminations was capable of being removed even though epithermal flux is unavoidably reduced.

Faghihi [5] also studied many materials as the filter and recommended Al(40%) + AlF_3(60%) as the suggested filter since it has low epithermal neutron absorption cross sections, high thermal absorption cross section and smaller radiation capture cross section.

2.4. Reflector

A reflector is used for directing the neutrons to the beam port. A proper reflector must have high elastic scattering cross section and low absorption cross section for entering the neutrons to the moderator again. For finding a cheaper alternative reflector, Uhlar et al. [20] tested a two-layer reflector; tungsten (W) as the inner part and Mo as the outer part. Approximate with the same performance, the mass of the W/Mo reflector is five times smaller than to the single component reflector made of W.

A research work by Burlon et al. [21] confirmed that Pb is a suitable choice as reflector. Pb has also shown better performance than graphite [15]. Rasouli et al. [3] also evaluated the possibility of using Pb and BeO as the reflector. Their different thicknesses were assessed to determine their reflective capabilities. It was concluded that lead was a suitable reflector to decrease the beam port size to a flat circular surface with a radius of 6 cm. Pb was also used as a reflector material for D–T neutrons [7], even though with a Bi collimator, spectra calculations gave similar results.

Mofakham et al. [12] investigated the use of graphite and boranyloxyboron (B_2O) reflectors to increase both relative and absolute thermal-epithermal neutron fluency. The best results in terms of thermal-epithermal fluency were obtained using a B_2O reflector. For a reflector thickness of 10 cm the relative thermal-epithermal fluency reaches a maximum value of 85%.

2.5. Collimator

A collimator is used to focus output neutron beam to the head phantom. Bi is a commonly used material for a collimator [7]. Reflector material such as Lithium Polyethylene in a trapezoidal shape is also occasionally used to collimate output neutron beam [5].

Uhlar et al. [22] utilized Pb supplemented with a Ni part as a collimator. In a further study, they also employed molybdenum (Mo) supplemented with a nickel part [20]. This is because at low neutron energies, nickel elastic cross section is higher than that of Mo and the Ni absorption cross section is lower than the Mo absorption cross section for a broad energy interval of moderated neutrons.

2.6. Gamma shielding
Gamma shielding is used to reduce gamma rays originated from the nuclear reactions and neutron capture in different BSA materials. Bi is normally used for this purpose. For instance, Rasouli et al. [3] employed Bi for gamma shielding. Results showed that using 2.6 cm Bi, D_γ/φ_{epi} was able to be decreased to lower than 2×10^{-13} Gy cm^2. Thus, it is suitable for BNCT.

3. BSA Models
In this section, three BSA models for BNCT are discussed. The first one is a BSA model proposed by Rasouli and Masoudi (figure 1) [23]. They used natural uranium as a neutron multiplier and thick layers of TiF_3 and Al_2O_3 as moderators. These materials were surrounded by a thick Pb reflector. Ni and Li-Poly were used as shield and collimator, respectively. This configuration showed low contamination of the fast and thermal neutrons and gamma at the beam port, which satisfies almost all IAEA criteria.

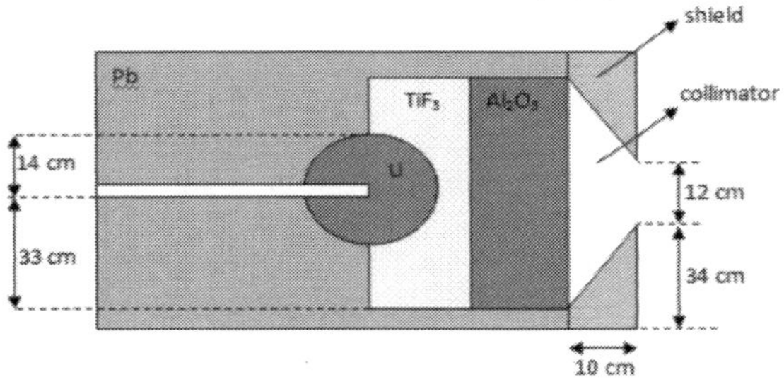

Figure 1. The BSA model proposed by Rasouli and Masoudi [23].

Rasouli et al. [3] also designed BSA for BNCT based on the use of 14 cm in radius metallic uranium as multiplier system for D–T neutron source and moderator/filter/reflector arrangement. TiF_3 with 23 cm in thickness and Fluental with 36 cm in thickness are used for the moderator to achieve a proper epithermal neutron flux. A configuration of 4 cm Fe, 1 mm Li, and 2.6 cm Bi are used for fast neutron, thermal neutron, and gamma ray filters, respectively, as illustrated in figure 2.

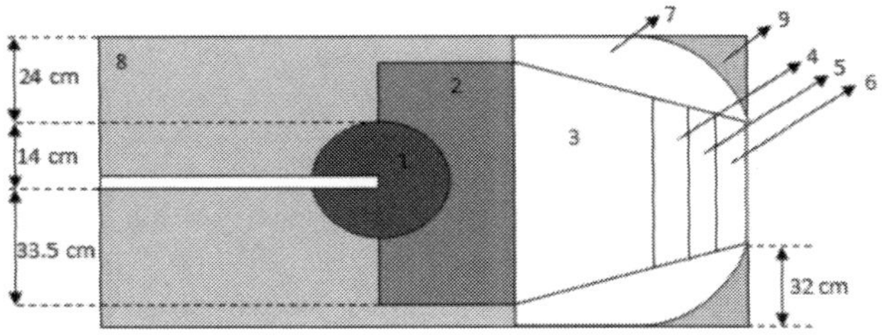

Figure 2. The BSA model proposed by Rasouli et al. [3]. Remarks: 1 is 14 cm uranium, 2 is 23 cm TiF_3, 3 is 36 cm Fluental, 4 is 4 cm Fe, 5 is 1 mm Li, 6 is 2.6 cm Bi, 7 is Pb as collimator, 8 is Pb as reflector, 9 is shield.

Rahmani and Shahriari [24] proposed a BSA model containing layers of Mg, Al, Fe, Pb, Bi, C, and Ni joined with their oxide and fluoride compounds as illustrated in figure 3. Collimators, neutron and gamma shields with various materials and sizes were also added to BSA configurations. The suitable thickness of these materials in cylindrical geometry was chosen to satisfy gamma and neutron dose requirements in the treatment position. In this model, 10 cm of Ni was selected as a collimator material in conical shape. Pb with 25 cm in thickness, 5% borated polyethylene, and 1 mm Cd were used as a reflector, neutron shield, and thermal neutron filter, respectively.

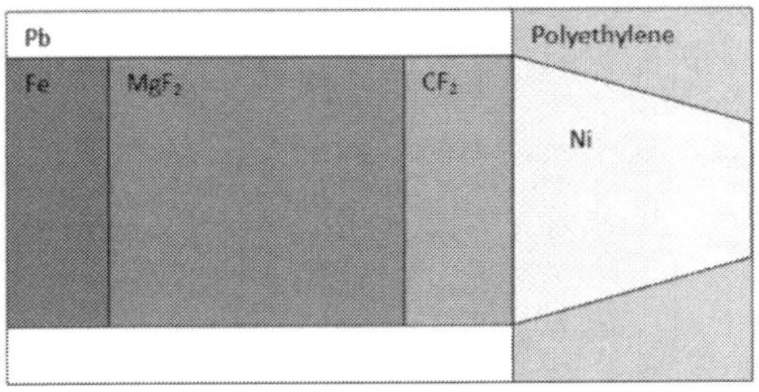

Figure 3. The BSA model proposed by Rahmani and Shahriari, [24].

It should be noted that as the suitable option for the treatment of deep-seated tumors, the epithermal neutrons should possess a minimum intensity of 5×10^8 n/cm^2 s for a reasonable time treatment. Therefore, selecting an appropriate neutron source able to provide this epithermal flux is pivotal before undertaking BSA design. The neutron yield of the available neutron sources capable of satisfying such criteria is in the range of $6.6 \times 10^{11} - 4.4 \times 10^{14}$ n/s [23].

The BNCT parameters of three configurations are presented in table 1. Based on the BNCT free beam parameters of those three BSA configurations, it was found that only the BSA designed by Rasouli et al. [3] satisfies all IAEA criteria. Although the BSA proposed by Rasouli and Masoudi [23] cannot satisfy all IAEA criteria, its neutron beam is effective for deep-seated brain tumor treatments even with D–T neutron generator yielding as low as 2.4×10^{12} n/s. It was found in this configuration that with increase in the neutron source yield, the treatment time decreases. Also, a shorter treatment time can be achieved with the cost of high neutron source intensity. Thus, considering the trade-off between reduction of treatment time and neutron source intensity is vital. In those three configurations, use of specific filters is very important for removing the fast and thermal neutrons and gamma contamination from neutron beam. The free beam parameters recommended for BNCT would be fulfilled by the use of filter. However, the epithermal neutron flux is decreased.

Table 1. BNCT parameters of the three BSA configurations .

BSA model	$\varphi_{epi} \times 10^9$ (n/cm^2 s)	$\varphi_{epi}/\varphi_{thermal}$	$\varphi_{epi}/\varphi_{Fast}$	$D\gamma/\varphi_{epi} \times 10^{-13}$ (Gy cm^2)	$D_{fn}/\varphi_{epi} \times 10^{-13}$ (Gy cm^2)
Rahmani and Shahriari, [24]	8.19	383	17.20	1.18	7.98
Rasouli et al. [3]	4.43	121.2	23.75	1.98	0.59
Rasouli and Masoudi [23]	1.04	20.21	-	5.79	0.67
IAEA criteria [2, 3, 23, 24]	>0.5	>100	>20	<2	<2

4. Conclusion

Progress has been made towards the development of BSA design for moderating primary neutrons into the epithermal energy range. It is found that the BSA configuration designed by Rasouli et al. meets all of the IAEA criteria. It consists of 14 cm uranium as multiplier, 23 cm TiF_3 and 36 cm Fluental as moderator, 4 cm Fe as fast neutron filter, 1 mm Li as thermal neutron filter, 2.6 cm Bi as gamma ray filter, and Pb as collimator and reflector. Use of specific filters for removing the fast and thermal neutrons and gamma contamination from neutron beam and use of a neutron source providing a proper epithermal flux are vital to obtain a BSA configuration satisfying the IAEA criteria.

Acknowledgments

This work is supported by the government research budget through Thammasat University. M Asnal thanks Sirindhorn International Institute of Technology for scholarship under the Excellent Foreign Students (EFS) program.

References

[1] Montagnini B, Cerullo N, Esposito J, Giusti V, Mattioda F and Varone R 2002 *Nucl. Instrum. Methods Phys. Res. A.* **476** 90
[2] Kiger III W S, Sakamoto S and Harling O K 1999 *Nucl. Sci. Eng.* **131** 1
[3] Rasouli F S, Masoudi S F and Kasezas Y 2012 *Ann. Nucl. Energy* **39** 18
[4] Eskandari M R and Kashian S 2009 *Ann. Nucl. Energy* **36** 1100
[5] Faghihi F 2012 *Proc. Inter. Conf. Ind. Eng. Operations Manag.* (Istanbul) p 719
[6] X-5 Monte Carlo Team 2003 *MCNP: A General Monte Carlo N-Particle Transport Code Version 5* LA-UR-03-1987 (Los Alamos: Los Alamos National Laboratory)
[7] Koivunoro H, Auterinen I, Kosunen A, Kotiluoto P, Seppälä T and Savolainen S 2004 *Phys. Med. Biol.* **48** 291
[8] Auterinen I and Hiismki P 1993 *Proc. Clinical BNCT Workshop* (Helsinki) p 14
[9] Verbeke J M, Costes S V, Bleuel D, Vujic J and Leung K N 1998 *Proc. Int. Conf. on the Physics of Nuclear Science and Technology* (New York) p 105
[10] de Boer J J 2008 *New Filter Design with Monte Carlo Calculation* (dissertation) (Delft: Delft University of Technology)
[11] Miyamuru H and Murata I 2011 *J. Nuc. Sci. Technology* **1** 533
[12] Mofakham N N, Payamil F, Eghbali A and Mansouri S 2014 *Switzerland Res. Park J.* **103** 474
[13] Briesmeister J F 2000 *MCNP–A General Monte Carlo N-Particle Transport Code.Version 4C* LA-13709-M (Los Alamos: Los Alamos National Laboratory)
[14] Yani S 2002 *Design and Benchmark Experiment for Cyclotron Based Neutron Source for BNCT.* CYRICA, Annual Report
[15] Burlon A A, Kreiner A J, Valda A A and Minsky D M 2004 *Appl. Rad. Isotopes* **61** 811
[16] Tyminska K 2009 *Pol. J. Med. Phys. Eng.* **15** 77
[17] Wei G 2005 *Lithium-6 Filter for Fission Converter-Based Boron Neutron Capture Therapy Irradiated Facility Beam* (dissertation) (Massachusetts: Massachusetts Institute of Technology)
[18] Venhuizen J R 1996 *INEL BNCT Research Program Annual Report 1995,* Idaho National Engineering Laboratory Report
[19] Briesmeister 1986 *MCNP - A General Monte Carlo Code for Neutron and Photon Transport* LA-7396 (Rev 2) (Los Alamos: Los Alamos National Laboratory)
[20] Uhlar R, Miroslava K, Peter A and Jaromir P 2014 *J. Radio. Anal. Nucl. Chem.* **300** 809
[21] Burlon A A, Kreiner A J, White S M, Blackburn B, and Yanch J C 2001 *Med. Phys.* **28** 796
[22] Uhlar R, Alexa P, Pistora J 2013 *Nucl. Instrum. Methods* **298** 81
[23] Rasouli F S and Masoudi S F 2012 *Appl. Rad. Isotopes* 70 2755
[24] Rahmani F and Shahriari M 2011 *Ann. Nucl. Energy* **38** 404

AUTHOR INDEX